Integrated Vehicle Health Management

The Technology

Other SAE books of interest:

Integrated Vehicle Health Management: Perspectives on an Emerging Field
By Ian K. Jennions
(*Product Code: R-405*)

Integrated Vehicle Health Management: Business Case Theory and Practice
By Ian K. Jennions
(*Product Code: R-414*)

Aircraft Maintenance: The Art and Science of Keeping Aircraft Safe
By Bruce R. Aubin
(*Product Code: T-115*)

Integrated Vehicle Health Management

The Technology

Edited by **Ian K. Jennions**

SAE INTERNATIONAL™

Warrendale, Pennsylvania
USA

SAE INTERNATIONAL™
400 Commonwealth Drive
Warrendale, PA 15096-0001 USA
E-mail: CustomerService@sae.org
Phone: 877-606-7323 (inside USA and Canada)
724-776-4970 (outside USA)
Fax: 724-776-0790

ISBN 978-0-7680-7952-4
SAE Order Number R-429
DOI 10.4271/R-429

Library of Congress Cataloging-in-Publication Data
Integrated vehicle health management : the technology / [edited by] Ian K. Jennions.
p. cm.
Summary: "This is the third book SAE has published on IVHM. The scope of this book was to return to the, now familiar, taxonomy and investigate each of the broad theme areas of IVHM in more detail. The aim was to provide a bridge between the introduction to the subject, covered in the first book, and technical papers describing new work. This book is basically in three parts: the technology chapters (2-9), then the Applications chapter (10), before the Emerging Technologies (11). The contributions made in these sections are now summarized in this order, before drawing the book to its conclusion"-Provided by publisher.
ISBN 978-0-7680-7952-4
1. Electronics in transportation. 2. Embedded computer systems. 3. Automatic test equipment. 4. Systems engineering. 5. Transportation--Technological innovations. I. Jennions, Ian K. II. Society of Automotive Engineers.
TA1235.I54 2013
629.25'49--dc23
2013016408

To purchase bulk quantities, please contact:

SAE Customer Service
Email: CustomerService@sae.org
Phone: 877-606-7323 *(inside USA and Canada)*
724-776-4970 *(outside USA)*
Fax: 724-776-0790

Visit the SAE International Bookstore at http://books.sae.org

Table of Contents

Acknowledgments

SAE International's series on the subject of Integrated Vehicle Health Management (IVHM) introduces its third volume, this time focusing on the technology that actually makes the implementation of IVHM possible.

Written by twenty-seven authors from the academic, government, and industry settings, the title coalesces the expertise gathered from many years of hands-on experience and intense research projects, in many different areas of the world—indeed a unique mix of talents and technical backgrounds.

As with the previous titles of the series, this one is also coordinated by Dr. Ian K. Jennions, Director of the IVHM Centre at Cranfield University (UK).

The focus of *Integrated Vehicle Health Management: The Technology* is to bring the reader closer to the technical aspects of a new and growing body of knowledge, a novel way to the future in commercial aviation's asset management and maintenance.

Here are the authors who collectively made this work happen. A list of their pictures and biographies can be found at the end of the book:

- Abhinav Saxena, NASA Ames Research Center
- Ashok N. Srivastava, NASA Ames Research Center
- Chaochao Chen, University of Maryland
- Charles E. Dibsdale, Optimized Systems and Solutions (OSyS)
- Dinkar Mylaraswamy, Honeywell International Inc.
- George Vachtsevanos, Georgia Institute of Technology
- Ian K. Jennions, Cranfield University
- Kai Goebel, NASA Ames Research Center
- Karl Reichard, Pennsylvania State University
- Keith Jackson, Meggitt PLC
- Kevin Swearingen, The Boeing Company
- Kirby Keller, The Boeing Company
- Marcos E. Orchard, University of Chile
- Michael Pecht, University of Maryland
- Michael H. Azarian, University of Maryland
- Michael J. Augustin, IVHM Inc.
- Michael J. Roemer, Impact Technologies LLC/Sikorsky Innovations Co.
- Neil Eklund, GE Global Research Center
- Peter Foote, Cranfield University
- Raj M. Bharadwaj, Honeywell International Inc.

- Ranjith-Kumar Sreenilayam-Raveendran, University of Maryland
- Ravi Rajamani, Meggitt PLC
- Robert W. Mah, NASA Ames Research Center
- Seth Kessler, Metis Design Corporation
- Sony Mathew, University of Maryland
- Sureshkumar Perinpanayagam, Cranfield University
- Tarapong Sreenuch, Cranfield University

And to the reviewers, whose time and dedication made the quality of this publication a reason to be proud of, again our immense gratitude:

- Andre Lafon, Airbus
- Antony Waldock, BAE Systems
- Chip Queitzsch, FAA
- Chris Pomfret, Treble One LLC
- David Followell, The Boeing Company
- George Vachtsevanos, Georgia Institute of Technology
- Ginger Shao, Honeywell International Inc.
- Ian Campbell, Meggitt PLC
- Ian K. Jennions, Cranfield University
- Joao Pedro Pinheiro Malere, Embraer
- Len Losik, Failure Analysis Co.
- Merv Floyd, Meggitt PLC
- Michael McCullough, BAE Systems
- Peter Smout, Aero Engine Controls
- Ravi Rajamani, Meggitt PLC
- Richard Greaves, Meggitt PLC
- Rhonda Walthall, Hamilton Sundstrand
- Seth Kessler, Metis Design Corporation
- Steve P. King, Rolls-Royce
- Steve Parker, Meggitt PLC
- Timothy Wilmering, The Boeing Company

Our commitment to publishing new developments in IVHM is unwavering. New titles are on their way, and we hope the readers will find them easy to understand and helpful well into the future.

SAE International

CHAPTER 1 Introduction

Ian K. Jennions, *IVHM Centre, Cranfield University, UK*

> "We are continually faced with a series of great opportunities brilliantly disguised as insoluble problems."
>
> —John W. Gardner

1.1 Background

SAE International formed an IVHM (Integrated Vehicle Health Management) Steering Group in the fall of 2010. The idea reflected the large number of groups, given below, within SAE's organization, that deal with health management of systems or subsystems (but without the integrated, holistic view of the vehicle asset or fleet).

- S-18: Aircraft and Systems Development and Safety Assessment
- E-32: Aerospace Propulsion Systems Health Management
- G-11: Reliability, Maintainability/ Supportability and Probabilistic Methods Group
- G-11 SHM: Structural Health Monitoring and Management
- AS-3: Fiber-Optics and Applied Photonics
- A-6: Aerospace Actuation, Control and Fluid Power Systems
- AE-5: Aerospace Fuel, Oil and Oxidizer Systems Steering Group
- A-5 Aerospace Landing Gear Systems

The newly formed group, through the IVHM HM-1 technical group, proceeded to organize an IVHM technical track for the SAE AeroTech meeting in Toulouse in October 2011 and also had the idea for a book on IVHM that could be handed out to participants to promote engagement in this exciting new field. The book: *IVHM Perspectives on an Emerging Field* [Jennions, ed. 2011] was the result and, surprisingly, it spawned a trilogy of books.

This first book explored the overall subject of IVHM at a very high level, sufficient for senior decision makers and technical people interested in the field, who were not familiar with IVHM, to be able to understand the overall subject and value proposition. What emerged during writing of the book, and the conversations that naturally ensued, was the realization that the major barrier to IVHM adoption lay not in the technology itself but in an ability to express a clear business case for its adoption. Such a case would lay out the cost of the engineering design, equipment, and development needed, as well as the organizational (cultural) cost of instilling service thinking along with training on new procedures and processes.

This resulted in the second book: *IVHM: Business Case Theory and Practice* [Jennions, ed. 2012]. It was predicated on the paradigm shift observed by the transformation of businesses from selling a product, and deriving future income from spare part sales, to selling a service in which regular income is received in return for effective maintenance of the asset. The resulting service business requires a much deeper understanding of how the asset is used and how parts degrade, and hence can be maintained, thus providing the rationale for IVHM. The objective of the book was to provide the tools and techniques for constructing a business case while also providing some of the context in which they are framed. It is essentially a book for the commercial, rather than the technical, side of the organization, but should prove readable by both.

1.2 Scope

The scope of the current book returns to the taxonomy introduced in the first book and replicated here as Figure 1.1. It was originally produced by the IVHM Centre and has been adopted by the SAE IVHM steering group; it is not intended to be all encompassing but rather to show the essential elements and how they fit together.

Each of the broad themes in the taxonomy (e.g., architecture—Systems Engineering), system's design (Design Tools and Toolkits),

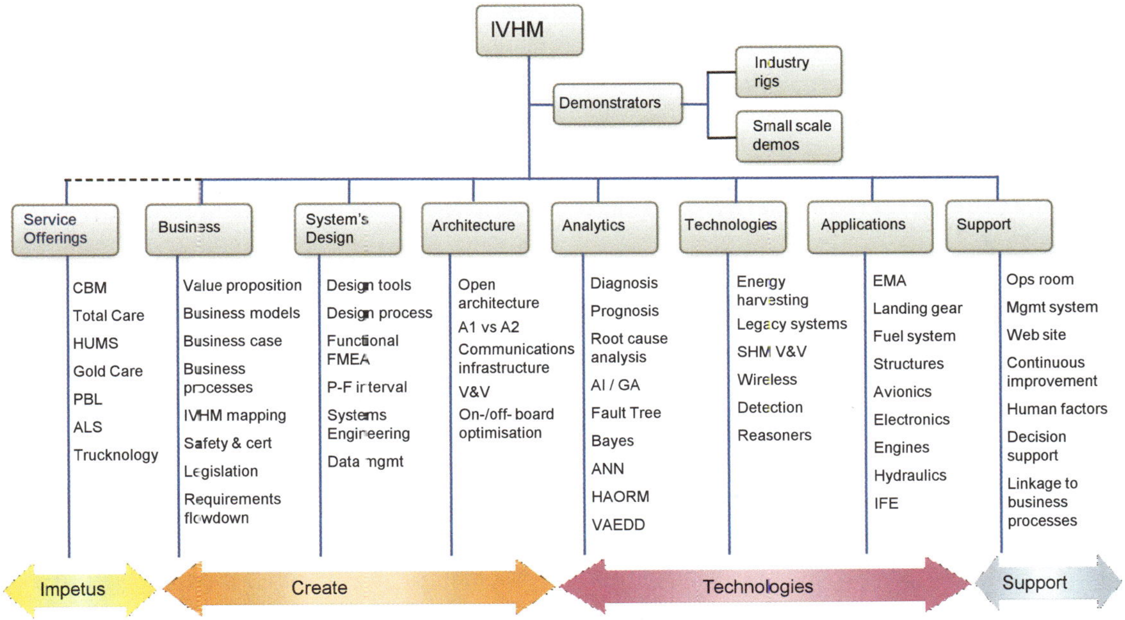

Figure 1.1 Generic IVHM taxonomy [Jennions 2011].

and analytics (Algorithms), were covered at a high level in the first book. It is the intention of this book to revisit these themes and add the detail necessary for an interested reader to go from the high-level view to the technical papers in each individual subject area. As such, this is an IVHM technology primer, all of which may not be easily understood and assimilated. If your background is in sensors, then you might not understand prognostics; if reasoning, then architecture may prove difficult. Either way, this book provides the technically literate a route to pursue and develop their knowledge in this broad subject area.

1.3 Book Structure

This book is broadly organized in three parts. The first eight chapters, after this, deal in detail with a number of the technologies at the center of this field. They are written by some of the leading practitioners in the world and offer a unique collection of technical material supplemented with real-world examples. Next is a chapter that looks at applications of IVHM technology in four very different fields: structures, rotating machinery, motorsport, and wind turbines. The third part is a chapter on disruptive technologies. Although IVHM is seen as being a disruptive technology in its own right, given its ability to radically change a business model, there are other technologies on the horizon that are thought to provide the next wave of disruption; these are examined here in some detail. Finally, a summary is given and some concluding remarks made.

1.4 References

Jennions, I. K., ed. 2011. *Integrated Vehicle Health Management—Perspectives on an Emerging Field*, ISBN 978-0-7680-6432-2.

Jennions, I. K., ed. 2012. *Integrated Vehicle Health Management—Business Case Theory and Practice*, ISBN 978-0-7680-6432-2.

Jennions, I. K. 2011. "The Story so Far, the Development of an IVHM Centre," 14th Australian International Aerospace Conference, March 2011 (http://www.cranfield.ac.uk/ivhm/).

CHAPTER 2 Sensors, Instrumentation, and Signal Processing

Sureshkumar Perinpanayagam, *IVHM Centre, Cranfield University, UK*

> "Installing sensors all over the world means we are creating a global nervous system."
>
> —Jason Silva

2.1 Introduction

Sensing, instrumentation, and signal processing constitute the foundation of Integrated Vehicle Health Management (IVHM). The diagnostic and prognostic algorithms developed in IVHM depend very much on what kind of sensor suite we employ, instrumentation we use for signal de-noising and digitization, and signal-processing techniques required to extract appropriate information for IVHM applications. From the sensor point of view, we need to decide on the type, number, and location of the sensors; and their size, weight, cost, dynamic range, and other characteristics. The data collected by these sensors are rarely useful in their raw form, so we need appropriate instrumentation and signal processing to convert the raw data into useful information or features that are indicative of a fault or failure. These features then can be used as a basis for the development of condition-based monitoring or prognostic prediction algorithms. Hence, sensing, instrumentation, and signal processing constitute the fundamental first steps in IVHM. These are also needed in an accurate and timely manner for maintenance scheduling.

2.2 Overview of Measurement Systems and Components

2.2.1 Introduction

Measurement systems are used in all aspects of IVHM and typically contain a number of basic building blocks. Figure 2.1 shows a typical measurement chain layout for an IVHM system, detailing some of the items that are usually found. The following sections outline the main elements of the measurement chain.

2.2.2 Sensors and Transducers

Sensors and/or transducers are devices that measure physical or chemical parameters and generally transform those parameters into electrical signals. The different types of sensors are discussed in section 2.3.

2.2.3 Preamplifier

This component is necessary to pre-amplify the signals from the sensors. Signal amplification performs the important functions of increasing the resolution of the input signal. For example, the output of an electronic

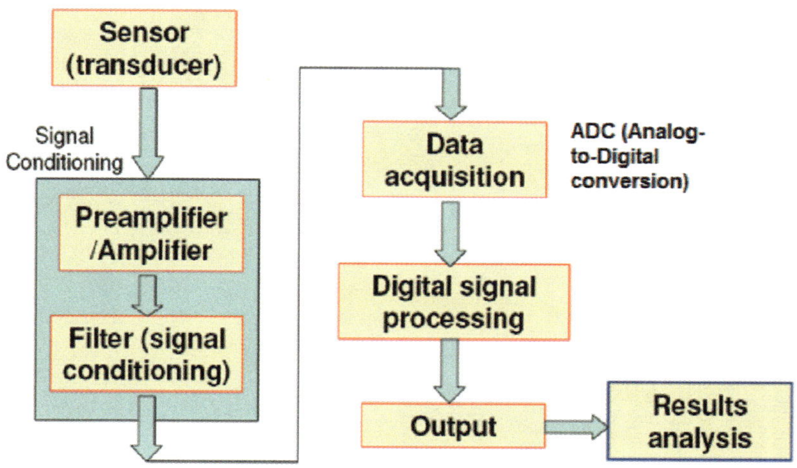

Figure 2.1 Measurement chain for an IVHM system.

temperature sensor, which is probably in the millivolts range, is too low for an Analog-to-Digital Converter (ADC) to process directly. In this case, it is necessary to bring the voltage level up to that required by the ADC. The choice of amplifier depends heavily on the type of transducer used and should, in effect, be regarded as part of it. In all cases, its role is to strengthen the (usually) small signals generated by the sensor-transducer configuration so that they could be fed into an ADC.

2.2.4 Filtering

Filtering is the most common signal-conditioning function, as usually not all of the signal frequency spectrum contains valid data. Signals coming from sensors can be very noisy, of low amplitude, biased, and dependent on secondary effects, such as temperature. Moreover, the physical quantity of interest may be related to measurements indirectly. The basic technique for improving the Signal-to-Noise Ratio (SNR) is low-pass filtering, as noise generally dominates the desirable signal at high frequencies.

2.2.5 Analog-to-Digital Converter

The ADC converts the analog input signals to digital values, and the processor (or other logic) records them as a set of N discrete values, evenly spaced in the period T during which the measurement is made. This topic is covered in detail in section 2.4.

2.2.6 Digital Signal Processing

Digital signal processing covers a wide variety of techniques that are used to extract useful information from the data obtained after converting analog signals. These techniques are discussed in more detail in section 2.7.

2.2.7 Data Storage and Output Display

The results of the measurement process must be displayed in a form that the human operator can perceive. The best form for the display may be numerical or graphical, discrete or continuous, permanent or temporary—depending on the particular measurand and how the operator will use the information.

The resulting data can then be used for Data-Driven Anomaly Detection and Diagnosis (chapter 3) and Prognostics (chapter 4).

2.3 Sensors and Transducers

2.3.1 Introduction

Generally, the terms sensor and transducer are used interchangeably but can be thought of in the following way. A sensor is the complete assembly required to detect and communicate a particular event, while a transducer is the element within that assembly that transforms energy from one form to another (from the Latin *transducere* to lead across). There are many other possibilities apart from the above examples for energy conversion by a transducer. These are represented in Figure 2.2.

As we move around the periphery of Figure 2.2, we find the various forms of energy.

Mechanical energy refers to the potential and kinetic energies of a mass of any material.

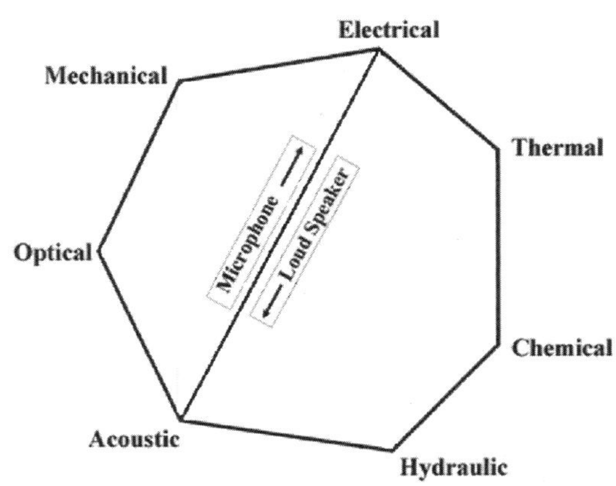

Figure 2.2 Chart of various energy formats.

Optical energy refers to energy in the form of light or electromagnetic radiation very similar to light, such as infrared or ultraviolet radiation. With the advent of the laser, this has become an important form of sensing.

Acoustic energy refers to the energy of sound waves, either in air or some other conducting medium.

Hydraulic energy refers to the energy contained in a fluid (liquid or gas). This energy can be in the form of kinetic energy of a flowing fluid, or it can be the potential energy of a fluid under pressure.

Chemical energy refers to the energy associated with the formation and reactions of various chemical compounds.

Thermal energy refers to the energy available in a material as a result of its temperature.

The last form of energy shown in Figure 2.2 is **electrical energy**. This is the energy that can be imparted to an electric charge and is the most useful means of conveying information to instrumentation, which is of particular interest to the IVHM community.

2.3.2 Requirements for a Sensor-Transducer System

There are three general requirements for a sensor-transducer system used for IVHM purposes:

1. Accuracy
2. Stability
3. Lack of interference with the physical variable being measured

It is obvious why an IVHM sensor must be accurate. It is essential to know how the output signal from the sensor is related to the input quantity. In most applications, it is desirable that this relationship be linear. However, this is not essential, and with today's ready access to computers and microprocessors, nonlinear characteristics can be "linearized" with little effort, provided the nonlinear characteristic is reproducible. This means that the output signal is directly proportional to the input quantity. In some cases, nonlinear relationships are desirable. Although linearity of a sensor's calibration characteristic is often a desired feature, sensors with nonlinear calibration characteristics are also very useful when (computer) microprocessors are a part of the measurement chain. It has always been possible to use these processors of the sensor output signal to "linearize" a nonlinear characteristic. There are also situations, as illustrated in the previous paragraph, when it may be desirable for a particular application to have a special functional relationship between the variable being sensed and a measured output; microprocessors can also make this possible. One of the roles of a microprocessor in a measurement chain is to convert the sensor's actual calibration characteristic to the sensor output that is desired for the application. The important feature of the sensor, in this case, is no longer linearity but stability and reproducibility. As long as the sensor's calibration characteristic is known and does not change, processing by a microprocessor can convert this to the characteristic that is desired.

A sensor must be stable if it is to be able to provide reproducible data. Changing of the sensor's properties with time, temperature, humidity, gravitational pull, etc. means that it cannot be used as the first stage of a measurement chain unless it is periodically used to measure known quantities considered to be calibration standards.

Finally, the presence of the sensor must not disturb the system being measured in any way if it is to provide accurate data. In many measurements, this cannot be achieved, and so the sensor is made to provide a minimum of disturbance to the system being measured. Since the sensor converts some of the energy of the variable being measured into a new form, it takes this energy away from the system being measured. If this is a substantial amount of energy, it can indeed result in an error in the measurement. It is, therefore, important to consider the effects of a given sensor on a measurement when selecting the sensor to make that particular measurement.

2.3.3 Mechanical Sensors

There are a number of sensors that are directly employed in dynamic systems to measure mechanical quantities. These include position, speed, acceleration, torque, strain, etc., and they are amply described in textbooks and technical publications. The operation of these sensors, including accelerometers and strain gauges, is described in textbooks such as de Silva [1989].

2.3.4 Optical Sensors

Fiber-optic sensors have been designed to measure strain, temperature, displacement, chemical concentration, and acceleration, among other material and environmental properties. Their main advantages include small size, light weight, immunity to electromagnetic and radio frequency interference, high- and low-level endurance, fast response, high sensitivity, and low cost. The basic physics governing fiber-optic sensors also make them relatively tolerant to high-temperature effects, high EMI (Electromagnetic Interference), and highly corrosive environments.

2.3.5 Acoustic Sensors

A number of sensors are used to measure acoustic quantities for monitoring the health of critical structures, such as airplanes, bridges, and buildings. These sensors are particularly useful for large structures, as sonic waves travel long distances and have the potential to measure large areas of structure for IVHM. These sensors include *acoustic sensors, guided wave ultrasonic sensors, bulk wave ultrasonic sensors*, and *diffuse wave ultrasonic sensors*. Several publications on these techniques are included in the references [Cook and Bertholet 2001; Leonard, Malyarenko, and Hinders 2000; Gao, Shi, and Rose 2005; Giurgiutiu and Bao 2004; Wilcox 2003].

2.3.6 Hydraulic Sensors

Hydraulic sensors measure mechanical properties of liquids or air, such as pressure or flow [Akers, Gassman, and Smith 2006; Manring 2005]. There are various kinds of flow sensors and flow meters, including some that have a vane that is pushed by the fluid, and can drive a rotary potentiometer, or ones that use magnetic fields.

2.3.7 Chemical Sensors

Chemical sensors typically measure variations in a chemical mixture and/or the concentration of the target chemical or the physical state of the chemical solution [Turner, Wilson, and Kaube 1987]. These could, for example, be used to measure the pH of a liquid, the presence of water in a solution, the conductivity of a liquid, etc.

2.3.8 Thermal Sensors

Thermal energy variations in many mechanical, electrical, and electronic systems are excellent indicators of impending failure conditions. Temperature in excess of control limits should be monitored and used in conjunction with other measurements to detect and isolate faults. Large temperature gradients and temperature in excess of specified limits are considered key indicators of incipient failure conditions for IVHM. There are a number of sensors available based on heat transfer principles; these include widely-used thermocouples [Williams 1977] and resistance temperature detectors.

2.3.9 Electrical Sensors

Electrical sensors track quantities such as voltage, current, power and their derivatives, and harmonic content [Jones 1988; Ramsden 2006]. They are primarily used with electromechanical and electrical or electronic systems in the transportation, utilities, communication, and computing sectors.

2.4 Analog-to-Digital Conversion and Associated Issues

2.4.1 Introduction

Most of the signals encountered in engineering applications are analog. To process analog signals using digital techniques, the analog signals must first be converted into digital signals. Digitization involves two processes: sampling (digitization in time) and quantization (digitization in amplitude), as shown in Figure 2.3. This whole process is called analog-to-digital conversion. In all analog-to-digital conversions, information from the original signal is lost. A digital signal is an approximation (usually very accurate) of the original analog signal and thus, it is extremely important to correctly design and implement the analog-to-digital conversion process.

In the next sections, sampling and quantization are discussed in depth, and the issues that need to be considered when converting data from an analog continuous form to digital are discussed.

2.4.2 Sampling

Sampling can be illustrated through the changing temperature on a single day, as shown in Figure 2.4.

Even though the temperature is changing continuously, an observatory that is measuring or sampling the temperature once every two hours will present information as shown in Table 2.1.

We could also say that the analog temperature signal is sampled once every T seconds. It should be noted that in our future discussion, we should assume that the sampling interval or space between the samples is constant, although irregular sampling is possible.

The most important parameter in the sampling process is the sampling period T or the sampling frequency f_s, which is defined in Eq. 2.1 as

$$f_s = \frac{1}{T} \tag{2.1}$$

Sampling frequency is given in units of "sample per second" or Hertz. If the sampling

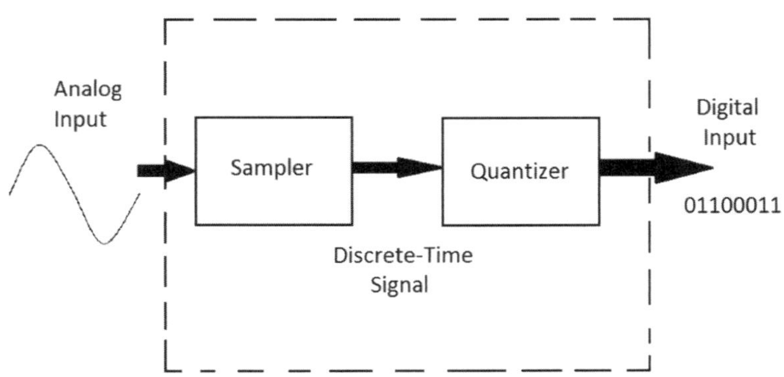

Figure 2.3 Analog-to-digital conversion.

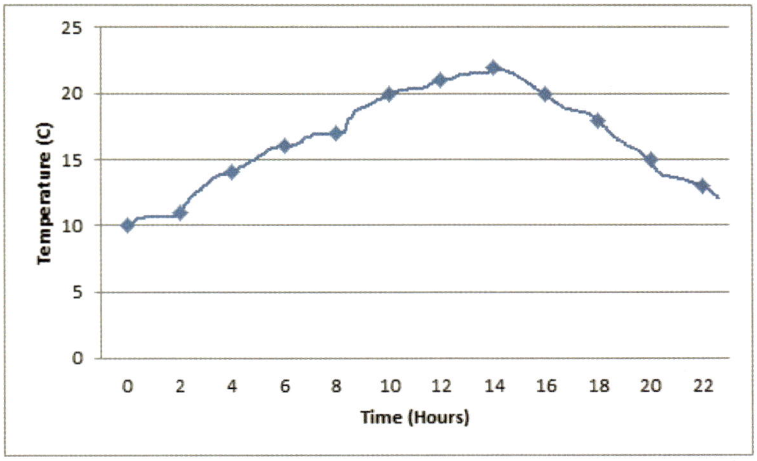

Figure 2.4 Temperature variation throughout a day.

is too frequent, then the algorithms that use these samples will have to process a large amount of data in the same time frame. If the sampling is too sparse, then important information might be missing in the sampled signal. The information content of a sampled signal is dictated by the Nyquist sampling theorem, discussed next.

Hour	Temperature
0	10
2	11
4	14
6	16
8	17
10	20
12	21
14	22
16	20
18	18
20	15
22	13

Table 2.1 Temperature Measured Every Two Hours of the Day

2.4.2.1 Minimum Sampling Rate—Nyquist-Shannon Theorem

The Nyquist-Shannon sampling theorem [Shannon 1949] provides the minimum sampling rate at which a continuous time signal needs to be sampled so that the original signal can be completely reconstructed by the samples alone.

If a continuous time signal contains no frequency components higher than ω Hz, then it can be completely determined by uniform samples taken at a rate of f_s samples per second, where, as in Eq. 2.2:

$$f_s \geq 2\omega \qquad (2.2)$$

Or in terms of sampling period, in Eq. 2.3:

$$T \geq \frac{1}{2\omega} \qquad (2.3)$$

The minimum sampling rate allowed by the sampling theorem ($f_s = 2\omega$) is called the Nyquist rate.

If the sampling frequency is less than twice the highest frequency component, as required by the sampling theorem, the replicated spectra overlap each other, causing distortion to the original spectrum. Under this circumstance, the original spectrum can never be recovered faithfully. This effect is known as aliasing, which is discussed in the next section.

2.4.2.2 Issues Associated with Sampling—Aliasing

The effect of aliasing on an input signal can be demonstrated by sampling a sine wave of frequency f_a using different sampling frequencies. Figure 2.5 shows such a sinusoidal function sampled at three different rates: $f_s = 4\ f_a$, $f_s = 2\ f_a$, and $f_s = 1.5\ f_a$.

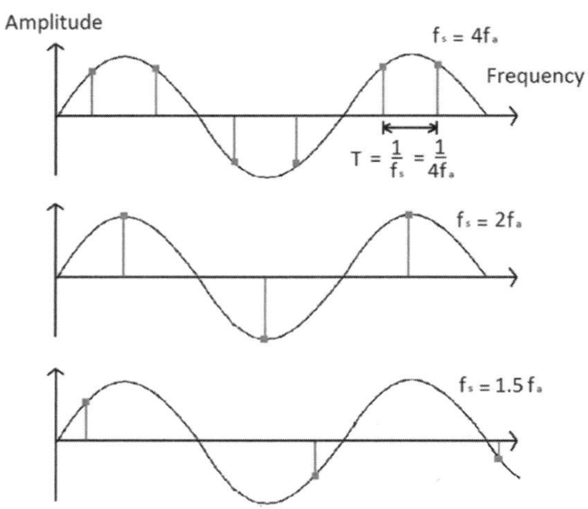

Figure 2.5 A sinusoid sampled at three different rates.

For the case with a sampling frequency 1.5 times the sine wave frequency, the sampling frequency is below the Nyquist rate. In this case, we would expect aliasing to occur. If we join the sampled points together, it can be observed that the rate at which the resulting function repeats itself differs from the frequency of the original signal. In fact, if we interpolate between the sample points, a smooth function with lower frequency is the result, as shown in Figure 2.6.

Therefore, it is no longer possible to recover the original sine wave from these sampled points. We say that the higher-frequency sine wave now has an "alias" in the lower-frequency sine wave, inferred from the samples. In other words, these samples are no longer representative of the input signal and, therefore, any subsequent processing will be invalid. Notice that the sampling theorem assumes that the signal is strictly band limited. In the real world, typical signals have a wide spectrum and are not band limited in the strict sense.

Anti-aliasing filters are always analog filters, as they process the signal before it is sampled. The sampling process incorporating an ideal low-pass filter as the anti-alias filter is shown in Figure 2.7, with the blue shaded area being the range through which the signal is admitted. The ideal filter has a flat pass band, and the cut-off is very sharp. Because the cut-off frequency of this filter is half of that of

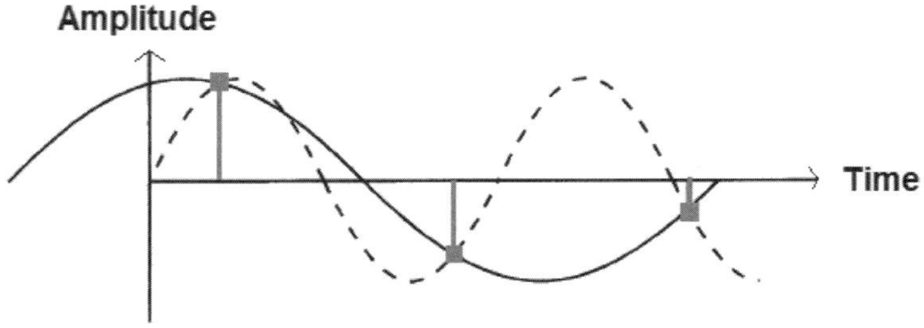

Figure 2.6 Effect of aliasing.

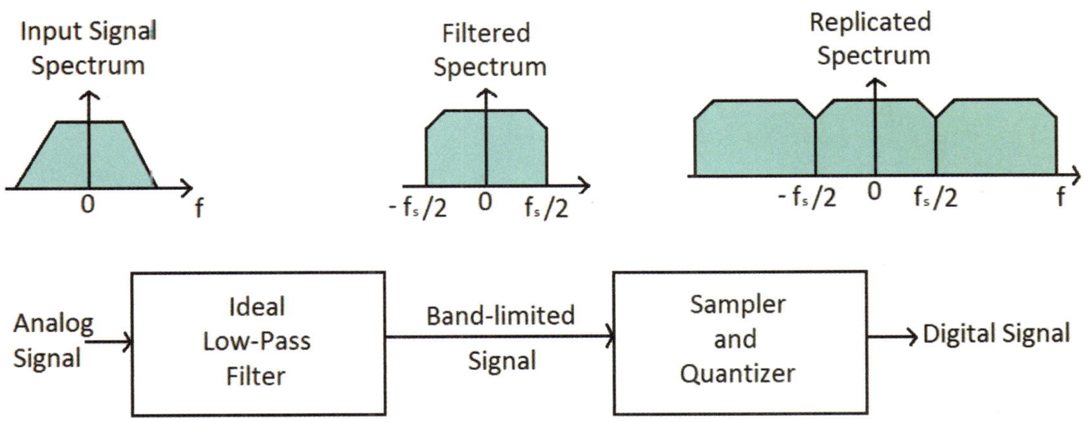

Figure 2.7 The analog-to-digital conversion process with anti-alias filtering.

the sampling frequency, the resulting replicated spectra of the sampled signal do not overlap each other. Thus, no aliasing occurs.

The practical choice of sampling rate is determined by two factors for a certain type of input signal. On one hand, the sampling theorem imposes a lower bound on the allowed values of the sampling frequency (Nyquist). On the other hand, the economics of the hardware imposes an upper bound. These economics include the cost of the ADC and the cost of implementing the analog anti-alias filter. A higher-speed ADC will allow a higher sampling frequency but may cost substantially more. However, a lower sampling frequency will put a more stringent requirement on the cut-off of the anti-aliasing filter, necessitating a higher-order filter and a more complex circuit, which again may cost more.

In real-time applications, each sample is acquired (sampled), quantized, and processed by a microprocessor. A higher sampling rate will mean that there are more samples to be processed within a certain amount of time. If T_{proc} represents the total microprocessor's processing time, then the time interval between samples T will need to be greater than T_{proc}, otherwise the processor will not be able to keep up. This means that if we increase the sampling rate, we will need a higher-speed microprocessor chip.

2.4.3 Quantization

Quantization is the discretization of the sampled signal amplitude when converting an analog signal into digital form.

The ADC assumes that the input values cover a full-scale range, say R. Typical values of R are between 1 and 15 V. Since the quantized sampled value is represented by B-bits, it can take only one of 2^B-1 possible quantization levels. If the spacing between these levels is the same throughout the range R, then we have a uniform quantizer. The spacing between quantization levels is called the quantization width or the quantizer resolution.

For uniform quantization, the resolution is given by Eq. 2.4.

$$Q = \frac{R}{2^B - 1} \tag{2.4}$$

In practice, the input signal x(t) must be preconditioned to lie within the full-scale

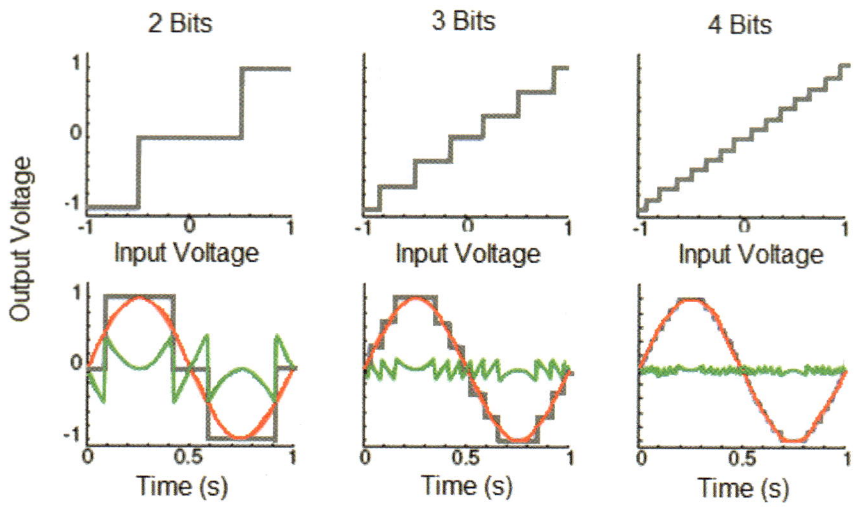

Figure 2.8 Quantization of a sine wave by 2-bit, 3-bit and 4-bit quantizers.

range of the quantizer. Figure 2.8 shows the quantization of a sine wave by 2-bit, 3-bit and 4-bit quantizers.

2.4.4 Leakage

Leakage is a problem that is a direct consequence of the need to take only a finite length of time history coupled with the assumption of periodicity for frequency-domain analysis. The problem is best illustrated by the two examples shown in Figure 2.9, in which two sinusoidal signals of slightly different frequencies are subjected to the Fourier transform from time to frequency domains.

In the first case (a), the signal is perfectly periodic in the sampling time window, and the resulting spectrum is quite simply a line, at the frequency of the sine wave. In the

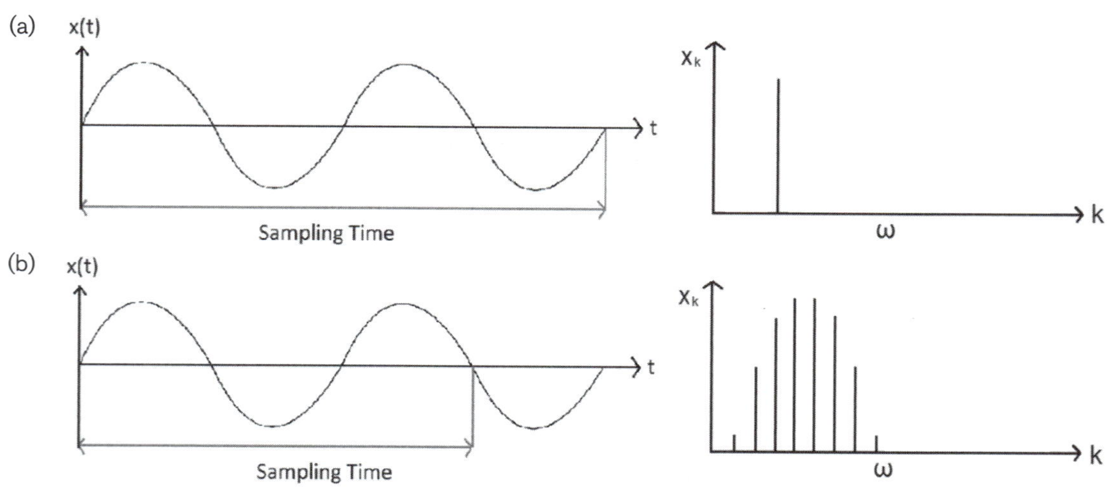

Figure 2.9 The effect of leakage.

second case (b), the periodicity assumption is not strictly valid, and there is a discontinuity implied at each end of the sample. As a result, the spectrum produced for this case does not indicate the single frequency from which the original time signal originates, and the frequencies in the chart are not actually represented in the specified lines of the spectrum. Energy has leaked into a number of the spectral lines close to the true frequency, and the spectrum is spread over several lines close to the true frequency. This problem becomes most acute when the signal frequencies are lower.

Leakage is a serious problem in many applications of digital signal processing, including IVHM, and there are several ways of avoiding or minimizing its effects. They include changing the duration of the measurement sample length to match any underlying periodicity in the signal, increasing the sampling time so that the separation between the spectral lines (the frequency resolution) is finer, adding zeros to the end of the measured sample, and modifying the signal sample using "windowing."

2.4.4.1 Windowing

In many situations, the most practical solution to the leakage problem involves the use of windowing, and there is a range of different windows for different classes of problems.

Windowing involves the imposition of a prescribed profile on the single time signal prior to performing the Fourier transform, and the profiles or windows are generally depicted as a time function. In Figure 2.10, the different windows are shown in the middle column, and the results of using the windows are shown in the right-hand column. The Hanning and Cosines Taper windows [Harris 1978] are used typically for continuous signals, such as steady periodic or random signals. The exponential window is used for transient signals, where much of the important information is concentrated in the initial part of the time record.

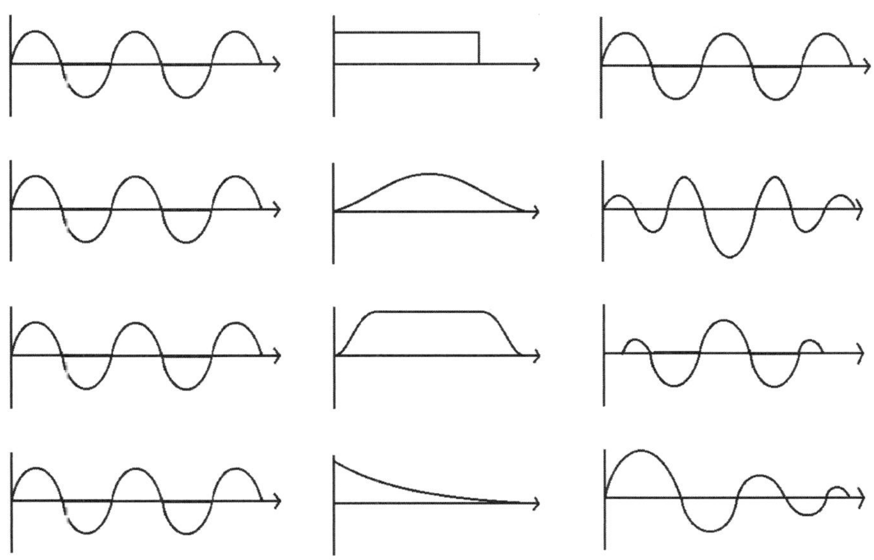

Figure 2.10 Effects of various window techniques on signals.

In all cases, rescaling is required to compensate for attenuation of the signal by application of the window.

2.4.5 Averaging

Averaging is very important for dealing with random signals. When analyzing random signals, it is not sufficient to compute Fourier transforms, as they do not exist for random processes, and we must instead obtain an estimate for the spectral densities and correlation functions that are used to characterize the random signal. Generally, it is necessary to perform an averaging process involving several individual time records or samples before the results obtained can be used with confidence. The number of averages required depends on the statistical reliability and the removal of spurious noise from the signals.

Figure 2.11 (a) shows m samples each of duration t. Averaging could be performed using the Fourier transform (covered later) on time record 1, 2, and 3. Alternatively, it is possible to perform the second transform as soon as possible using the most recent N points, even though some of these points are used in the previous transform. This form of averaging is pictorially depicted in Figure 2.11 (b) and is called overlap averaging.

2.5 Instrumentation Choice for an IVHM System

In designing instrumentation systems for use in IVHM systems, several considerations must be taken into account. One must first be concerned with the overall accuracy and precision of the system. The accuracy of a system is the degree of closeness of measurement of a quantity to that quantity's true value. The precision of a measurement system can be expressed in a number of alternative forms. For example, a meter that

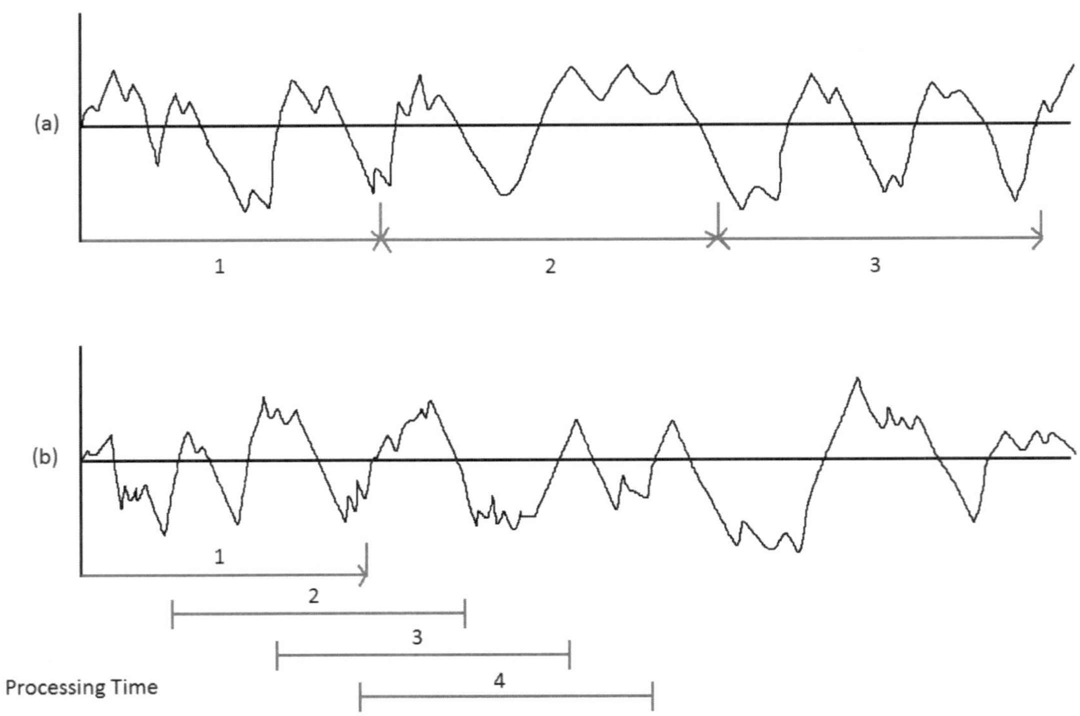

Figure 2.11 Illustration of averaging (a) and overlap averaging (b).

displays a reading of 2.434 V is more precise than one that displays a reading of 2.43 V. High-precision measurements do not imply high accuracy, because precision makes no comparison to the true value.

In IVHM, the requirements of the specific variable being measured determine the necessary accuracy for the instrument. If it is necessary to know the sound intensity of an engine to within ± 5% of its true value, it is important to use a sound intensity meter that has an accuracy that yields errors of less than ± 5%. Sometimes, it is not necessary to know the value of a variable accurately, but it is important to be able to correctly measure very small changes in this variable. To do this, then, a high-precision instrument is required. Even if the accuracy of our sound intensity meter is only ± 5%, if it has a precision of 1% of the total intensity, then we can measure changes in the sound intensity much more accurately than we can measure the total sound intensity. The smallest change in a measured variable that can be detected by the instrument is known as the resolution of the instrument. The resolution tells us how close two different measurements of a variable can be while still being able to tell that these measurements are different.

To understand the overall accuracy and precision of an instrument, we sometimes look at a quantity known as its transfer function. This may be illustrated by the generalized instrument of Figure 2.12.

This instrument has some time-varying input: namely, the variable being measured,

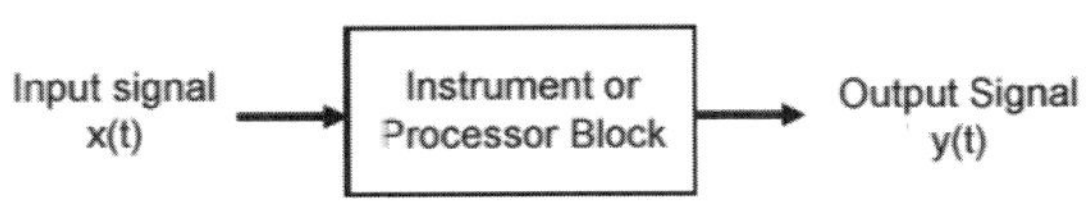

Figure 2.12 General instrument block for defining the transfer function.

which we shall denote by x(t) and some output that we shall denote by y(t). In the case of our example of the sound intensity meter, the input x would be the sound in the room being measured, and our output would be the indication on the display device. The transfer function of the instrument describes what happened to the input x(t) to get the output y(t). This can be described mathematically by Eq. 2.5.

$$y(t) = T[x(t)] \tag{2.5}$$

where T is the transfer function of the instrument. For an instrument with good accuracy and precision, the transfer function must be well defined and stable.

Another factor that is essential in describing the overall accuracy and precision of an instrument is its fidelity. For an instrument to measure a variable accurately, it must be able to follow changes in this variable over the course of time. As in a high-fidelity home music system, the instrument must be able to faithfully reproduce the variable that it is measuring. This means that the maximum rate of change of the variable to be measured and its frequency spectrum must be known, so that the proper instrument to do this job can be specified.

Another important consideration in determining the accuracy of an instrumentation system is the method of its calibration. A highly stable system need not be calibrated very often; however, for the highest accuracy, it must be calibrated with an established standard. The procedure of calibration consists of measuring the standard with the instrument and then adjusting the instrument so that it reads the known value of the standard. The value of the standard should be known to at least the same accuracy as that of the instrument to be calibrated. If the measurement is to be taken immediately after the instrument is calibrated, a standard having a known value to at least the rated precision of the instrument may be used to improve the accuracy of the instrument. In the case of this measurement, the instrument will be as "accurate" as its "precision."

An important consideration in determining the amount of calibration required for a particular instrument is its stability. An instrument that presents a great amount of drift will have to be calibrated several times throughout the period of time in which it is used, while one that is relatively stable need only be calibrated once.

Another factor that is closely related to the accuracy and precision of an instrument is its signal-to-noise ratio. We can define the signal as the information about the actual variable being measured by the instrument, and the noise as all other signals that are observed on the instrument display. Thus, if we can make our signal zero (make sure that the variable being measured by the instrument is zero), then everything that is displayed by the instrument will be noise. When there is a signal present, this noise can interfere with the signal to reduce the accuracy and precision of the instrument in observing that signal. The signal-to-noise ratio is a useful quantity to tell us how much noise is present with respect to the signal being measured.

We can define the signal-to-noise ratio by means of Eq. 2.6.

$$\text{Signal-to-noise ratio} = \frac{\text{Signal power}}{\text{Noise power}} \tag{2.6}$$

For highly accurate measurements, an instrument must have a signal-to-noise ratio that is a large number such as 10,000. As a means of presenting this number in a more readable format, we often present signal-to-noise ratios using a logarithmic scale. Electrical engineers have borrowed the logarithmic scale used to measure sound intensity to do this. Signal-to-noise ratios are usually presented using the decibel scale, as shown in Eq. 2.7.

$$\text{Signal-to-noise ratio} = 10\log_{10}\frac{\text{Signal power}}{\text{Noise power}} \tag{2.7}$$

In instrumentation, we often speak of signals in terms of the voltage of that signal. Since in electrical terms, we can determine power by Eq. 2.8:

$$\text{Power} = \frac{V^2}{R} \tag{2.8}$$

we could substitute Eq. 2.8 into Eq. 2.7 to obtain Eq. 2.9, the signal-to-noise ratio in terms of voltages that we might measure in an actual instrumentation system:

$$\text{Signal-to-noise ratio} = 20\log_{10}\frac{\text{Signal voltage}}{\text{Noise voltage}} \tag{2.9}$$

This is generally the way that instrumentation engineers present signal-to-noise ratio.

Voltage is much easier to measure than power, and so when an instrumentation engineer refers to signal-to-noise ratio, they are usually referring to Eq. 2.9. Signal-to-noise ratio is a very important concept in instrumentation systems. Although in the past we were primarily concerned with the sensitivity of an instrument, now that high-quality amplifiers are available to boost the strength of very weak signals, this is no longer a major concern. Instead, we need to know what the relationship between our signal and the noise might be. We can always amplify a weak signal, but if it is buried in noise, we will also amplify the noise. The signal-to-noise ratio will remain the same. Amplifiers in general introduce some noise themselves, so the signal-to-noise ratio might even be lower after the amplification process.

2.6 Feature Extraction Techniques

2.6.1 Introduction

Feature or condition indicator selection and extraction constitute the cornerstone of IVHM. In the feature selection and extraction, the signal data are converted into information and knowledge that can be used to identify or track a fault or failure. Fault diagnosis depends mainly on extracting a set of features from the sensor data that can distinguish between the fault classes of interest and detect and isolate the fault at its early incipient stage. Figure 2.13 provides a schematic view of feature extraction and selection and how they are related to other IVHM functions.

Fault diagnosis depends mainly on extracting a set of features from the sensor data that can be used to distinguish between different fault classes of interest. These features also could be used to isolate the area of a system where the fault is initiated. Good features related to fault diagnosis should have the following attributes: computationally inexpensive to measure, mathematically definable, explainable in physical terms, characterized by large mean distance from the other features and small variance, insensitive to extraneous variables and uncorrelated with other features.

Feature selection still relies primarily on expertise, observation, past historical evidence, and our understanding of the fault signature characteristics.

Feature extraction predominantly depends on signal processing techniques. Signal processing techniques enable this to happen by converting measurement data from the sensors to information and knowledge. In the following sections, the different feature extraction techniques for extracting useful information for IVHM are briefly reviewed. For further reading, please refer to the references.

2.6.2 Time Domain Feature Extraction Techniques

Time-domain data are measured typically at successive points in time, spaced at uniform intervals. Time-domain analysis [Engel et al. 2000] consists of methods for analyzing time-series data to extract meaningful statistics and other characteristics of the data. A simple form of analyzing time-domain data is regression analysis. This technique is used for estimating the relationships among variables with time. Other time-domain feature extraction processing treats signals as stochastic processes, dealing with their statistical properties (e.g., mean, covariance, etc.). Autocorrelation and cross-correlation

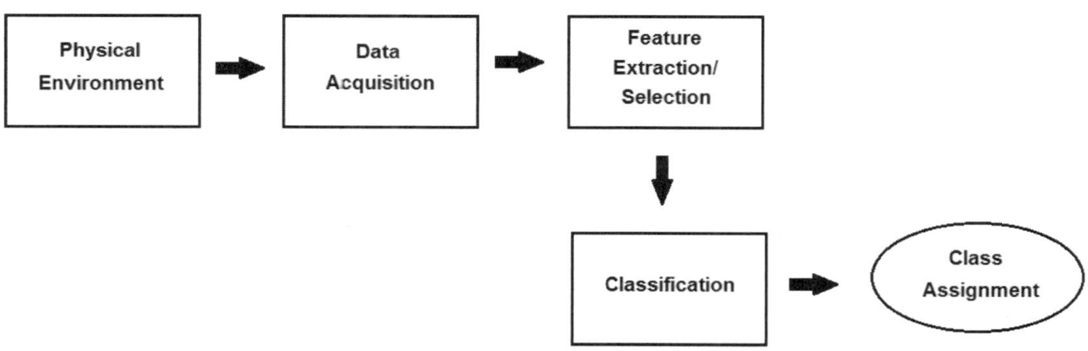

Figure 2.13 The schematic view of feature extraction/selection and relation to IVHM.

are used with time-domain data for finding repeating patterns, such as the presence of a periodic signal that has been buried under noise [Matlab 2004].

2.6.3 Frequency Domain Feature Extraction Techniques

In several IVHM applications, the frequency spectrum of a measured time signal contains considerable information that is useful for diagnosis [Eisenmann and Eisenmann 1998]. The engineer can study the spectrum to determine which frequencies are present in the input signal and which are missing.

In addition to frequency information, phase information is often needed. This can be obtained from the Fourier transform. With some applications, how the phase varies with frequency can be a significant consideration. A time-domain graph shows how a signal changes over time, whereas a frequency-domain graph shows how much of the signal lies within each given frequency band over a range of frequencies. A frequency-domain representation can also include information on the phase shift that must be applied to each sinusoid to be able to recombine the frequency components to recover the original time signal.

Signals are converted from the time to the frequency domain usually through a Fourier transform. The Fourier transform converts the signal information to a magnitude and phase component of each frequency. Often, the Fourier transform is converted to the power spectrum, which is the magnitude of each frequency component squared. The "spectrum" of frequency components is the frequency-domain representation of the signal [Matlab 2004]. The inverse Fourier transform converts the frequency-domain function back to a time function.

Fourier transformation is even more important for rotating machinery. For example, if the rotating machine is rotating with a base frequency of 100 Hz and has a four-bladed fan, then it will have a frequency component at 400 Hz. If this machine has a 12-toothed gear, then it will have a frequency component at 1200 Hz.

2.6.4 Time Frequency Extraction Techniques

Time-frequency analysis [Peng and Chu 2004] consists of those techniques that study a signal in both the time and frequency domains simultaneously, using various time-frequency representations. Rather than viewing a one-dimensional signal (a function, real or complex-valued, whose domain is the real line) and some transform (another function whose domain is the real line, obtained from the original via some transform), time-frequency analysis studies a two-dimensional signal—a function whose domain is the two-dimensional real plane, obtained from the signal via a time-frequency transform.

The mathematical motivation for this study is that functions and their transform representations are often tightly connected, and they can be understood better by studying them jointly, as a two-dimensional object, rather than separately. The practical motivation for time-frequency analysis is that classical Fourier analysis assumes that signals are infinite in time or periodic, while many signals in practice are of short duration, and change substantially over their duration.

One of the most basic forms of time-frequency analysis is the Short-Time Fourier Transform (STFT), but more sophisticated techniques have been developed: notably, wavelets.

The STFT is a Fourier-related transform used to determine the sinusoidal frequency and phase content of local sections of a signal as it changes over time.

In practice, the STFT is computed as a succession of Fast Fourier Transforms (FFTs) of windowed data frames, where the window "slides" or "hops" forward through time.

2.6.5 Wavelet Transforms

A wavelet is a mathematical function used to divide a given function or continuous-time signal into different scale components [Donald and Walden 2000]. One can usually assign a frequency range to each scale component. Each scale component can then be studied with a resolution that matches its scale. A wavelet transform is the representation of a function by wavelets. The wavelet transform is often compared with the Fourier transform, in which signals are represented as a sum of sinusoids. The main difference is that wavelets are localized in both time and frequency, whereas the standard Fourier transform is only localized in frequency. Wavelet transforms have advantages over traditional Fourier transforms for representing functions that have discontinuities and sharp peaks, and for accurately deconstructing and reconstructing finite, nonperiodic, and/or nonstationary signals. The STFT is more similar to the wavelet transform, in that it is also time and frequency localized, but there are issues with the frequency/time resolution trade-off. Wavelets often give a better signal representation using multi-resolution analysis [Mallat 1989], with balanced resolution at any time and frequency.

2.7 Case Study—Machine Fault Simulator Unbalance Localization

In this case study, a Machine Fault Simulator (MFS) has been used to detect and localize unbalance faults in rotating machinery. Unbalance as a fault is among the most commonly occurring faults in rotating machinery. All rotating machines contain an inherent amount of unbalance, and therefore they operate after balancing has been performed to bring the unbalance to within a given tolerance. The fault "unbalance" can therefore be considered as an unbalance occurring outside of this given tolerance. Such conditions can be caused by a wide variety of phenomena. Taking the example of a gas turbine, unbalance may be caused by a buildup of material on compressor or turbine disks, blade damage (ranging from small cracking/chipping to full blade-off), and incorrect/inaccurate maintenance/balancing procedures—to name but three examples.

Modern-day rotating machines operate with a high level of reliability, and yet the drive for ever-increased operation and decreased unscheduled maintenance is providing additional challenges for industry. The airline industry provides a current example of this desire, with airlines pushing manufacturers to enable shorter turnaround times and to keep aircraft in the air longer, increasing cost-benefit. Despite the high level of reliability, unbalance, along with other rotor-dynamic faults, remains an aspect that requires consideration in this drive for increased reliability and improved maintenance procedures.

The MFS can be seen in Figure 2.14.

To investigate unbalance, a single accelerometer has been used, with a system set up for data collection and classification. The flow

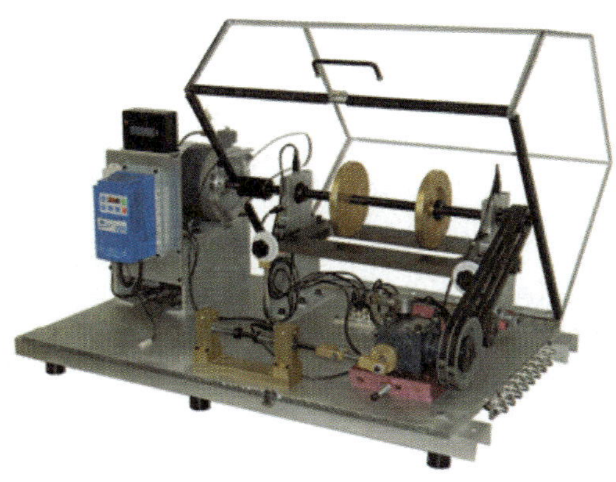

Figure 2.14 Spectraquest MFS.

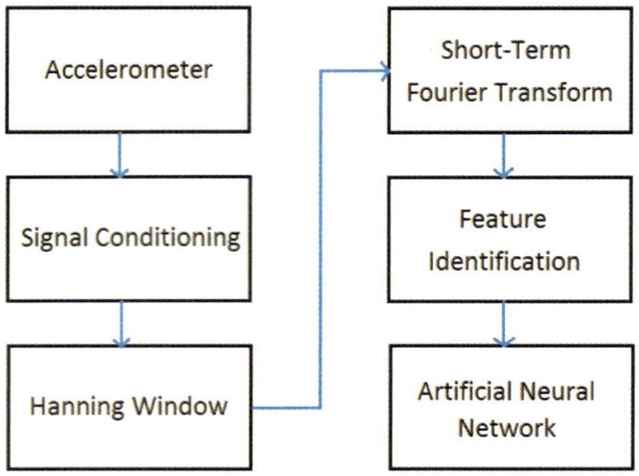

Figure 2.15 Flow of IVHM data.

of data can be described in a broad sense by referring to Figure 2.15.

This process can be further described as follows:

Sensor: A single accelerometer has been used in this case to identify and localize unbalance. The sensor is rated at 10 mV/*g*, and the calibration curve can be seen in Figure 2.16.

Signal Conditioning: The next stage involves the conversion of data from analog to digital, with signal amplification from the single accelerometer also occurring.

Sampling: Vibration data typically requires a high sampling rate to extract relevant features. In this case, a sampling rate of 25 kHz has been used. This ensures that the relevant features can be identified from the system.

Low-Pass Filter: At this point in the analysis, a low-pass frequency has been implemented. This enables the data range to be restricted to that which is to be studied. Numerous low-pass filters are available. However, for the case of diagnosing and localizing unbalance in the MFS, a fourth-order Butterworth filter [Butterworth 1930] has been applied, due to the smooth "roll off" rate of this filter.

STFT: The application of an STFT at this stage in the cycle enables the conversion from the time domain to the frequency domain. An example set of frequency-domain data can be seen in Figure 2.17. In this, the dominant "1X" (the speed of rotation) can be seen to occur at 5 Hz.

Averaging: Time-synchronous averaging enables a certain amount of variation to be removed, accounting for unwanted peaks and troughs caused by noise in the processed signal. It is for this reason that this stage has been implemented.

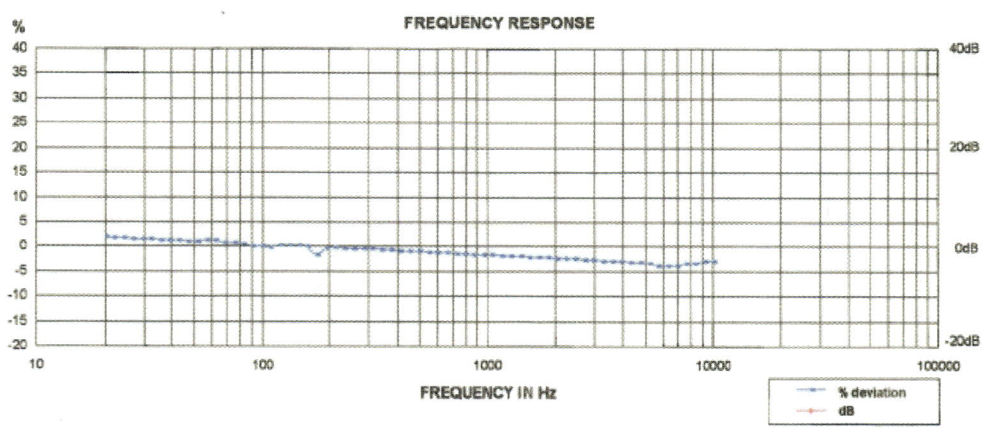

Figure 2.16 Accelerometer calibration chart for 98.76 mV/*g*.

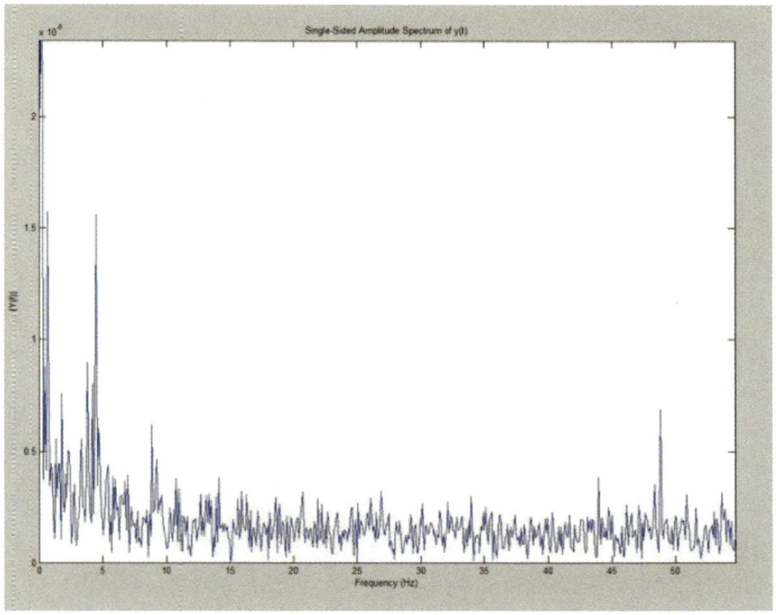

Figure 2.17 STFT applied frequency-domain data.

Normalization: In the case of faults in rotating machinery, the ability to look at the relative strengths of features in the frequency domain is important; the absolute values of amplitudes is less so. It is for this reason that a normalization stage has been implemented, as this allows for the relative peaks across the frequency spectrum to be studied for faulty conditions, providing unique information that may be disguised when looking at absolute amplitudes. Post averaging and normalized data can be seen in Figure 2.18.

Figure 2.18 also displays the differences between unbalance faults placed upon four different discs on the MFS. The processed

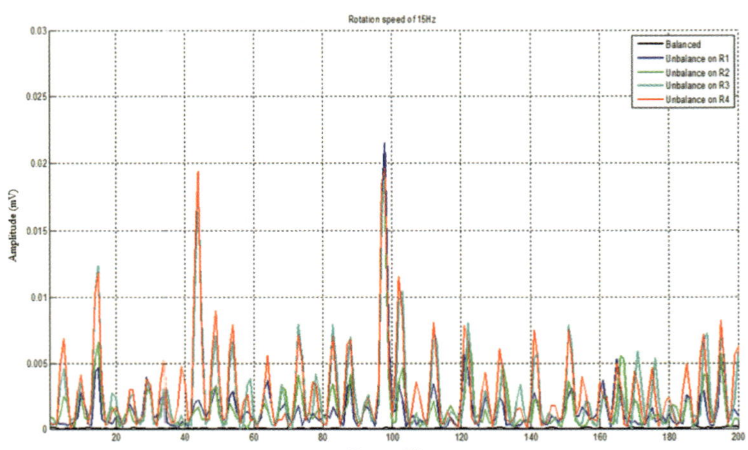

Figure 2.18 Prenormalized frequency spectrum for unbalance localization cases.

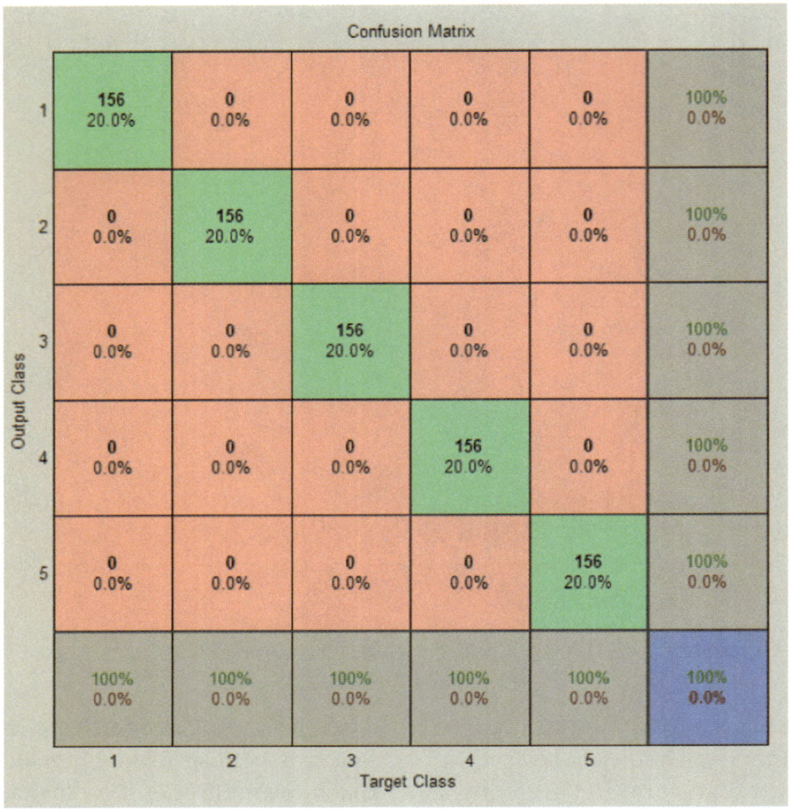

Figure 2.19 ANN applied for automated fault diagnosis and localization.

spectrum displays obvious differences between the cases, thus enabling fault diagnosis and localization to be performed.

Classification: A wide range of classification techniques is available. For this study, an Artificial Neural Network (ANN) has been applied, which enables automated fault diagnosis and localization. The ANN is trained using a set of known data, against which others can be tested (see chapter 3). An example matrix that details the success of training an ANN can be seen in Figure 2.19.

Through this process, unbalance faults can be accurately identified and localized across four locations on the MFS. This simple case study displays the importance of each step in the process, as without each section, 100% accuracy in fault detection and localization would not be possible. For further reading on this case study, refer to Walker, Perinpanayagam, and Jennions [2013].

2.8 References

Akers, A., M. Gassman, and R. Smith. 2006. *Hydraulic Power System Analysis*, Taylor & Francis, New York.

Butterworth, S. 1930. "On the Theory of Filter Amplifiers," *Experimental Wireless & the Wireless Engineer*, 536–541, Oct. 1930.

Cook, D. A., and Y. H. Berthelot. 2001. "Detection of Small Surface-Breaking Fatigue Cracks in Steel Using Scattering of Rayleigh Waves," *NDT&E International* 34: 483–492.

de Silva, C. W. 1989. *Control Sensor and Actuators*, Englewood Cliffs, NJ: Prentice-Hall.

Eisenmann, Sr., R. C., and R. C. Eisenmann, Jr. 1998. *Machinery Malfunction Diagnosis and Correction*, Englewood Cliffs, NJ: Prentice-Hall.

Engel, J., B. J. Gilmartin, K. Bongort, and A. Hess. 2000. "Prognostics, the Real Issues Involved with Predicting Life Remaining," *Aerospace Conference Proceedings*, Vol. 6, March 2000, pp. 457–469, Big Sky, MT.

Gao, H., Y. Shi, and J. L. Rose. 2005. "Guided Wave Tomography on an Aircraft Wing with Leave in Place Sensors," D. O. Thompson and D. E. Chimenti (eds.), *Review of Progress in Quantitative Nondestructive Evaluation*, Vol. 760, Golden, Colorado, American Institute of Physics, pp. 1788–1794.

Giurgiutiu, V., and J. Bao. 2004. "Embedded-Ultrasonic Structural Radar for in Situ Structural Health Monitoring of Thin-Walled Structures," *Structural Health Monitoring* 3(2): 121–140.

Harris, Fredric J. 1978. "On the Use of Windows for Harmonic Analysis with the Discrete Fourier Transform," *Proceedings of the IEEE* 66(1): 51–83, Jan. 1978.

Jones, R. V. 1988. *Instruments and Experiences*, John Wiley and Sons, London.

Leonard, K. R., E. V. Malyarenko, and M. K. Hinders. 2000. "Ultrasonic Lab Wave Tomography," *Inverse Problem* 18(6): 1795–1808.

Mallat, S. G. 1989. "A Theory for Multiresolution Signal Decomposition: The Wavelet Representation," *IEEE Transactions on Pattern Analysis and Machine Intelligence*, 11(7):674–693.

Manring, Noah D. 2005. *Hydraulic Control Systems*, John Wiley & Sons, Inc., Hoboken, NJ.

MATLAB Reference Guide. 2004. Cambridge, MA: MathWorks, Inc.

Peng, Z. K., and F. L. Chu. 2004. "Application of the Wavelet Transforming Machine Condition Monitoring and Fault Diagnosis: A Review with Bibliography," *Mechanical Systems and Signal Processing* 18: 199–221.

Percival, Donald B., and Andrew T. Walden. 2000. *Wavelet Methods for Time Series Analysis*, Cambridge University Press, ISBN 0-521-68508-7.

Ramsden, E. D. 2006. *Hall-effect sensors: theory and applications*, Second Illustrated Edition, Elsevier.

Shannon, C. E. 1949. "Communication in the presence of noise," *Proc. Institute of Radio Engineers*, Vol. 37, No. 1, pp. 10–21, Jan. 1949. Reprint as classic paper in: Proc. IEEE, Vol. 86, No. 2, Feb. 1998.

Turner, Anthony, George Wilson, and Isao Kaube. 1987. *Biosensors: Fundamentals and Applications*, Oxford University Press.

Walker, R., S. Perinpanayagam, and I. K. Jennions. 2013. "Localizing Unbalance Faults in Gas Turbines," *ASME Turbo Expo 2013*, San Antonio, TX, USA.

Wilcox, P. 2003. "Omni-Directional Guided Wave Transducer Arrays for the Rapid Inspection of Large Areas of Plate Structures," *IEEE Transactions on Ultrasonics, Ferroelectrics, and Frequency Control*, 50(6): 699–709.

Williams, J. 1977. "Designer's Guide to Temperature Sensors," EDN, May 5, 1977.

CHAPTER 3 Data-Driven Anomaly Detection and Diagnosis

Neil Eklund, *General Electric Global Research*

> "It is of the highest importance in the art of detection to be able to recognize, out of a number of facts, which are incidental and which vital."
>
> —Sir Arthur Conan Doyle

3.1 Introduction

An IVHM (Integrated Vehicle Health Management) system performs three major functions: anomaly detection, diagnosis, and prognosis. *Anomaly detection* (also known as *fault detection*) is the detection of unusual system behavior. *Diagnosis* (also known as *fault isolation*) is the process of determining the root cause of the unusual behavior. Fault *prognosis* (discussed in depth in the next chapter) estimates the remaining time before the fault causes the system or subsystem to fail.

The goal of anomaly detection is to find, as early as possible, patterns in data indicating deviation from normal system behavior. Given that anomalous system behavior has been detected, it may be diagnosed as benign, as one or more underlying fault states, or as unknown (Figure 3.1). The "anomalous but benign" state may arise, for example, as a result of system maintenance (e.g., a gas turbine water wash), and it is often desirable to suppress reporting of these events. Sometimes a very specific fault state is indicated (e.g., "hydromechanical fuel control unit bypass valve stuck closed"). However, note that occasionally several fault states may be indistinguishable based on the existing data, and the system may require physical inspection to really isolate the fault (e.g., vibration in rotating machinery may come from any number of root causes). The "unknown" state acts as the catch-all for when the system is known to be behaving unusually but there is not enough evidence to classify the anomaly as benign, or one of the faulty states. This might arise when machinery is exposed to conditions that are novel and were never anticipated when developing the IVHM system. Without the unknown state, many classifier systems will incorrectly allocate the state to a known fault. When an unknown diagnosis arises, engineers must examine the data closely and apply domain knowledge to understand the nature of the fault.

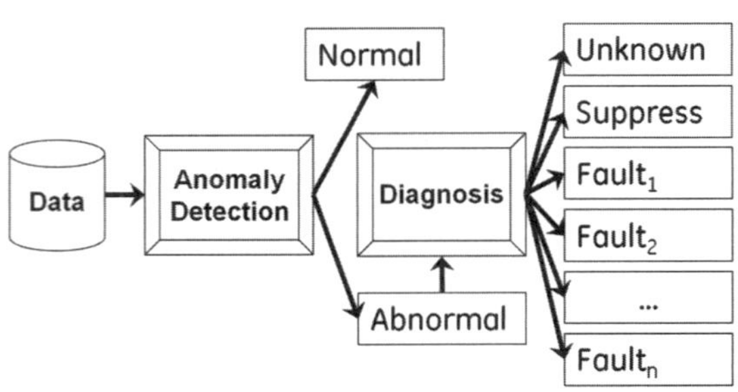

Figure 3.1 Schematic of anomaly detection and diagnosis.

This chapter is concerned with data-driven fault detection and isolation. Practical issues around the use of machine learning techniques to build models and classify systems are discussed. Additionally, several useful machine learning techniques are outlined.

3.2 System Modeling and Residuals

Operating mode and environmental conditions often have a pronounced effect on the behavior of systems. For example, consider starting a car on a very cold day; in comparison to a warm day, the oil is much more viscous, and the chemical reaction that produces current in the battery is slower, reducing the amps that can be delivered to the starter. Data from cold-day starts and hot-day starts will be very different, and the spread on a statistic such as "time between initial key turn and engine catching" will be enormous.

Models of system performance help to account for environmental effects and operating mode. Performance can be estimated using a high-fidelity physics-based model (e.g., a gas turbine cycle deck), an entirely data-driven model (e.g., a neural network), or a hybrid model incorporating elements of both techniques. Models use as inputs environmental data (e.g., ambient temperature) and command data (e.g., throttle setting) to predict system performance (e.g., temperatures, pressures, speeds, and flows). Data from "upstream" sensors may also be incorporated to predict "downstream" performance (e.g., high-pressure compressor discharge temperature can be one of the features used to predict exhaust gas temperature in a gas turbine engine). Non-parametric data can sometimes be useful for increasing the fidelity of the model. For example, consider a fleet of assets operated by two different owners with dramatically different maintenance programs. By encoding the asset owner—and, by proxy, maintenance policy—into the model, additional variance can be accounted for in the data.

Rather than operating on the raw data, *residuals*, the difference between the raw data and the modeled output of the system, are typically used for anomaly detection and fault isolation. Using a model of the system to remove sources of variance in the data that can be accounted for by environmental effects and command settings makes unusual system behavior easier to detect.

3.3 Machine Learning

3.3.1 Introduction to Machine Learning

Machine learning is concerned (among other things) with building models that associate a certain label to particular input data. One useful example is *classification*, where the label is of two or more classes (e.g., normal or abnormal). Another example is *regression*, where a real-valued output is assigned to each input. The goal of all machine learning models is to generalize well (i.e., while the model was trained on a particular set of data, it should perform accurately on previously unseen data).

The inputs to machine learning models are called *features* (often called "variables" or "parameters"). Features may be raw data, or data that has been transformed to make it more informative. For example, body mass index is the ratio of a person's mass in kilograms to the square of their height (in meters). This ratio has been demonstrated to be much more useful in predicting heart disease and other medical outcomes than height and weight alone. Another example is converting sampled data from the time domain to the frequency domain via the Fourier transform.

Another reason to transform data is to convert the data to be better able to be "understood" by an algorithm (there are not many algorithms that will take the text input directly—most require numeric input). Consider an automobile model with three levels of engine—economy, base, and sport. One might represent the three levels numerically with two binary features, for which [0 0] represents the economy, [0 1] represents base, and [1 1] represents sport.

Note that some care needs to be put into numeric transformations. One might be tempted to convert, say, months into integers directly (January is 1, February is 2, etc.). However, June is not twice March! For cases in which there is a cyclic relationship (mostly, time), encoding it as

$$\left[\sin\left(\frac{2\pi(time)}{\max(time)}\right) \quad \cos\left(\frac{2\pi(time)}{\max(time)}\right)\right] \tag{3.1}$$

the order and distance properties are preserved, but inappropriate magnitude information is removed.

Frequently, there are irrelevant or redundant features in the data. Eliminating these extraneous features makes models more interpretable, reduces training time, and makes models generalize better to unseen data. There are numerous feature selection approaches in the literature [Guyon and Elisseeff 2003]. Additionally, features can be selected based on domain knowledge, but in many instances (particularly for complex systems), models built with algorithmic feature selection outperform models built from features selected using domain knowledge alone.

3.3.2 Supervised and Unsupervised Learning

Machine learning algorithms can be classified as *supervised* or *unsupervised*. Supervised approaches rely on having well-labeled training data, while unsupervised models do not require any labels—instead, they make the assumption that the bulk of the training data is "normal."

Diagnosis requires supervised models. This can be problematic for several reasons. First, engineered systems are typically quite reliable—for the most part, ships stranded at sea or aborted aircraft takeoffs are quite rare—so fault data can be very difficult to come by. Often, there are millions of records of normal data (known as the *majority class*), and only a handful of faulty records (known as the *minority class*); this problem is known as *class imbalance*. The problem with class imbalance is that it tricks the models into labeling everything as "normal," as that is the correct answer most of the time.

An effective heuristic for addressing class imbalance is to down-sample the majority (nominal) class, such that the number of normal and faulty instances is approximately the same in the training data. An additional level of sophistication is to perform stratified sampling of the majority class such that the entire operational space is represented proportionately. Finally, one can often partially train a model on one realization of stratified sampling, and then train it further on another realization, and so on, such that the train model "sees" a greater fraction of the data in different combinations.

Another common issue is that labeling the fault cases in "real life" can be difficult. When exactly does a fault initiate? Given multiple time series of recorded data, with each series ending in a failure (or a known fault that did not result in failure), it is often difficult to determine the exact onset of a fault. Ask three engineers, you may get five answers. To overcome this problem, it is often effective to label n samples at the end of a series as faulty, where n is agreed upon by domain experts. Then n-1 to n-b data points can be discarded, where b is some number that is larger than the possible onset time by a wide margin. The remaining points can be labeled "normal" (Figure 3.2) for training a classifier.

Both supervised and unsupervised methods can be used for fault isolation, and unsupervised methods can be used directly for anomaly detection. In the absence of labels, supervised methods can be used *for anomaly detection only*. Given operational data that is known to be largely nominal, minority class data can be created either by sampling the marginal distribution of the majority class, or by uniformly sampling the range of the majority class (Figure 3.3). The sampled data is labeled as a different class (the minority class) from the real data, a classifier is trained to distinguish the two, and the sampled data is discarded. An anomaly is detected when the resulting trained classifier indicates that it detects the minority class.

3.3.3 Unsupervised Methods

Unsupervised methods operate without labels. There might be faults and outliers in the data, but their location is unknown. There are two fundamental tasks for unsupervised learning: clustering—grouping objects such that objects in the same cluster are more similar to each other than to objects in other clusters; and dimensionality reduction—projecting a high dimensional space onto a smaller dimension (often two).

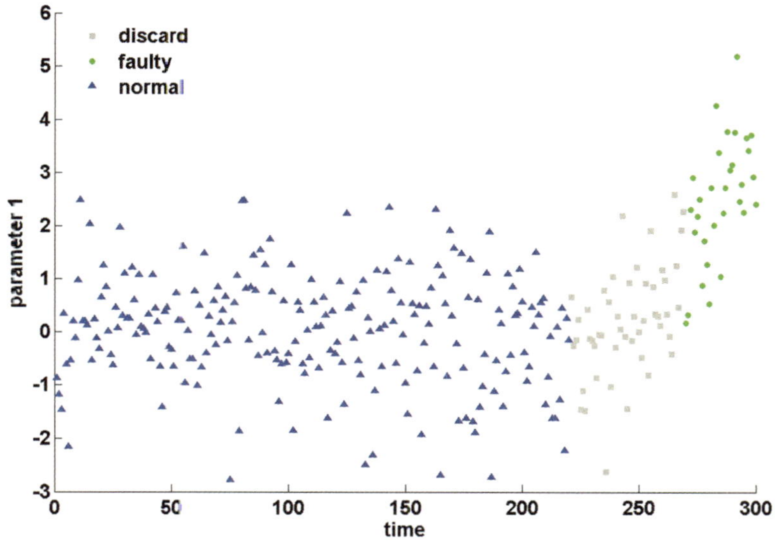

Figure 3.2 Heuristic for labeling data for classifier training when the onset of the fault is unclear.

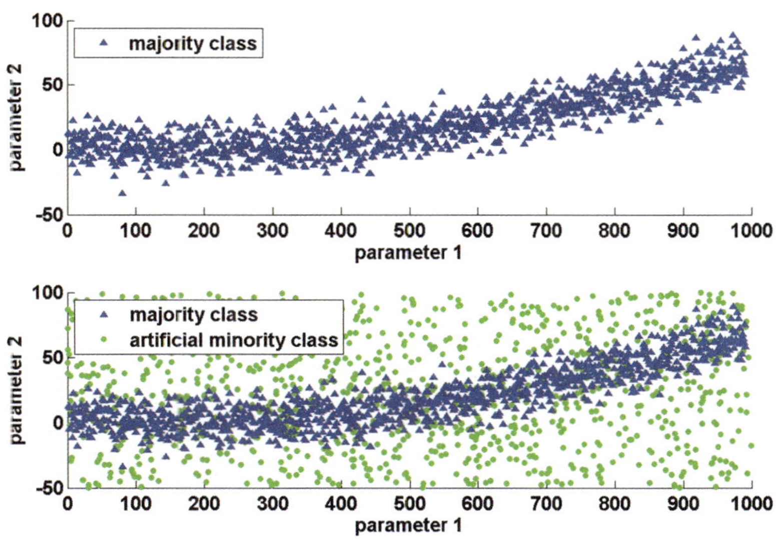

Figure 3.3 Using artificial minority class data to train a supervised algorithm to do anomaly detection. Top panel: raw data. Bottom panel: raw data with artificial minority class.

There are *many* ways to do clustering and dimensionality reduction. An example of each method—k-means clustering and self-organizing maps—is presented in the following paragraphs.

Clustering is often used in an IVHM context for identifying different operational modes of a system. For example, Figure 3.4 is a three-dimensional plot of some real data from complex thermomechanical machinery,

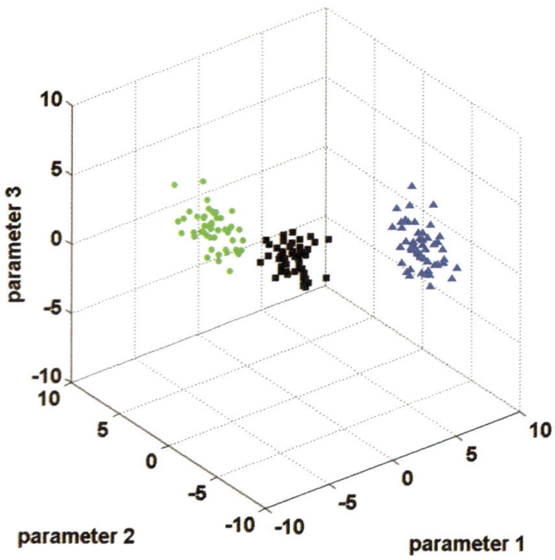

Figure 3.4 K-means clustering for operational data. Note the three clear operational modes.

colored by cluster, and found using the k-means algorithm (described in section 3.3.3.1). Note that there are three clear operational modes in the data. One might wish to model system performance of each mode (using a supervised method) separately. This is a toy example, as it might be solved visually by just plotting the data in only three dimensions. However, in the case where it takes, say, eight parameters to characterize the operational state of a system, automated clustering becomes critical.

Dimensionality reduction is useful for visualizing high-dimensional data. For example, Figure 3.5 shows the result of a self-organizing map (described in section 3.3.3.2) applied to some real data from complex thermomechanical machinery. On its own, the data is difficult to interpret. But projected down to two dimensions, it is easy to see that there is some structure in the data. The top left plot shows the u-matrix (distance between nodes), and the subsequent plots show the distribution of different parameters. Note how easy it is to assess the distribution of parameters within each region—correlations are clear to see. Additionally, the u-matrix can be used to pick out potential clusters of similar behavior for further investigation.

3.3.3.1 K-Means Clustering

K-means clustering is a method of data exploration—a way of looking for patterns or structure in the data that are of interest, without having labels in advance. Clusters

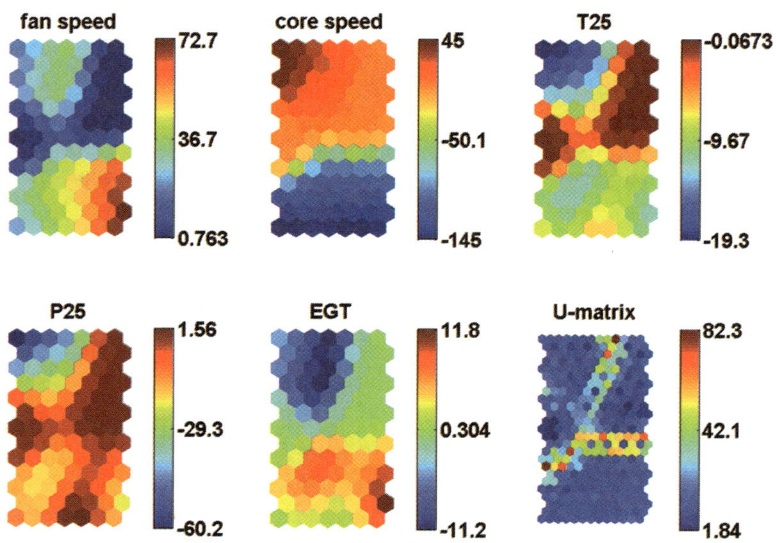

Figure 3.5 A self-organizing map for operational data.

are a collection of objects similar to one another within the same cluster, and dissimilar to the objects in other clusters. Typical applications of clustering are as a stand-alone tool to get insight into data distribution, or as a preprocessing step for other algorithms.

For comparing different ways of clustering the same data, some kind of cost function for the clustering algorithm is necessary. The goal of clustering is to try to minimize such cost functions. Generally, cost functions depend on some quantities, such as the centers of the clusters, the distance of each point in a cluster to the cluster center, or the average degree of similarity of all points in a cluster. Cost functions are algorithm specific, so comparing the results across different clustering algorithms is difficult. One common metric is Euclidian distance.

The k-means algorithm:

1. pick k random points to be cluster centers (centroids)
2. compute the distances between data and centroids
3. assign data to nearest centroid
4. move centroid to mean of cluster
5. If cluster's centroids are stable or some other stopping criterion is achieved, stop algorithm; otherwise, repeat steps 2–5.

Figure 3.6 depicts the first six iterations (after which stability is reached) of k-means clustering for k = 3. Note that there is an element of chance to this algorithm. If poor cluster centers are chosen by chance, unusual clusters can result (Figure 3.7).

K-means has problems when clusters are of differing sizes, densities, or of non-globular shapes; the technique also has problems with outliers. One solution is to use many clusters. Although each cluster has less meaning, the pattern of data is better represented.

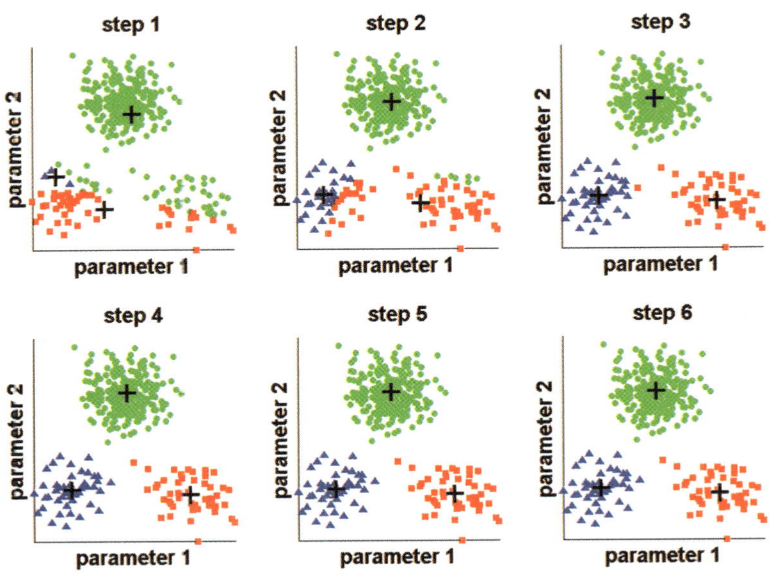

Figure 3.6 The first six iterations of k-means clustering for k = 3. The + symbols represent the cluster centers, and the colors represent the c uster assignment.

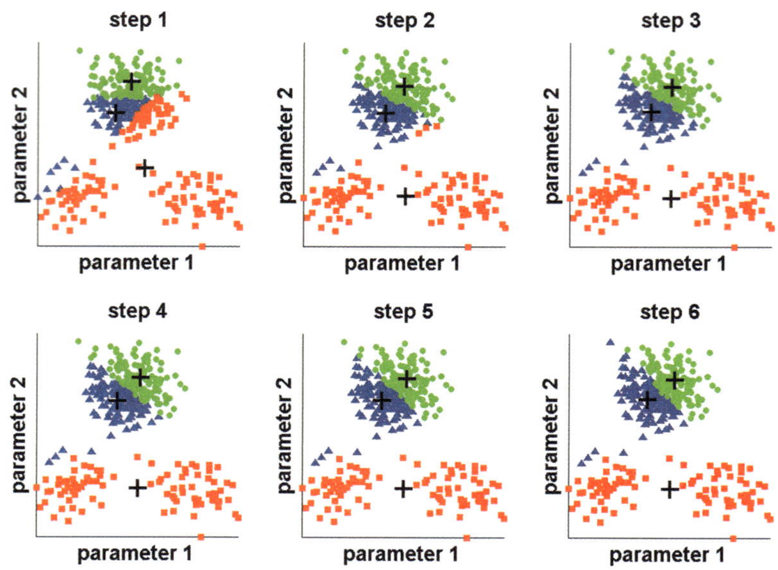

Figure 3.7 Another realization of k-means, with an unfortunate choice of initial cluster centers.

The strengths of k-means are:

- It is relatively efficient.
- It often terminates at a local optimum.
- A global optimum may be found using evolutionary algorithms or simulated annealing.

The weaknesses of k-means are:

- Requires interval or ratio data
- Need to specify k, the number of clusters, in advance
- Susceptible to noisy data and outliers
- Not suitable to discover clusters with non-convex shapes
- Works best when the clusters in the data are of approximately equal size
- Attribute significance cannot be determined

3.3.3.2 Kohonen Self-Organizing Maps

The Self-Organizing Map (SOM) was invented by Tuevo Kohonen [1990], inspired by microelectrode studies with cats that demonstrated that brain cells tuned to similar orientations tend to be physically located in proximity with one another. SOMs are an artificial neural network mapping high-dimensional data into (typically) a two-dimensional representation space, with similar cases in neighboring regions. SOMs use artificial processing units known as nodes, which are connected to form networks that mimic, at an abstract and *very* simplified level, the function of the brain.

The SOM algorithm performs vector quantization by adjusting the weights from common input nodes to M output nodes. Continuous valued input vectors are presented without specifying the desired output (i.e., there is no supervised learning). After the learning, weight will be organized such that topologically close nodes are sensitive to inputs that are physically similar. The algorithm is:

1. initialize network
2. get input
3. find winning neuron

4. update winning neuron
5. update neighborhood
6. adjust neighborhood size
7. goto step 2

Initialization consists of creating a set of units i in a two-dimensional grid; each unit is assigned a weight vector m_i, which is the same dimension as the input data (Figure 3.8). The initial weight vectors are assigned random values.

Input vectors $x(t)$ are selected from the training set at random. Output neurons are ranked by activity level $c(t)$,

$$c(t): \min_i \{\|x(t) - m_i(t)\|\} \tag{3.2}$$

Initially, some output nodes will randomly be a little closer to some particular type of input; these nodes become "winners," and their weights are adjusted to move them even closer to the inputs. For example, in a toy case of two output nodes, with weights [2 4] and [5 3], competing for the input vector [1 2], $c(t)$ is 2.2 and 4.1, respectively, so the winning node would be [2 4]. Over time, nodes in the output layer become representative prototypes for examples in the input. The magnitude of the neighborhood adaptation is controlled via a learning parameter α, which decays over time:

$$m_i(t+1) = m_i(t) + \alpha(t) \cdot h_{ci}(t) \cdot [x(t) - m_i(t)] \tag{3.3}$$

Using (typically) a Gaussian neighborhood-kernel, h_{ci},

$$h_{ci}(t) = \exp\left(-\frac{\|r_c - r_i\|^2}{2 \cdot \delta(t)^2}\right) \tag{3.4}$$

where $\|r_c - r_i\|^2$ denotes the distance between the winner node c and input vector i. The time-varying parameter δ enables formation of large clusters in the beginning and fine-grained input discrimination toward the end of the learning process, while the gradual lowering of the learn rate ensures stability.

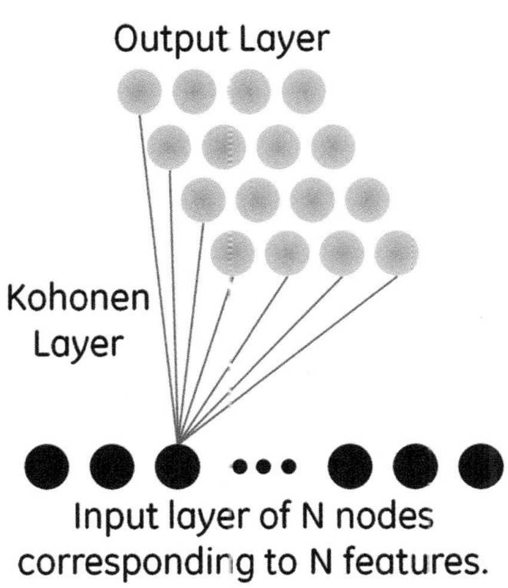

Figure 3.8 Cartoon of SOM. Input nodes are at the bottom, output nodes on the top. Lines represent m_i.

Given a trained SOM, new input vectors are allocated to the winning node. Figure 3.9 shows the weight vectors for an untrained and (one realization of) a trained SOM using the red, green, and blue (RGB) components of randomly generated color data. Note that the initial output is perfectly homogeneous, due to the random nature of initializing m_i. However, the trained weights are clustered by color, and similar colors (e.g., purples) are adjacent, while dissimilar colors (e.g., red and green) are further apart on the map.

The strengths of SOMs are:

- Reduction of dimensionality makes data easier to interpret.
- Can accommodate very large data sets.
- Fast to train and execute, low computational overhead.
- Demonstrated on large, complex data sets.

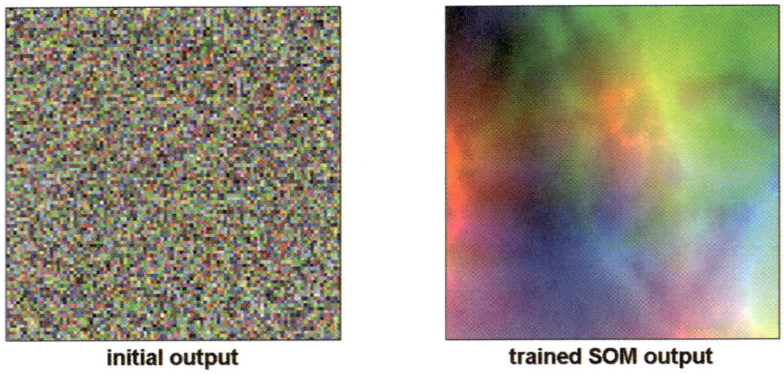

Figure 3.9 Output nodes for color data. Left is untrained SOM; right is trained SOM. Note the clustering of colors, and the proximity of like colors.

The weaknesses of SOMs are:

- Requires sufficient data to form clusters, so rare events may not be well clustered.
- Unimportant features add noise to clustering; often making different SOMs for different subsystems is an effective approach to overcome this.
- The orientation of the weights changes randomly from run to run, making it difficult (not impossible) to compare even two SOMs trained on the same data.

3.3.4 Supervised Methods

3.3.4.1 Introduction

Except in the case of anomaly detection, as just noted, supervised methods require labels. There are two fundamental tasks for supervised learning: classification—where objects are grouped into two or more classes (discrete labels), and regression—mapping inputs to continuous labels.

Figure 3.10 is a typical application of regression. The ambient conditions and command data (*X*) and plant output (*Y*) are recorded in a database. When sufficient data is available, a model can be trained using the historic

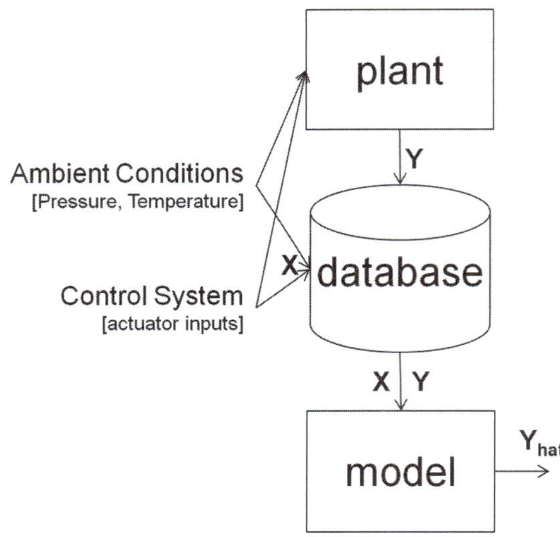

Figure 3.10 Supervised learning for regression.

*X*s and corresponding *Y*s. The output of the model—an estimate of *Y*—is known as Y_{hat}.

Given a trained model, the difference between *Y* and Y_{hat} are known as residuals. If residuals are close to zero, then the plant is running in a manner similar to how it has in the past. If residuals are non-zero, then something is anomalous—possibly a fault.

Supervised learning is used as a classifier, typically for fault isolation. If a high-fidelity physics-based model of the system is available, and engineers can agree how to model a fault (often the greatest challenge), then artificial data can be created to simulate fault events. However, more typically, a small amount of real fault data for a limited number of fault modes collects in the database (Figure 3.10), and that is used to train a classifier using data labeled with the approach outlined in section 3.3.2.

3.3.4.2 Decision Trees

Decision trees [Breiman et al. 1984] are a simple modeling method where data is partitioned through a sequence of decisions to arrive at a class assignment. Decision trees can perform either classification or regression.

Figure 3.11 is a decision tree for heart disease risk. It is evaluated from the top down. At the root node, all cases are split depending on the value of minimum systolic blood pressure. If it is less than 91, they are classified as low risk; otherwise, the cases are evaluated in the next node, looking at age. The cases continue to be partitioned on yes/no questions until all of them have been classified.

The algorithm for creating a decision tree is:

1. for each feature, find the value that minimizes the Gini impurity, I_G, for all cases in the node:

$$I_G(f) = 1 - \sum_{i=1}^{m} f_i^2 \tag{3.5}$$

   where there are m classes, and f_i is the fraction of cases labeled class i in the data set
2. choose the combination of feature and split value that minimizes impurity, and split the data accordingly
3. until some stopping criterion is met (e.g., max depth of tree), goto 1

The strengths of decision trees are:

- Very simple to interpret—it's easy to track the manner in which a decision was arrived

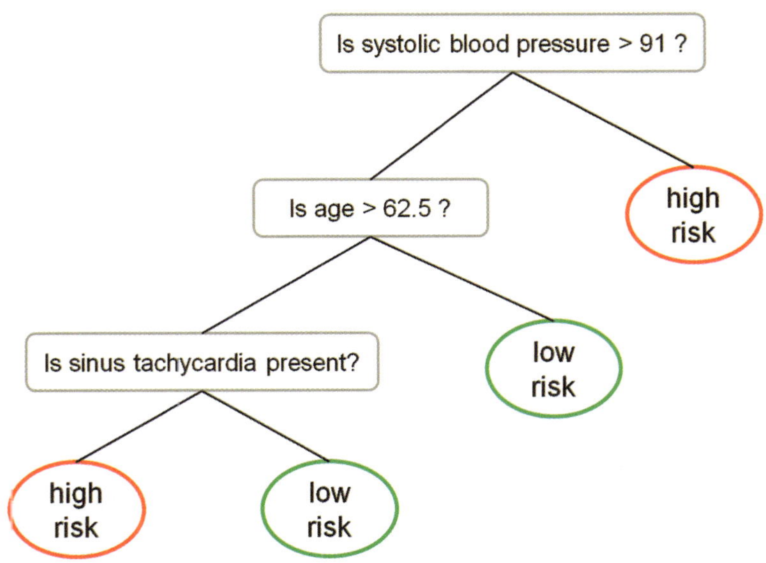

Figure 3.11 Sample decision tree for heart disease risk.

- Accommodates categorical data, unlike many methods
- Very fast to train and evaluate

The weaknesses of decision trees are:

- It is a greedy algorithm, so not necessarily optimal
- Prone to overfitting (modeling the training data too closely) if care is not taken; a technique called pruning can address this (see [Breiman et al. 1984])
- Not complex enough to represent some problems

3.3.4.3 Random Forests

Random forests (RF) are tree-based algorithms invented by Leo Breiman [2001]. To this day, they remain one of the most accurate supervised learning approaches. RF can be used for classification or regression.

RF works by partitioning the data through multiple, uncorrelated weak classifiers. Averaging across all weak classifiers minimizes model variance, resulting in very accurate predictions. The RF algorithm is:

1. Given N cases and M features, choose the number of features to sample at each node, $m << M$ (a good default value for m is $\sqrt{M}$).
2. while the number of trees is less than the desired number
   a. create a bootstrap sample (sample $n = N$ cases from all N available cases, with replacement).
   b. using the bootstrap sample, select m features randomly and calculate the best split as with a decision tree.
   c. goto step b until all the cases have been allocated to a leaf node (grow the tree all the way out—*do not* use pruning).

Note that there are two sources of randomness built into this approach. First, at each node in every tree, *m random* features are chosen to find the best split. Second, a random sampling of data is used to build each tree within the forest. Thus, each tree within the forest will be poorly correlated to all of the other trees.

As a byproduct of creation, random forests produce an estimate of generalization error, eliminating the need to use cross validation *for non time-series data only*. It does this by applying each tree to the samples that were not selected to be used in the bootstrap sample. This estimate is known as "out of bag error." Additionally, RF reports a metric of variable importance (determined by permuting the features) which can be used very effectively for feature selection.

The strengths of RFs are:

- It is one of the most accurate algorithms available for classification and regression [Caruana, Karampatziakis, and Yessenalina 2008].
- It takes relatively little time to train compared to other algorithms with similar performance, and is embarrassingly parallel (i.e., no effort is required to separate tree construction into parallel tasks).
- It can accommodate very large feature sets without feature selection.
- It provides an estimate of feature importance and generalization error.
- Some implementations can accommodate missing data.
- It accommodates class imbalance.
- It can provide a measure of proximity that accounts for categorical data; this measure can be used for outlier identification and clustering.

The weaknesses of RFs are:

- As with many other modeling approaches, RF is a "black box"—it is virtually impossible to interpret exactly in what manner decisions were arrived.

- Care has to be taken when using feature importance scores for categorical features with many levels; good implementations of the random forest algorithm account for this.
- In the presence of highly correlated features, the feature ranking can sometimes be misleading [Tolosi and Lengauer 2011].

3.3.4.4 Kernel Regression

Kernel Regression (KR) is a nonparametric technique that uses the similarity of an input pattern to reference patterns to generate an estimate via weighted average [Nadaraya 1964; Watson 1964; Atkeson et al. 1997]. KR can be used for both regression and classification problems.

Although KR can be performed using an unfiltered reference data set, it is computationally more efficient to algorithmically select exemplars (e.g., using the technique outlined by Hines and Garvey [2006]). Also, the data for KR must be mean centered (i.e., subtract the mean from each feature) and scaled (divide each mean-centered feature by its standard deviation).

Given n observations of p features ($\mathbf{X}$) and r outputs ($\mathbf{Y}$) for use as exemplars,

$$\mathbf{X} = \begin{bmatrix} X_{1,1} & \cdots & X_{1,p} \\ \vdots & \ddots & \vdots \\ X_{n,1} & \cdots & X_{n,p} \end{bmatrix} \quad \mathbf{Y} = \begin{bmatrix} Y_{1,1} & \cdots & Y_{1,r} \\ \vdots & \ddots & \vdots \\ Y_{n,1} & \cdots & Y_{n,r} \end{bmatrix} \tag{3.6}$$

and an observation of the p inputs, to get an estimate of the r outputs:

$$\mathbf{x} = \begin{bmatrix} x_1 & \cdots & x_p \end{bmatrix} \tag{3.7}$$

To quantify the similarity of the input pattern to each exemplar, Euclidean distance is typically used. To calculate the distance d_i of an input pattern to the ith exemplar:

$$d_i = \sqrt{\left(X_{i,1} - x_1\right)^2 + \cdots + \left(X_{i,p} - x_p\right)^2} \tag{3.8}$$

This distance must then be converted to a similarity metric (i.e., large distances must become small weights and small distances must correlate to large weights). The Gaussian kernel function [Fan and Gijbels 1996] is frequently used to calculate weight (or similarity) w_i of the input pattern to the ith exemplar:

$$w_i = e^{-\frac{d_i^2}{2h^2}} \tag{3.9}$$

where h is the kernel bandwidth—a measure of how close the input has to be to an exemplar to be deemed similar. Relatively large bandwidths are more liberal about what is considered similar and assign non-trivial weights to more exemplars, while relatively small bandwidths do the opposite. Thus, large bandwidths produce a very smooth input-output mapping (although may miss some subtleties); small bandwidths result in a more jagged mapping. For centered and scaled data, a bandwidth of 0.5 is often a good place to start.

Once weights are calculated, estimating outputs is straightforward. For kernel regression, the estimates are simply weighted averages of the exemplar patterns. For n exemplars, and the sum of all weights a, the estimates of the r outputs are calculated by

$$\hat{y} = \begin{bmatrix} \frac{1}{a}\sum_{i=1}^{n} w_i Y_{i,1} & \cdots & \frac{1}{a}\sum_{i=1}^{n} w_i Y_{i,p} \end{bmatrix} \tag{3.10}$$

For classification rather than regression, the exemplar outputs are simply binary encodings of the classes (e.g., in a two-class classifier, the exemplar outputs would be a matrix with two columns, the first with a 1 for every exemplar of the first class and 0 for every exemplar of the second class, and the second with a 0 for every exemplar of the first class and 1 for every exemplar of the second class). The class of the input vector is assigned to the majority class of the weighted output.

The strengths of KR are:

- Easy to optimize performance, as there is only one parameter to adjust, kernel bandwidth
- Good at estimating nonlinear data

The weaknesses of KR are:

- The performance of KR is sensitive to selecting good exemplars.
- Can take a lot of computing resources, particularly if there are a lot of exemplars.
- Sensitive to outliers, particularly during the centering and scaling step.

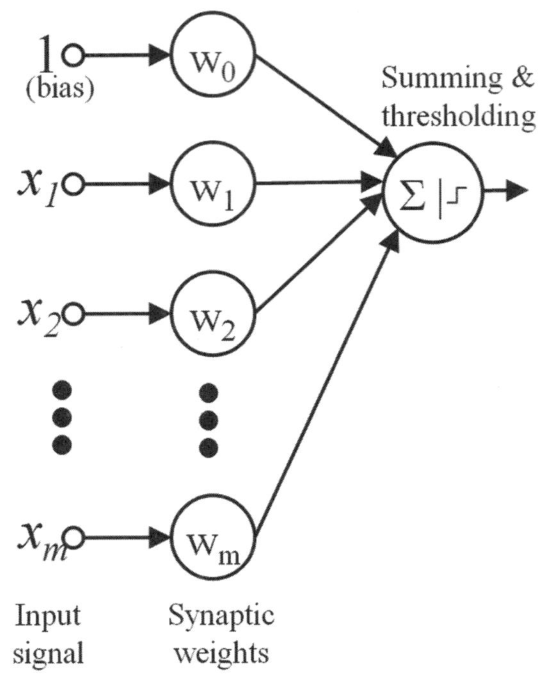

Figure 3.12 Schematic of an artificial neuron.

3.3.4.5 Feed-Forward Artificial Neural Networks

Feed-forward artificial neural networks, or just neural networks (NN) for short, are a biologically inspired modeling approach. McCullough and Pitts [1943] originally conceived of the idea, but it was not a practical approach for over four decades until Rumelhart, Hintont, and Williams [1986] invented a practical technique for training them.

Like SOMs, NNs use artificial processing units known variously as "neurons," "nodes," or "processing elements," which mimic, at a *very* simplified level, the function of the brain. NNs can be used for both regression and classification, and like kernel regression, benefit from centering and scaling the data beforehand.

Each neuron is quite simple (Figure 3.12). A neuron has $m + 1$ inputs, for signals x_0 through x_m, and weights w_0 through w_m. The x_0 input is fixed at one, and w_0 is known as "bias." The input to a neuron, u, is

$$u = \sum_{i=0}^{m} w_i x_i \tag{3.11}$$

and the output, y, of a neuron uses a sigmoid transfer function, typically

$$y(u) = \frac{1}{1+e^{-u}} \tag{3.12}$$

Neurons are arranged in *layers*. There is an input layer, one (typically, or at most two for very complex problems) hidden layers, and an output layer (Figure 3.13). The number of nodes in the hidden layer is selected, essentially, via trial and error—so long as there are enough nodes, performance is not *too* sensitive to this parameter, particularly if some sort of regularization is employed when tuning the weights. A simple but effective heuristic for the number of hidden nodes is "prime numbers smaller than the number of input nodes." If performance of a trained network *on the training data* is poor, try adding more neurons. If performance of a trained network *on the testing data* is poor, try removing some neurons.

The classic way to train the weights is using the backpropagation algorithm, or some variation on it:

1. initialize all weights to small random values
2. choose each case in turn; apply it to the input layer (one series of all cases is known as an *epoch*)
3. propagate the signal forward through the network
4. compute the deltas—the difference between the current output and the desired signal—for the output layer
5. compute the deltas for the preceding layers by propagating the error backward
6. update all weights in layer j and node i such that deltas would be diminished
7. while the error rate is unacceptable, goto 2

Each software package will have different options for the training algorithm. The tradeoffs are generally on speed and computing resources (i.e., memory).

Neural networks are prone to *overfitting* the training data, in which the network fits the training data near perfectly but generalizes poorly to novel data. There are two fundamental approaches to preventing overfitting. *Early stopping*, in which part of the training set is segregated into a *validation set*, and training is stopped when the performance on the validation set begins to decrease; and *regularization*, in which the weights are penalized for becoming large during training, which ensures smooth output. If regularization is used, a network is much less sensitive

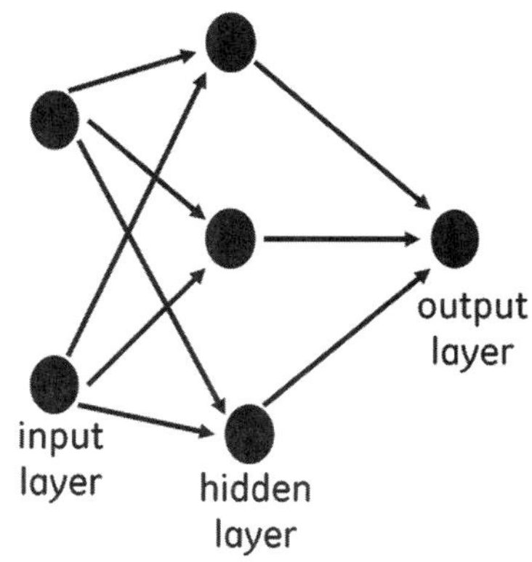

Figure 3.13 Layers in a feedforward network. Circles are nodes; lines are connections.

to the number of hidden nodes, provided it is sufficient.

The strengths of NNs are:

- Good at solving complex, nonlinear problems with interactions between the features.
- Very compact representation—low storage requirement for a trained network.
- Can approximate *any* function.

The weaknesses of NNs are:

- As with many other modeling approaches, NN is a "black box"—it is impossible to interpret the exact manner in which decisions are arrived.
- Care has to be taken to prevent overfitting the training set.
- Large computational cost to training.

3.3.5 Performance Evaluation Metrics

The performance of diagnostic systems can be analyzed using a *confusion matrix*. Figure 3.14 is a cartoon confusion matrix that shows the four possible outcomes for a two-class problem (such as anomaly detection). When the predicted positive class matches the

True class	Predicted class	
	positive	negative
positive	true positive	miss
negative	false alarm	correct rejection

Figure 3.14 A cartoon two-class confusion matrix showing the jargon used to describe the four outcomes.

actual positive class, it is known as a *true positive* (or a *hit*); when the predicted positive class matches the true negative, it is known as a *false alarm*. When the predicted negative class matches the actual positive class, it is known as a *miss*; when the predicted negative class matches the true negative, it is known as a *correct rejection*.

The same approach can scale up to multiple faults, even if a series of binary (two-outcome) classifiers is used. For example, Table 1.1 is a confusion matrix for a five-class system. As with the two-class case, it is ideal to be on the diagonal—all true positives. Off-diagonal elements provide insight as to how the system is making mistakes. For example, in Table 1.1, Fault 2 is often confused with "normal" when misdiagnosed.

Often the cost of making a misclassification can be calculated (or even estimated), and is frequently asymmetrical (e.g., in the case of cancer diagnosis, the cost of a miss can be much higher than a false alarm). If the cost of making correct classifications and errors is available, a confusion matrix can be used to optimize classifier thresholds by multiplying the cost (or gain, for correct classifications) of each cell by the count, and summing across the entire matrix. Using a non-gradient optimization method such as an evolutionary algorithms (or, realistically, often via trial and error), the overall cost can be minimized.

Receiver Operating Characteristic (ROC) curves, developed during the second World War to analyze performance of radar operators, are a common way to visualize the performance of a two-class classifier (e.g., an anomaly detection system). An ROC curve is constructed by varying the threshold on a classifier from the minimum to the maximum, and plotting the resulting relationship between true positive rate against the false alarm rate (Figure 3.15). A diagonal line indicates chance performance, while proximity to the upper left corner (0.0 false alarm rate, 1.0 true positive rate) indicates progressively better performance.

ROC curves allow the performance of a classifier to be assessed quickly. For example, classifier 2 in Figure 3.15 can detect about 40% of faults at about a 1% false alarm rate, and to detect 90% of faults at a 40% false alarm rate. ROC curves also make it possible to compare multiple classifiers. In the same figure, classifier 2 is superior to classifier 1 below about 35% false alarm rate; however, if the cost of false alarms was low, and it was imperative

	Normal	Fault 1	Fault 2	Fault 3	Fault 4
Normal	**749**	18	54	6	1
Fault 1	5	**170**	3	5	13
Fault 2	84	16	**623**	6	19
Fault 3	32	19	7	**787**	12
Fault 4	0	63	8	3	**505**

Table 3.1 Confusion Matrix for a More-Complex Fault Isolation System

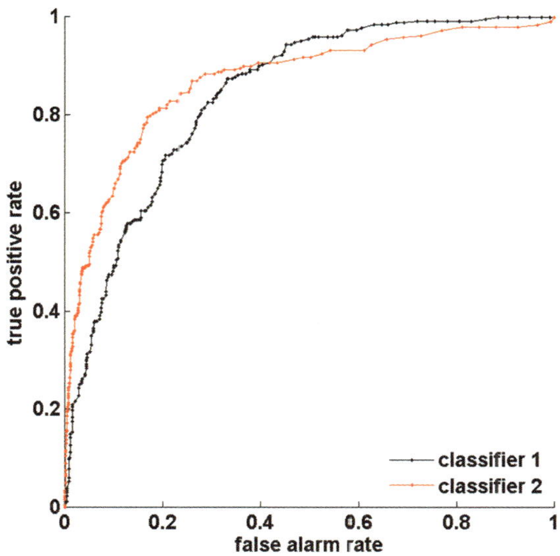

Figure 3.15 ROC curves. Classifier 2 is superior below a 35% false alarm rate.

to maximize detections, classifier 1 is a better choice above 35% false alarm rate. The integrated area under the curve (AUC) of an ROC curve is a useful metric if it is necessary (e.g., for optimization) to summarize performance as a single numeric metric. AUC has a minimum value of 0.5 (indicating chance performance) and a maximum of 1.0 (indicating perfect performance).

Note that ROC curves are only appropriate for assessing a single instant—"Is the fault present or not at this time?" However, IVHM data sets are typically time-series, and detecting faults with as much lead time as possible to allow for reallocating assets, scheduling maintenance, and minimizing interruptions is critical. Thus, it is desirable to have an approach that accounts for the ability of a classifier to detect a fault quickly as well as accurately. The time Receiver Operating Characteristic, or tROC [Xue et al. 2010], curve accounts for both classification accuracy and detection latency.

Time To Failure (TTF) is defined [Xue et al. 2010] as the time between when a classifier output first exceeds the alerting threshold and when the fault occurs (Figure 3.16; if the score never exceeds the threshold, TTF is set to zero). Given M_+ failure cases and M_- normal cases out of a total of M cases, average time to failure is calculated as

$$TTF_D = \frac{1}{M_+} \sum_{i=1}^{M_+} TTF \tag{3.13}$$

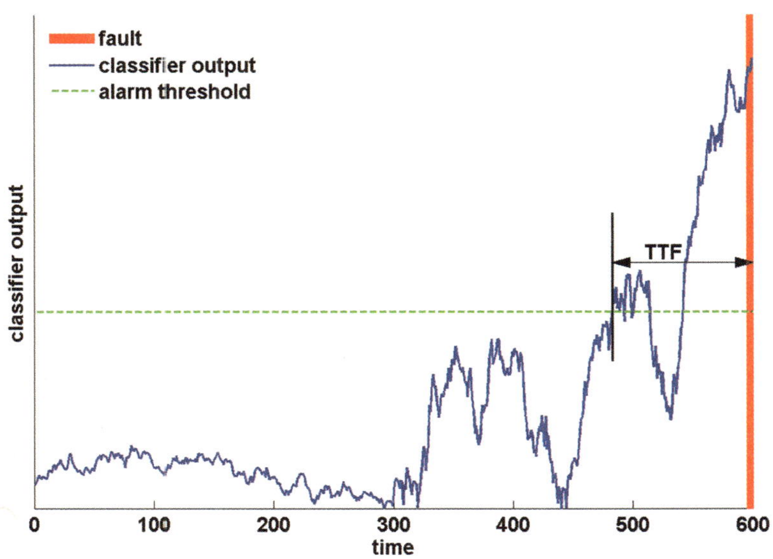

Figure 3.16 Definition of Time To Failure (TTF).

Normalized average TTF, $NTTF_D$, is defined as

$$NTTF_D = \frac{TTF_D}{t_E} \qquad (3.14)$$

where range t_E is the time considered "unrealistically (or unbelievably) early" for the detection of a particular fault modality, typically defined by a domain knowledge expert. Thus, for very short warning times (i.e., low TTF), $NTTF_D$ is near zero, and for $TTF = t_E$, $NTTF_D$ is one; for $TTF > t_E$, $NTTF_D$ is greater than one (however, this is not a valid region for tROC).

The false positive rate p_{FA} for the M_- (normal) cases is defined as

$$p_{FA} = \frac{FP}{M} \qquad (3.15)$$

where FP is the score for any normal case that exceeds the threshold. Note that the tROC curve is unaffected by class imbalance, as p_{FA} and $NTTF_D$ are calculated with information from within only one class (the majority or minority), which is important for IVHM applications.

Figure 3.17 is a sample tROC curve with some illustrative points marked on it. Point A represents a very conservative strategy that gives very little warning time but produces few false alarms (i.e., the threshold is likely very high). Point B corresponds to a low threshold for alarm, resulting in many false alarms, but relatively early detection. Point C represents a compromise between the two strategies. As with ROC curves, curves with higher $NTTF_D$ and lower p_{FA} dominate, so, for example, point D represents better performance than point C. Finally, point E represents perfect performance: alarms are raised as early as possible, and no normal cases are ever mistaken for faulty cases.

3.3.6 Cross-Validation

Imagine if a class of students was given a week to prepare for a standardized test in algebra, and as part of that preparation, they were permitted to train with *the exact test* they were going to take. Students could work out the problems very carefully at their leisure,

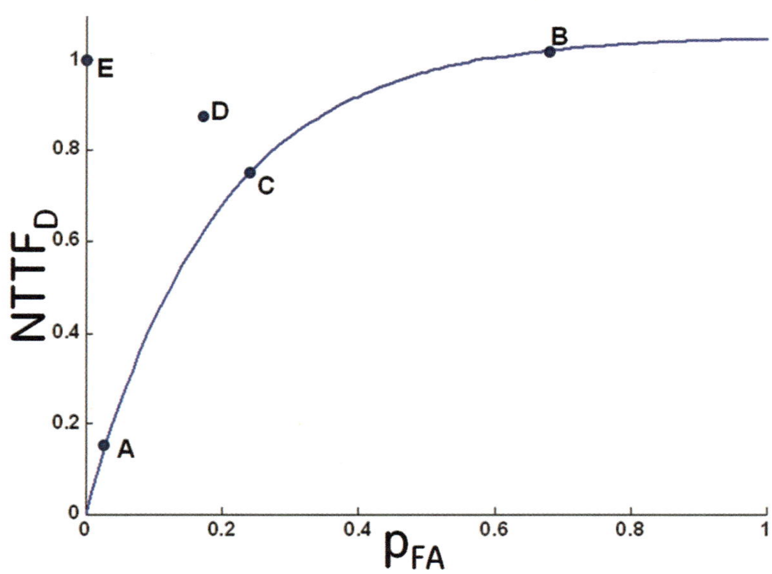

Figure 3.17 A sample tROC curve.

memorize the answers, and then do extraordinarily well on the actual test. However, the students would perform poorly on real-world applications of algebra; certainly less well than the standardized test performance would lead you to believe. Machine learning algorithms suffer from the same limitation.

Data-driven models need to be *trained* to learn the coefficients that relate the input data to the output labels. Training is accomplished (at the most abstract level) by exposing normal and abnormal cases to the algorithm. However, it is desirable to assess "real world" performance of a diagnostic system in a fair way. If we use the performance of the diagnostic system on the training data as a metric, we are virtually guaranteed to overestimate the ability of the system. Thus, we need separate training and test data.

In IVHM applications, labeled anomaly or fault data is very rare—often there will be only a handful of cases of the fault class, which makes it very difficult to set aside enough data to properly test on, and still have enough to train. In this circumstance, an approach known as Leave-One-Out (LOO) cross validation is the most effective way to assess performance. In LOO, each case is omitted in turn, and a model is trained on the remaining data. The trained model is then used to evaluate the left-out case. This process is repeated until all cases have been left out (Figure 3.18). Estimated error, $\hat{e}$, is calculated using

$$\hat{e} = \frac{1}{N}\sum_{i=1}^{N} e_i \tag{3.16}$$

where N is the number of cases, and e_i is the individual case error.

It is important to note that *cross validation is done at the case level*, not at the measurement level. For example, Figure 3.2 shows 300 points, of which some at the end are faulty. This entire time series would be left out at one time, not each of the 300 points.

3.4 Putting It All Together

3.4.1 Introduction

Data-driven anomaly detection and diagnostic systems are created in three phases: preparation, model building, and deployment. The following list is intended to prompt focused analysis of data-driven systems. Note that the solution to a problem at one stage is frequently found a few steps back. For example, class imbalance can sometimes be addressed by collecting more data,

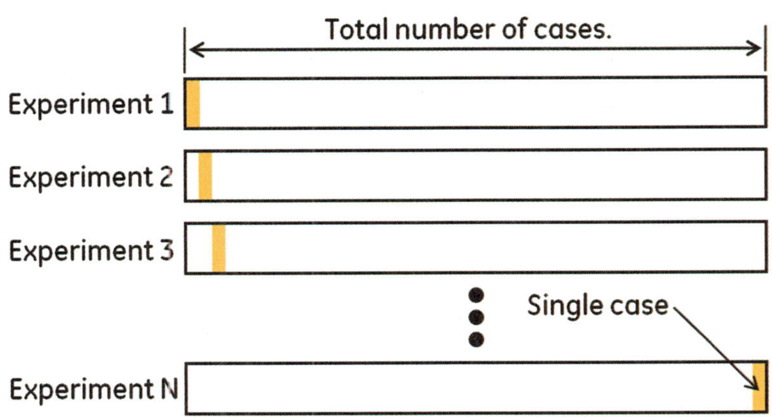

Figure 3.18 Schematic of leave-one-out cross validation.

or poor model performance might require revisiting stratification.

3.4.2 Preparation

1. **Platform identification.** What system are you trying to model? What subsystems is it composed of?
2. **Identifying sensors.** What raw sensor data is available? How much is available? Will it be sufficient to detect faults?
3. **Identifying fault cases.** Is there data from any known fault cases? Is the data labeled?
4. **Collecting data.** Can the data be assembled? Are there any commercial, legal, or political (e.g., International Traffic in Arms Regulations) limitations on the data? Is there any non-sensor data to be had (e.g., shop visit data)?
5. **Defining the feature space and transforming the data.** Is there any informative domain expertise? Can more informative features be constructed (e.g., via FFT)?

3.4.3 Model Building

6. **Method selection.** What form of modeling approach is most suitable to your data?
7. **Cross validation.** What approach for cross validation will you use? How can you be sure you are not "contaminating" your model with the test data?
8. **Sampling and stratification.** Is there significant class imbalance? Is there a good way to stratify the data for sampling? What is a good approach for sampling the majority class?
9. **Model training.** Do you have enough computing power to build models in a reasonable time? Do you need to downsample your data just to fit it into memory?
10. **Model assessment.** Is the model performance suitable? Does it detect faults far enough in advance? Do thresholds need to be tuned to optimize performance?

3.4.4 Deployment

11. **Operational constraints.** How quickly do you need the data processed? How many computing resources are required? What form should the results be delivered in?
12. **Robustness.** Is there a mechanism to accommodate for missing data? What if the data arrives asynchronously? How do you know if the model is "working"?
13. **Model maintenance.** What mechanism is in place to track model performance over time? Is there a plan to re-train models as they become stale? Are "unknown" faults being tracked and identified?

3.5 References

Atkeson, Christopher G., Andrew W. Moore, and Stefan Schaal. 1997. "Locally Weighted Learning," *Artificial Intelligence Review*, Vol. 11, pp. 11–73, 1997.

Breiman, Leo, J. H. Friedman, R. A. Olshen, and C. J. Stone. 1984. *Classification and regression trees*, Monterey, CA: Wadsworth & Brooks/Cole Advanced Books & Software.

Breiman, Leo. 2001. "Random Forests," Machine Learning 45 (1): 5–32.

Caruana, R., N. Karampatziakis, and A. Yessenalina. 2008. "An empirical evaluation of supervised learning in high dimensions," In *Proceedings of the 25th International Conference on Machine Learning*, July 2008, pp. 96–103.

Fan, J., and I. Gijbels. 1996. *Local Polynomial Modeling and Its Applications*, Chapman & Hall/CRC, New York, NY.

Guyon, I., and A. Elisseeff. 2003. "An Introduction to Variable and Feature Selection," *Journal of Machine Learning Research*, 3, 1157–1182.

Hines, J. Wesley, and Dustin Garvey. 2006. "Traditional and Robust Vector Selection Methods for Use with Similarity Based Models," 5th International Topical Meeting on Nuclear Plant Instrumentation, Control and Human-Machine Interface Technologies, Albuquerque, NM: November 12–14, 2006.

Kohonen, Teuvo. 1990. "The self-organizing map," Proceedings of the IEEE 78.9 (1990): 1464–1480.

McCulloch, W. S., and W. Pitts. 1943. "A logical calculus of the ideas immanent in nervous activity," *The Bulletin of Mathematical Biophysics*, 5(4), 115–133.

Nadaraya, E. A. 1964. "On Estimating Regression," *Theory of Probability and Its Applications*, Vol. 10, pp. 186–190, 1964.

Rumelhart, D. E., G. E. Hintont, and R. J. Williams. 1986. "Learning representations by back-propagating errors," *Nature*, 323(6088), 533–536.

Tolosi, L., and T. Lengauer. 2011. "Classification with correlated features: unreliability of feature ranking and solutions," *Bioinformatics*.

Watson, G. S. 1964. "Smooth Regression Analysis," *Indian Journal of Statistics*, Series A, Vol. 26, pp. 359–372, 1964.

Xue, Ya, N. H. W. Eklund, Xiao Hu, and Weizhong Yan. 2010. "Time receiver operating characteristic (tROC) curves: A new tool for evaluating the performance of diagnostic systems," The 2010 International Joint Conference on Neural Networks (IJCNN), pp.1–7, July 18–23, 2010.

CHAPTER 4 Prognostics

Kai Goebel, *NASA Ames Research Center*
George Vachtsevanos, *School of Electrical and Computer Engineering, Georgia Institute of Technology*
Marcos E. Orchard, *Department of Electrical Engineering, University of Chile*

> "With silver spears you may conquer the world."
>
> —Oracle at Delphi

4.1 Introduction

The methods for performing prognostics have changed considerably since the days when Philip II of Macedon consulted the Oracle at Delphi. Knowledge discovery, statistical learning, and—more specifically—an understanding of the system evolution in time when it undergoes fault conditions (note that we distinguish here between fault and failure, the former an undesirable out-of-spec condition that may worsen toward a failure, the complete loss of function), are critical for adequate implementation of successful prognostic systems. Prognosis may be understood as the generation of long-term predictions describing the evolution in time of a particular signal of interest or condition indicator, with the purpose of estimating the Remaining Useful Life (RUL) of a failing component/subsystem. Predictions are made using a thorough understanding of the underlying processes and factoring in the anticipated future usage.

In the past, IVHM practitioners have used the terms "trending" and "prognosis" interchangeably. Operators then may have followed the evolution of a single variable, temperature for example, and performed mentally a linear extrapolation as to when the variable might reach a specified threshold requiring equipment maintenance. Failure prognosis, on the other hand, involves fault detection and tracking/prediction of the fault evolution (growth) via process modeling, data analysis, and the utility of estimation algorithms, such as Kalman filtering, particle filtering, and nonlinear regression, among others. We use the term "prognosis" in this chapter in deference to trending, as suggested above.

A fundamental challenge in prognosis stems from the "large-grain" uncertainty inherent in the prediction task. Long-term prediction of the fault evolution, to the point that may result in a failure, requires means to represent and manage the inherent uncertainty. Indeed, data uncertainty, process (system) uncertainty, load uncertainty, and measurement and modeling uncertainties are potential contributors to prognostic uncertainty.

The prognosis scheme should consider critical state variables (such as condition indicators) as random processes in such a way that, once their probability distributions are estimated, other important attributes (such as confidence intervals) may be computed.

An important distinction must be drawn between two major categories of prognostic algorithms: health-based vs. usage-based prognostics. The former refers to prognostic approaches developed and applied online in real time as the system/component at hand is monitored and data are streaming into a processor for fault detection and failure prognosis. In this case, an incipient failure or fault is detected first with specified confidence, and then the prognostic algorithm is initiated to predict the time evolution. The final fault state acts as the initial condition for prognosis.

In contrast, usage-based prognosis considers the past, current, and assumed future usage or stress patterns of the system to estimate the system's remaining useful life. Such prognostic methods do not presuppose the existence of fault or incipient failure modes, in contrast to health-based prognostics. The remaining useful life may be estimated at any time in the system's operating history in the absence of a fault. The prediction may be continuously updated as new evidence accumulates. Life-cycle management tools for critical systems take advantage of usage-based prognostic routines to arrive at times needed for maintenance, repair, and overhaul of such systems. The enabling technologies include neuro-fuzzy systems, response surface methodologies, similarity-based methods, and regression analysis techniques, among others.

There are some basic ingredients that are common to all prognostic approaches. These are a model that describes both the system under investigation and damage propagation, a quantification of the damage threshold, an algorithm that handles the propagation of the damage into the future, and a mechanism to deal with uncertainty. These elements will be discussed in more detail in the next few sections.

4.2 System Model

The system model describes the characteristics of the system under nominal conditions. Ideally, such a model should be able to factor in the effects of operational and environmental conditions as well as any other conditions that cause different system response under nominal conditions. The system model could integrate domain expertise and be implemented using either knowledge-based rules, a physical description of the system under investigation, or a structure that could learn the system behavior from examples (for instance, using machine learning techniques).

4.3 Damage Propagation Model

A damage propagation model describes how the damage is expected to grow in the future. It should, similar to the system model, account for operational and environmental conditions as well as any other conditions that have an impact on the damage. While one often thinks of damage as a monotonically increasing phenomenon, it is possible for the domain in which damage is evaluated to have non-monotonic attributes. These could be either intrinsic attributes (for example, recovery effects in batteries' capacity or power semiconductors) or extrinsic effects, such as partial maintenance actions.

Depending on the fault mode, damage propagation may exhibit different symptoms, and it may be necessary to consider dedicated damage propagation models for different fault modes.

4.4 Prognostic Algorithm

The role of the prognostic algorithm is applying the damage propagation model into the future. The damage propagation algorithm must properly consider the effects of environmental and operational conditions, and, possibly, healing phenomena.

Depending on the implementation, it is sometimes not completely easy to separate the damage propagation model and the prognostic algorithm. For the general case, we will treat the damage propagation model and the prognostic algorithm as separate.

4.5 Damage Threshold

Damage thresholds define the condition representing the end-of-life, or failure. An end-of-life threshold in prognostics must be a measurable condition, although it should be noted that this condition is not always the complete destruction of a component. In fact, to be effective there should always be some margin between this threshold and complete destruction. Often, thresholds are performance specifications that represent a graceful degradation in performance that constrains the operation of the component, or system. Thus, it is typical that the system may continue to operate beyond the end-of-life threshold conditions (but outside of the specifications for maintenance).

4.6 Prognosis and Uncertainty Characterization

Prognostics is not really useful unless the uncertainties in the predictions are accounted for. Uncertainty management tools seek to improve the signal (fault) to noise (uncertainty) ratio. They begin by determining the uncertainty sources in terms of an uncertainty tree and then exploiting filtering or kernel-based methods for uncertainty management [Orchard et al. 2010]. A remaining life estimate that has no quantification of the uncertainty bounds leaves the user with little practical information. Accounting for the various sources of uncertainty and rigorously combining them will allow decision makers to justify the action taken. Uncertainties need to be managed carefully, because a haphazard stacking of the various uncertainties may lead to wide bounds that wipe out the benefits of estimating remaining life. Sources of uncertainty arise from a wide range of influences including the models (both their structures and their parameters), the current state estimate, the future load and environmental conditions, and sensor noise, just to name a few.

Figure 4.1 summarizes the range of possible prognostic approaches as a function of the applicability to various systems and their relative implementation cost. The pyramid starts at the base with generic, statistical life usage and experience-based prognostic models, migrates to basically data-driven techniques employing evolutionary or trending models, while the top of the pyramid is occupied by physics-based models employed for prognostic purposes. Prognostic technologies typically use measured or inferred features, in combination with data-driven and/or physics-based models, to predict the condition of the system at some future time [Orchard and Vachtsevanos 2009; Engel et al. 2000]. Prognostic techniques combining data-driven and physics-based models are sometimes referred to as 'hybrid' prognostics. Model-based prognostics, at the top of the pyramid, is expected to guide the future development of reliable and verifiable prognostic algorithms for complex systems such as aircraft.

Inherently probabilistic or uncertain in nature, prognosis can be applied to failure modes governed by material condition or by functional loss. Prognosis algorithms can be generic in design but specific in terms of

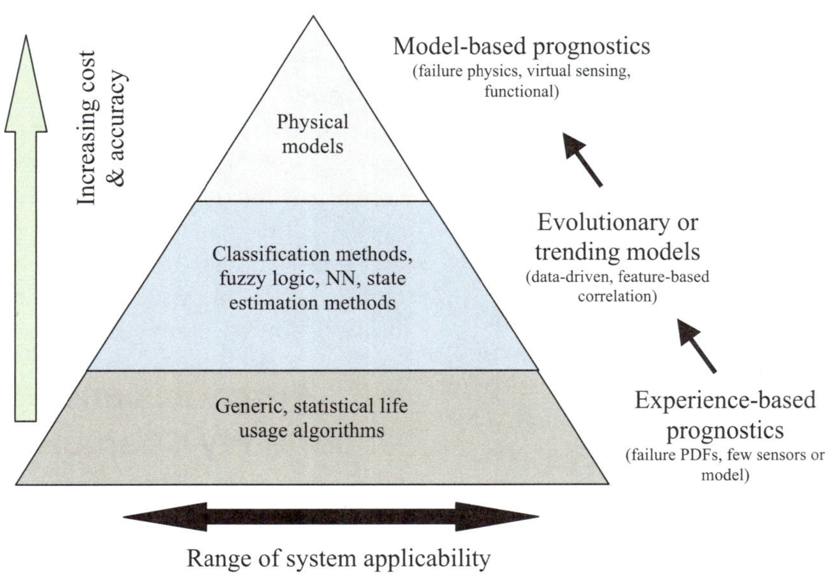

Figure 4.1 A taxonomy of prognostic approaches.

application. Prognosis system developers have implemented various approaches and associated algorithmic libraries for customizing applications that range in fidelity from simple historical/usage models to approaches that use advanced condition indicator analysis or physics-of-failure models.

Depending on the complexity and criticality of the component/system being monitored, various levels of data, models, and historical information are needed to develop and implement the desired prognostic approach.

In the engineering disciplines, fault prognosis has been approached via a variety of techniques ranging from Bayesian estimation and other probabilistic/statistical methods to artificial intelligence tools and methodologies based on notions from the computational intelligence arena. Specific enabling technologies include multistep adaptive Kalman filtering [Lewis 1986], autoregressive moving-average models [Pham and Yang 2010], stochastic autoregressive integrated-moving-average models [Jardim-Gonçalves et al. 1996], Weibull models [Groer 2000], forecasting by pattern and cluster search [Frelicot 1996], and parameter estimation methods [Ljung 1999]. From the artificial intelligence domain, case-based reasoning [Aha 1997], intelligent decision-based models, and min-max graphs have been considered as potential candidates for prognostic algorithms. Other methodologies, such as Petri nets, neural networks, fuzzy systems, and neuro-fuzzy systems [Studer and Masulli 1996], have found ample utility as prognostic tools. Physics-based fatigue models [Tangirala 1996] have been employed extensively to represent the initiation and propagation for structural life and failure prediction.

4.7 Prognostic Techniques

4.7.1 Model-Based Techniques

Model-based prognostic schemes include those that employ a dynamic model of the process being predicted. These can include physics-based models, Autoregressive Moving-Average (ARMA) techniques, Bayesian filtering algorithms, and empirical-based methods. Model-based methods provide a technically comprehensive approach that has been used traditionally to understand component failure mode progression.

Figure 4.2 depicts a model-based prognostic scheme. Input from the diagnostic block is combined with stress profiles and feeds into the fault growth model. An estimation method (in this case particle filtering) is called upon to propagate the fault model initially, one step at a time, while model parameters are updated online in real time as new sensor data become available.

Eventually, the model is allowed to perform long-term prognosis of the remaining useful life of the failing component/system with confidence bounds. The fault model PDF is convolved with the hazard zone PDF when the former reaches the threshold bounds and the resultant PDF is projected along the time axis (which is usually measured in "cycles" of operation) depicting the system's remaining life statistics.

Implementing physics-based models can provide a means to calculate the damage to critical components as a function of operating conditions and assess the cumulative effects in terms of component life usage. By integrating physical and stochastic modeling techniques, the model can be used to evaluate the distribution of remaining useful life as a function of uncertainties in component strength/stress properties and loading conditions for a particular fault.

Bayesian estimation techniques can satisfy these requirements. In particular, particle filtering and learning strategies [Orchard and Vachtsevanos 2009, 2007] employ a state dynamic model and a measurement model

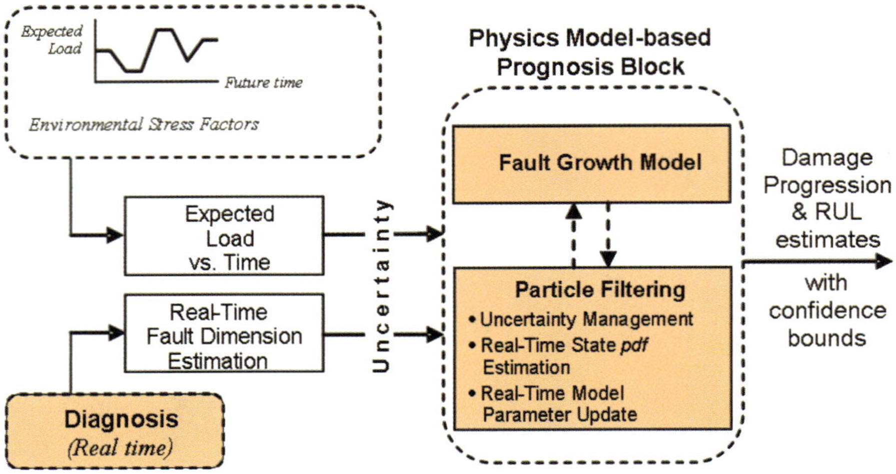

Figure 4.2 A scheme for model-based prognosis.

to predict the posterior probability density function of the state to predict the time evolution of a fault or fatigue damage. Particle filters avoid the linearity and Gaussian noise assumption of Kalman filtering, and provide a robust framework for long-term prognosis while accounting effectively for uncertainties. Correction terms are estimated in a learning paradigm to improve the accuracy and precision of the algorithm for long-term prediction.

Particle filtering methods assume that the state equations that represent the evolution of the fault mode in time can be modeled as a first order Markov process with additive noise and conditionally independent outputs [Arulampalam et al. 2002]. Let

$$x_k = f_{k-1}\left(x_{k-1}\right) + \omega_{k-1} \tag{4.1}$$

$$z_k = h_k\left(x_k\right) + v_k \tag{4.2}$$

where the state vector x_k includes a set of parameters that characterizes the evolution in time of the fault condition; the process noise ω_{k-1} represents the model uncertainty; z_k is the observation (measurement); and v_k is the measurement noise (uncertainty associated to sensors specifications and feature computation processes).

While there are several flavors of particle filters, the focus here is on algorithms based on the concept of *Sampling Importance Resampling* (SIR), in which the posterior filtering distribution denoted as $\pi(x) = p(x_k|z_k)$ is approximated by a set of N weighted particles $\left\{\langle x_k^i, w_k^i\rangle; i = 1,\ldots, N\right\}$ sampled from an arbitrarily proposed distribution $q(x)$ that intends to be "similar" to $\pi(x)$ (i.e., $\pi(x) > 0 \Rightarrow q(x) > 0$ for all $x \in R^{n_x}$), The *importance weights* w_k^i are proportional to the likelihood $p(z_k \mid x_k^i)$ associated to the sample x_k^i, and normalized as in Eq. 4.3:

$$w_k^i = \frac{\pi\left(x_k^i\right)/q\left(x_k^i\right)}{\sum_{j=1}^{N} \pi\left(x_k^j\right)/q\left(x_k^j\right)} \tag{4.3}$$

such that $\sum_{j=1}^{N} w_k^i = 1$, and the posterior distribution (a.k.a the *target* distribution) can be approximated as

$$p\left(x_k | z_k\right) = \sum_{i=1}^{N} w_k^i \delta\left(x_k - x_k^i\right) \tag{4.4}$$

Thus, as in any Bayesian processor, the filtering stage is implemented in two steps: the computation of the a priori state density estimate (prediction step), and the update of the estimate according to the information presented by new measurements. Using the model in Eq. 4.1, the prediction step becomes

$$p(x_k|z_{k-1}) \approx \sum_{i=1}^{N} w_{k-1}^i \delta\left(x_k - f_{k-1}\left(x_{k-1}^i\right) - \omega_{k-1}^i\right) \quad (4.5)$$

The update step, on the other hand, modifies the particle weights according to the relation

$$\overline{w}_k^i = w_{k-1}^i \frac{p(z_k \mid x_k^i)p(x_k^i \mid x_{k-1}^i)}{q(x_k^i \mid x_{k-1}^i, z_k)} \quad (4.6)$$

$$w_k^i = \frac{\overline{w}_k^i}{\sum_{j=1}^{N} \overline{w}_k^j} \quad (4.7)$$

It is possible that all but a few of the importance weights degenerate such that they are close to zero. In that case, one has a very poor representation of the system state (and also wastes computing resources on unimportant calculations). To address that, *resampling* of the weights can be used [Arulampalam et al. 2002]. The basic logical flowchart is shown in Figure 4.3.

Prognosis, and thus the generation of long-term predictions, is a problem that goes beyond the scope of filtering applications because it involves future time horizons. Hence, a particle-filtering-based prognostic approach requires proposing a procedure to project the current estimate of the state Probability Density Function (PDF) in time. The simplest implementation that can be used to solve this problem uses Eq. 4.1 recursively to propagate the posterior PDF estimate defined by $\{\langle x_p^i, w_p^i \rangle; i = 1,\ldots, N\}$ in time, until x_p^i fails to meet the system specifications at time t_{EOL}^i. The RUL PDF—i.e., the distribution $p(t_{EOL}^i - t_p)$—is given by the distribution of w_p^i. Figure 4.4 shows the flow diagram of the prediction process.

4.7.2 Data-Driven Techniques

In some cases involving complex systems, it may be difficult or impossible to derive dynamic models based on all the physical processes involved. In such cases, it is possible to assume a certain form for the

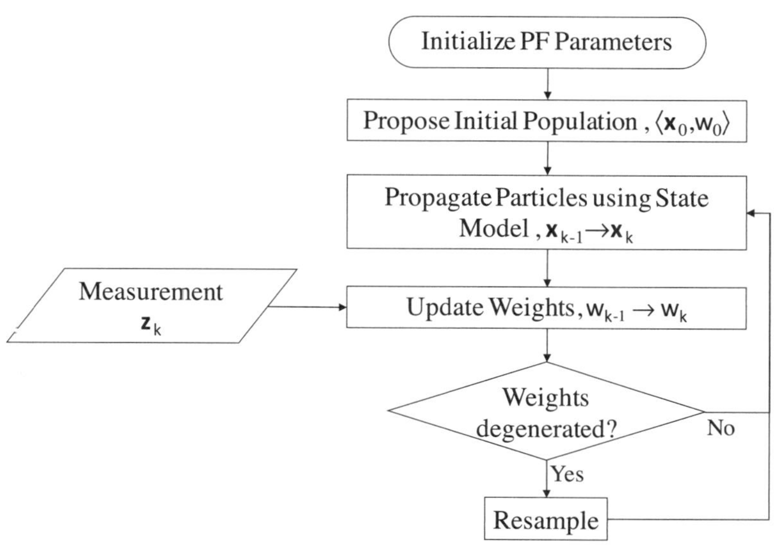

Figure 4.3 Particle filtering flowchart.

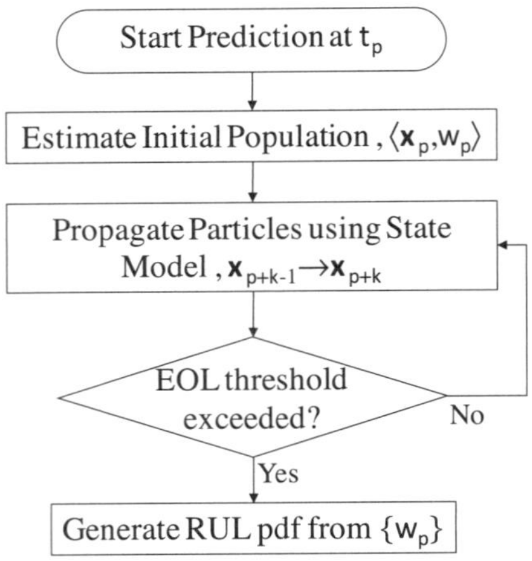

Figure 4.4 Prediction flowchart.

dynamic model and then use observed inputs and outputs of the system to determine the model parameters needed so that the model indeed serves as an accurate surrogate for the system. This is known as model identification.

For fault diagnosis and failure prognosis, a variety of input-output mappings have been employed as surrogate system models. Specifically, one may have historical fault/failure data in terms of time plots of various signals leading up to failure, or statistical data sets. In such cases, it is very difficult to determine any sort of model for prediction purposes. In such situations, one may use nonlinear network approximators that can be tuned using well-established formal algorithms to provide desired outputs directly in terms of the data. They provide structured nonlinear function mappings with very desirable properties between available data and desired outputs.

In prediction, Artificial Neural Networks (ANNs), fuzzy systems, and other computational intelligence methods, based on the linguistic and reasoning abilities of humans, have provided an alternative tool for both forecasting researchers and practitioners [Sharda 1994]. Werbos [1988] reported that ANNs trained with the backpropagation algorithm outperform traditional statistical methods such as regression and Box-Jenkins approaches. In a forecasting competition organized by Weigend and Gershenfeld [1993] through the Santa Fe Institute, all winners of each set of data used ANNs. Unlike the traditional model-based methods, ANNs are data-driven and self-adaptive, and they make very few assumptions about the models for problems under study. ANNs learn from examples and attempt to capture the subtle functional relationship among the data. Thus, ANNs are well suited for practical problems, for which it is easier to have data than knowledge governing the underlying system's fault behavior. Generally, they can be viewed as one of many multivariate nonlinear and nonparametric statistical methods [Cheng and Titterington 1994]. Data-driven approaches to failure prognosis also take advantage of recurrent neural networks, dynamic wavelet neural networks, neuro-fuzzy systems, and a variety of statistical tools. For training and validation, they use the current and past history of input data and feedback outputs via unit delay lines. The main problem of ANNs is that their decisions are not always evident. Nevertheless, they provide a feasible tool for practical prediction problems.

4.7.3 Statistical Techniques

For situations in which sophisticated prognostic models are not or cannot be utilized (perhaps because a high investment in such models was not justified by the business case, low failure-criticality rates, or where there is an insufficient sensor network to assess condition), a statistical reliability or usage-based prognostic approach may be the only alternative. This form of prognostic algorithm is the least complex and requires that a history of component failure or operational usage profile data is available. One typical approach would be to fit a Weibull distribution (or other statistical failure distribution) to such failure or inspection data [Groer 2000; Schömig and Rose 2003]. Despite the obvious loss of information (compared to condition-based approaches), a statistical reliability-based prognostic distribution can still be used to drive interval-based maintenance practices which can then be potentially revised by the information obtained from maintenance. A regularly updated maintenance database is critical for this approach.

4.8 Measuring Prognostics Performance

Chapter 5, IVHM Performance Metrics, lists a number of measures that can be used to evaluate prognostic performance. Some metrics of particular interest are listed here.

4.8.1 Prognostic Horizon

Prognostic Horizon (PH) fulfills two roles: first, it identifies whether an algorithm predicts within a specified error margin (specified by α, a statistical confidence parameter) around the actual End-Of-Life (EOL, the time index for actual end of life, according to the defined failure threshold); and, second, it indicates how much time the algorithm provides for any corrective action to be taken. In other words, it assesses whether an algorithm yields a sufficient prognostic horizon; if it does not, it may not be useful or meaningful to compute other metrics. PH is defined as the difference between the EoP (End-of-Prediction) and the current time index i; it is calculated using data accumulated up to the time index i, provided that the prediction meets a minimum set of desired specifications. These specifications may be defined in terms of an allowable error bound (α) around true EOL. It is expected that PHs are determined for an algorithm-application pair offline during the validation phase. These numbers can then be used as guidelines when the algorithm is deployed in test applications when the actual EOL is not known in advance. While comparing algorithms, an algorithm with a longer prediction horizon (H) would be preferred. The prediction horizon is computed as $H = \text{EoP} - i$, where $i = \min\{j \mid (j \in l) \wedge (r_*(1-\alpha) \le r^l(j) \le r_*(1+\alpha))\}$, $r_*^l(i)$ is the true RUL at time i given that data are available up to time i for the lth UUT (Unit Under Test), and $r^l(i)$ is the RUL estimate for the lth UUT at time i as determined from measurement and analysis.

For instance, a PH with error bound of $\alpha = 0.05$ identifies when a given algorithm starts predicting estimates that are within 5% of the actual EOL. Other specifications may be used to derive PH as desired.

4.8.2 α-λ Performance

α-λ Performance identifies whether the algorithm performs within desired error margins (specified by the parameter α) of the actual RUL at any given time instant (specified by the parameter λ) that may be of interest to a particular application. This presents a more stringent requirement of staying within a converging cone of error margin as a system nears EOL. The time instances may be specified as percentage of total remaining life from the point the first prediction is made or a given absolute time interval before EOL is reached. For instance, we define α-λ accuracy as the prediction accuracy to be within $\alpha \cdot 100\%$ of the actual RUL at specific time instance t_λ expressed as a fraction of time between the point when an algorithm starts predicting and the actual failure. For example, this metric determines whether a prediction falls within 20% accuracy (i.e., $\alpha = 0.2$) halfway to failure from the time the first prediction is made (i.e., $\lambda = 0.5$). The metric is visualized in Figure 4.5. An extension of this metric based on other performance measures is straightforward:

$$[1-\alpha] r_*(t) \le r^l(t_\lambda) \le [1+\alpha] r_*(t) \qquad (4.8)$$

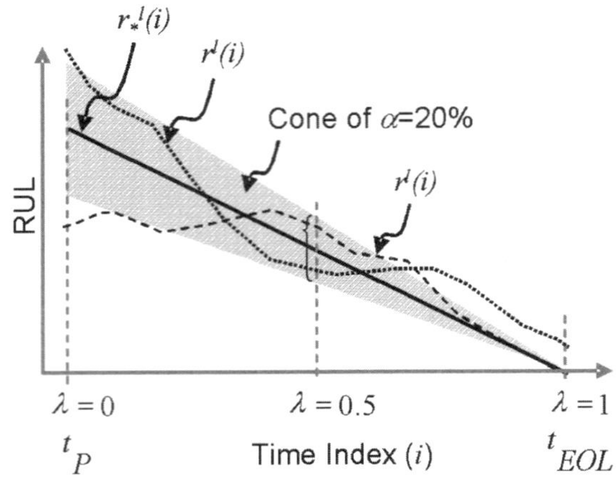

Figure 4.5 α-λ Performance visualization [Saxena et al. 2009].

where α is the accuracy modifier, λ is the time window modifier, $t_\lambda = P + \lambda(\text{EOL} - P)$, and P is the time index at which the first prediction is made by the prognostic system. Note that α-λ Performance and Prognostic Horizon can also be computed as precision metrics.

4.8.3 Prognostic Dynamic Standard Deviation (DSTD)

Other performance measures intend to quantify the volatility of generated predictions, something that can be achieved by computing the standard deviation of the expected EOL over a sliding window:

$$\text{DSTD} = \varphi \left(\sqrt{Var\left(E\left\{EOL \mid y_{1:j}\right\}\right)_{j=k_{pred}-\Delta:k_{pred}}} \right)_{\forall k_{pred} \in [1, EOL]} \tag{4.9}$$

where k_{pred} is the cycle in which the prognostic algorithm is executed, Δ is the number of samples considered in the sliding window, φ is the logistic function that aims to scale the results in the range [0,1]. For this measure, the better DSTD, the closer to zero is the measure (DSTD = 0 indicates perfect null volatility and the fact that new measurements do not alter the output of the prognostic algorithm).

4.8.4 Critical-α Index

Decision-making support-systems cannot depend solely on information about the expectation of random variables, because the tails of PDFs contain critical information about the risk that is associated to process operation. The *critical-α* index is a measure based on the concept of the JITP (Just In Time Point) that helps to quantify this point. The *critical-α* index is a measure of risk aversion (a significant factor to be considered when using implementations that overestimate the remaining useful life of a system) and is defined as the maximum $\alpha \in [0,100]$ that guarantees that the $\text{JITP}_{\alpha\%}(k_{pred})$ value is smaller than the ground truth value of the EOL time instant, for all $k_{pred} \in [1, \text{EOL}]$:

$$\alpha_{crit} = \arg\max_{\alpha} \left\{\text{JITP}_{\alpha\%}(k_{pred}) \le EOL\right\}_{\forall k_{pred} \in [1, EOL]} \tag{4.10}$$

Decision-making support systems that consider prognostic algorithms with larger critical-α values in their design are capable of implementing more aggressive strategies. This is based on the fact that these prognostic routines are conservative; thus it is possible to accept the risk of accumulating larger failure probability mass before recommending a corrective action. However, a large critical-α value is also an indicator that the variance of the predicted EOL PDF is large (i.e., less precise estimates of the EOL). For this reason, a good design should try to lessen this problem by selecting prognostic algorithms that allow not only the use of large critical-α values, but also minimize—over time—the difference between the ground truth EOL and the JITP values computed for the corresponding $\alpha_{crit}\%$.

4.9 Electro-Mechanical Actuator Case Study

4.9.1 Introduction

To illustrate the application of the principles outlined in this chapter, a detailed case study of an electro-mechanical actuator is presented. This case study describes the modeling and simulation steps, the experimental evaluation, and the development of fault diagnosis and failure prognosis algorithms for brushless DC motor winding insulation faults in the context of an aircraft EMA application. Electro-Mechanical Actuators (EMAs) are finding extensive utility as drives for modern aircraft systems, in addition to classical hydraulic devices. An EMA is configured as a closed-loop system consisting of a controller, motor(s), and sensing apparatus such as a resolver. Figure 4.6 depicts the main modules of a typical EMA, including the control loops.

A thorough Failure Modes, Effects, and Criticality Analysis (FMECA) study identified turn-to-turn winding faults as the primary mechanism, or mode, of failure. Physics-of-failure mechanisms were used to develop a model for the identified fault. The model, implemented in Simulink, simulates the dynamics of the motor with a turn-to-turn insulation winding fault.

An experimental test procedure was devised and executed to validate the model. After a fault identification step (not reported here), a condition indicator extraction routine preprocesses monitoring parameters and passes the resulting condition indicators to a particle filter, as a model-based algorithm for fault diagnosis and failure prognosis.

This case study highlights the challenges faced by the IVHM designer in addressing the unique and complex issues posed by such systems. It points toward the necessity for rigorous tools and methods that are based on physics-of-failure mechanisms of such devices: modeling and simulation studies in conjunction with experimental data and a thorough approach to data processing, condition indicator extraction, and novel diagnostic and prognostic routines with guaranteed performance (i.e., customer-specified confidence level with given false alarm rate, detection accuracy, and prescribed RUL prediction). Figure 4.7 shows a flow chart of the motor RUL prediction approach.

4.9.2 Actuator Model

An EMA high-fidelity 5th order state-space model was developed to represent higher-order dynamics for a closed-loop actuator position controller [Brown et al. 2009]. The model, which can be expressed by the linear

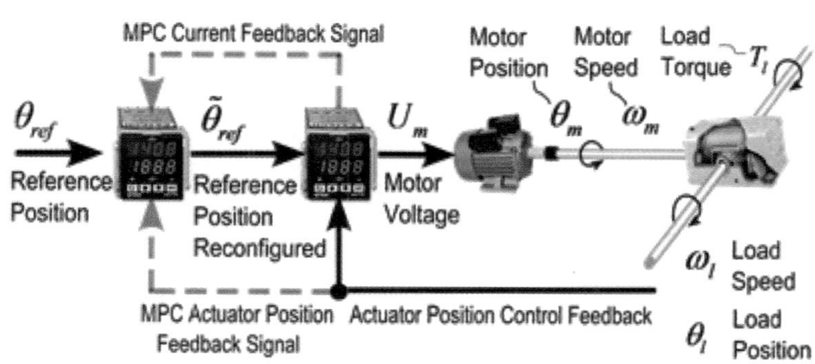

Figure 4.6 An EMA configuration with associated control loops.

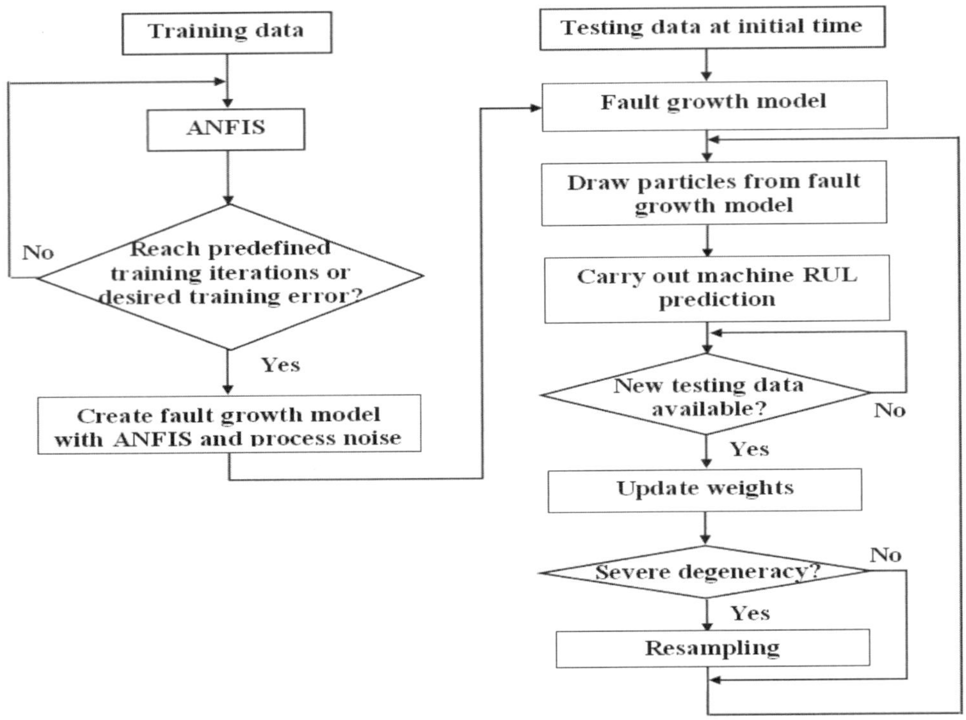

Figure 4.7 Flow chart of motor RUL prediction approach.

state-space system (A_m, B_m, C_m) is employed to relate the control inputs and measured outputs of the actuator to the internal system states of the brushless DC motor:

$$\dot{\tilde{x}}_m = A_m\tilde{x}_m + B_m u_m \qquad (4.11)$$
$$y_m = C_m\tilde{x}_m$$

where $\tilde{x}_{m0} = \tilde{x}_m(0)$. The internal state $\tilde{x}_m = \left[\tilde{i}_m \tilde{\theta}_m \tilde{\omega}_m \tilde{\theta}_l \tilde{\omega}_l\right]^T \in R^5$ is defined by the motor current, motor position and speed, and load position and speed; the control input $u_m = \left[\tilde{\theta}_{ref}\, T_{load}\right]^T \in R^2$ is defined by the reference position and external load disturbance; and the control output $y_m = \left[\theta_l\, i_m\right]^T \in R^2$ is defined by the load position and motor current.

The derivation of the motor dynamic equations begins by considering the physical layout of the stator windings and, using Kirchoff's second law, one obtains three differential equations, expressing the phase currents as functions of the motor voltages, the motor parameters, the winding inductances, and the mutual fluxes. The final dynamic model is expressed as follows:

$$\begin{bmatrix} \frac{di_{as}}{dt} \\ \frac{di_{bs}}{dt} \\ \frac{di_{cs}}{dt} \end{bmatrix} = \begin{bmatrix} L_{11}^* & L_{12}^* & L_{13}^* \\ L_{21}^* & L_{22}^* & L_{23}^* \\ L_{31}^* & L_{32}^* & L_{33}^* \end{bmatrix} \left\{ \begin{bmatrix} (U_m - \omega_r \psi_{am})\cos(\theta) \\ (U_m - \omega_r \psi_{bm})\cos(\theta - \frac{2\pi}{3}) \\ (U_m - \omega_r \psi_{cm})\cos(\theta + \frac{2\pi}{3}) \end{bmatrix} - \begin{bmatrix} r_{aa} i_{as} \\ r_{bb} i_{bs} \\ r_{cc} i_{cs} \end{bmatrix} \right\} \qquad (4.12)$$

4.9.3 Failure Modes, Effects, and Criticality Analysis

A FMECA revealed that gradual deterioration of the insulation due to thermal, electrical, mechanical, and environmental stresses is the leading root cause for motor failures [Lee, Younsi, and Kliman 2005]. Insulation can break down due to many causes, including voltage surge, over temperature, shock, vibration, excessive moisture, contamination, and metallic dust. A FMECA study for the EMA under test was carried out to identify the primary failure mode of interest, turn-to-turn winding insulation faults for a DC brushless motor. The major findings of this study are listed in Table 4.1.

To focus attention on the most critical actuator components, the criticality number and severity class of each failure mode and the major contributors to actuation system failure were identified. The severity class of a failure mode was assessed according to the definition of

I. A failure that would cause the actuator to flutter or disconnect and lead to mission loss
II. A failure that would cause the actuator to be held in a fixed position or be driven to a hardover condition
III. A failure that would result in marginal performance degradation
IV. A failure that has no effect on the system performance

According to MIL-STD-1629A *Procedure for Performing a Failure Mode, Effects and Criticality Analysis*, the critical number of a failure mode is defined as $C_m = \beta \alpha \lambda_p t$, where β is conditional probability for failure mode, α is failure mode ratio, λ_p is part failure rate, and t is duration of mission in hours.

Components with a high criticality number for the actuator include motor stator, motor resolver, bearing, and planetary gear head. They account for 95% of the criticality number sum. Furthermore, because motor

Item / Component	Failure Rate (hr^{-1})	Failure Mode Mechanism
Bearings	1.78E-05	Increased friction due to: • bearing wear • galling
Position Sensor	1.70E-05	Loss/incorrect feedback signal from resolver: • Turn-to-turn short • Turn-to-ground short • Open circuit
Brushless DC Motor	1.03E-05	Breakdown of stator assembly insulation due to: • Turn-to-turn short • Phase-to-phase short • Turn-to-ground short • Open circuit
Power Connector	6.64E-07	Electrical open due to: • Open pin • Open leadwire

Table 4.1 FMECA for EMA

stator and motor resolver are in severity class II, while bearing and planetary gear head are in severity class III, failures of motor stator and motor resolver will have a more severe impact on vehicle safe operation. Therefore, they were selected for the technology development and demonstration.

4.9.4 Component Phenomenology and System Performance Modeling

To evaluate the phenomenon of a motor winding short and the observed effects, a winding fault model was developed that represents the physics of the winding degradation. The schematic representing a winding fault for any single winding is shown in the left of Figure 4.8, where *k* out of *N* winding turns experience a short, and the extent is modeled by a shorting resistor R_f. When R_f approaches zero, the *k* winding turns are completely shorted. When R_f approaches infinity, the *k* winding turns do not have any short. By using Thévenin circuit transformation, an equivalent circuit is obtained in which w^f (between 0 and 1) represents the dimension of this turn-to-turn winding fault. That is, $w^f = 1$ represents no winding short at all while $w^f = 0$ represents an entire winding that is completely shorted. By applying this notion to all three windings, the modified three-phase wye-connected electrical diagram for a brushless DC motor, which is consistent with the test actuator, is shown in the right of Figure 4.8.

The modeling of a turn-to-turn fault caused by wearing of winding insulation within a single phase of the motor results in a connection between individual windings, as illustrated in the left of Figure 4.9.

A schematic representing a winding fault for any single winding is provided in Figure 4.10. The symbols L_s, R_s, and U_s represent the total winding inductance, resistance, and back-emf (electromotive force) voltage of the winding during normal conditions (R_f is open). The

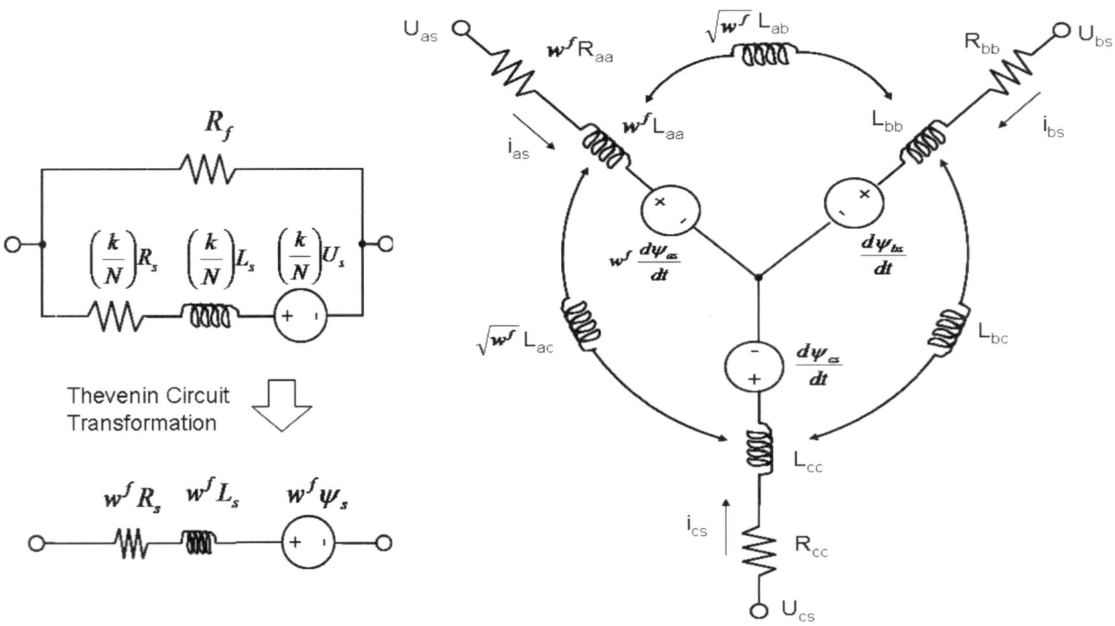

Figure 4.8 Turn-to-turn winding fault model.

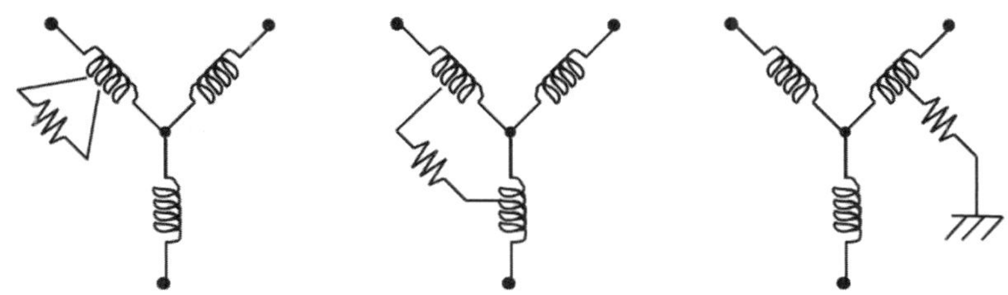

Figure 4.9 Motor winding fault modes.

symbols N, k, and R_f represent the number of total winding turns, number of winding turns between the fault, and the resistance of the insulation fault, respectively.

The circuit network on the left side can be reduced to a single resistor, inductor, and voltage source, as illustrated in Figure 4.11, by again applying the Thévenin circuit transformation.

4.9.5 Data Processing/Condition Indicator Extraction and Selection

Condition indicators computed from the preprocessed data are used to develop indications of a fault condition and operating conditions of the DC motor. Four condition indicators are extracted from the filtered voltage and current matrices. The voltage condition indicators f_1 and f_2 are related to

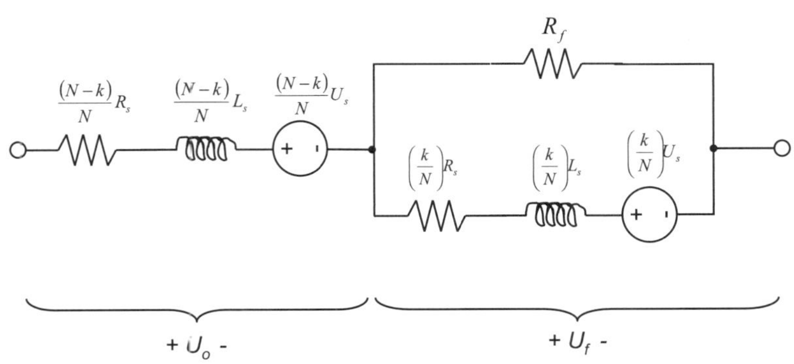

Figure 4.10 Schematic of insulation fault model.

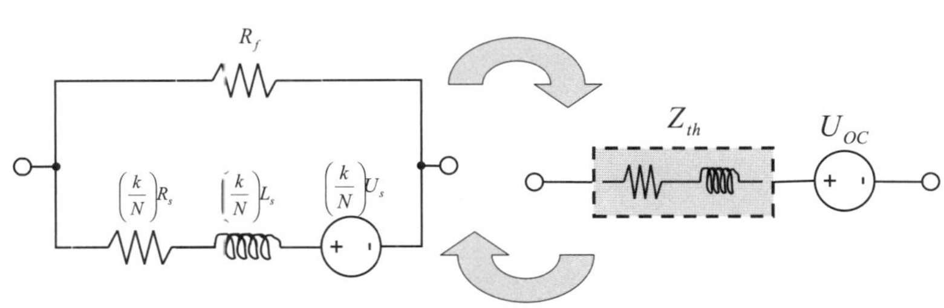

Figure 4.11 Thévenin circuit transformation of the winding fault model.

the operating speed of the motor and are expressed as

$$f_1 = \sum_{n=1}^{3} \frac{1}{3} RMS\left(U_{abc}^{1}(n)\right) \qquad (4.13)$$

$$f_2 = \sum_{n=1}^{3} \frac{1}{3} RMS\left(I_{abc}^{1}(n)\right)$$

where U_{abc}^{1} and I_{abc}^{1} are, respectively, obtained by filtering terminal-to-terminal voltage and current matrices around the frequency that defines the motor operating speed. Similarly, the current condition indicators f_3 and f_4 are computed as

$$f_3 = \sum_{n=1}^{3} \frac{1}{3} RMS\left(U_{abc}^{2}(n)\right) \qquad (4.14)$$

$$f_4 = \sum_{n=1}^{3} \frac{1}{3} RMS\left(I_{abc}^{2}(n)\right)$$

where U_{abc}^{2} and I_{abc}^{2} are, respectively, obtained by filtering terminal-to-terminal voltage and current matrices around the natural lag frequency of the resolver. The principle current feature is related to the asymmetry of the motor windings.

The final step is to map the computed condition indicators to the known fault values inserted during the seeded fault test. A neural network fitting tool in MATLAB is used to map condition indicators into fault dimensions. The neural network was trained appropriately and used for this mapping.

4.9.6 Motor Winding Short Model

Depending on where the short occurs at the stator's three windings, there are three winding fault modes, as shown in Figure 4.9. Turn-to-turn winding fault is when the short occurs across one winding. Phase-to-phase winding fault is when the short occurs between two windings. Phase-to-ground winding fault is when the short occurs between the winding and case. Because the insulation between windings, stator core, and case is normally well protected with a polymer and polymer/mica combination, while turn-to-turn insulation can only be applied with a thin film of polyamide on the surface of the magnet wire, turn-to-turn winding fault is the most prevalent winding fault mode and therefore was selected to be the focus of our program.

The schematic representing a turn-to-turn winding fault is shown in the left of Figure 4.9, where k out of N winding turns experience a short, and the extent of the short is modeled by a shorting resistor R_f. The symbols L_s, R_s, and U_s represent the total winding inductance, resistance, and back-emf voltage of the winding during normal conditions (i.e., $R_f = \infty$). The equivalent circuit that represents the reduced effectiveness of the winding is

$$w^f = 1 - \frac{k}{N}\left(1 - \left[1 + \frac{k}{N}\left(\frac{R_s}{R_f}\right)\right]^{-2}\right) \qquad (4.15)$$

Note that w^f always takes a value between 0 and 1. When $k = 0$, $w^f = 1$ and there is no winding short at all. When $k = N$ and $R_f = 0$, $w^f = 0$ and the entire winding is completely shorted.

4.9.7 Prognosis

The nonlinear dynamic state model [Brown et al. 2009; Zhang et al. 2008] for use with the particle filter is

$$\begin{cases} x_d(t+1) = f_b\left(x_d(t), \mathrm{n}(t)\right) \\ x_c(t+1) = f_t\left(x_d(t), x_c(t), w(t)\right) \\ f_p(t) = h_t\left(x_d(t), x_c(t), v(t)\right) \end{cases} \qquad (4.16)$$

where f_b, f_t, and h_t are nonlinear mappings, x_d is a collection of Boolean states associated with the presence of a particular operating condition in the system (normal operation, fault type #1, #2, etc.), x_c is a set of continuous-valued states that describe the evolution of the system given those operating conditions, f_p is a feature measurement, w and v are non-Gaussian distributions that characterize the process and feature noise signals, respectively. The function h_t is a mapping between the feature value $f_p(t)$ and the fault state $x_c(t)$. In the case of a turn-to-turn winding insulation short circuit fault, the values of $f_p(t)$ and $L(t)$ are related by the operating parameters, motor speed ω_m and motor current i_m. For simplicity, $n(t)$ may be assumed to be zero-mean, independent and identically distributed, uniform white noise. At any given instant of time, this framework provides an estimate of the probability masses associated with each fault mode, as well as a PDF estimate for meaningful physical variables in the system. PDF estimates for the system continuous-valued states (computed at the moment of fault detection) may be used as initial conditions in failure prognostic routines. As a result, a swift transition between the two modules (fault detection and prognosis) may be performed and reliable prognosis can be achieved within a few cycles of operation after the fault is declared. This characteristic is one of the main advantages of the particle-filter-based framework.

The fault model employed for fault diagnosis and failure prognosis is described next.

Historically, the rate of degradation is dependent on temperature:

$$L\,(k+1)=C_0\exp\left(-\frac{E_a}{k_b T_{wa}(k)}\right)$$

$$\text{where } C_0>0 \text{ (Arrhenius law)} \tag{4.17}$$

The relationship between winding current and temperature is expressed as

$$T_{wa}(k+1)=\left[\frac{1}{R_{wa}C_{wa}}\right]\left[\tau^2 R_0 R_{wa}-T_{wa}(k)\right] \tag{4.18}$$

T_{wa} = Difference between winding and ambient temperatures [°K]

R_{wa} = Thermal resistance [W/°K]

C_{wa} = Thermal conductance [°K/W]

I_m = Motor current [A]

R_0 = Winding resistance [Ω]

E_a = Activation energy [J]

k_b = Boltzman's constant [J/°K]

C_0 = Constant of proportionality

L = Fault dimension (winding degradation)

τ = Thermal time constant

An adaptation law is used to identify the growth model. The winding resistance depends on the temperature but changes slowly with time, while thermal measurements, T_{wa}, are used to update the model.

A discrete-time linear approximate model is written as

$$T_{wa}(k+1)=\hat{a}_m T_{wa}(k)+\hat{b}_m u(k),$$

$$\text{where } u=\sum_{n=1}^{3} i_n^2(k)$$

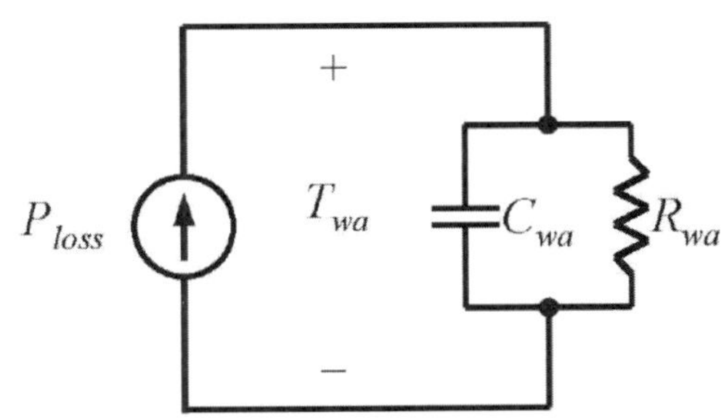

Figure 4.12 Simplified fault model.

The adaptation law is then

$$\hat{a}_m(k+1) = \gamma_a T_{wa}(k)e(k)$$

$$\hat{b}_m(k+1) = \begin{cases} \gamma_b\, u(k)e(k) & \text{if } \hat{b}_m(k) > \bar{b} \\ \gamma_b u(k)e(k) + \dfrac{\bar{b}-\hat{b}_m}{\hat{b}_m - \bar{b} + \varepsilon} & \text{if } \hat{b}_m(k) < \bar{b} \end{cases}$$

$$e(k) = y(k) - T_{wa}(k)$$

$$\text{where} \begin{cases} \bar{b}, \gamma_a, \gamma_b, \varepsilon > 0 \\ a_m(0) = a_{m0} \\ b_m(0) = b_{m0} \end{cases}$$

As a result, T_{wa} can be computed as

$$\begin{cases} T_{wa}(k+1) = \hat{a}_m T_{wa}(k) + \hat{b}_m u(k) \\ L(k+1) = L(k) + \beta(k)\exp\left[-\dfrac{E_a}{k_b T_{wa}}\right] + \omega_1(k) \\ \beta(k+1) = \beta(k) + \omega_2(k) \end{cases}$$
$$\text{Feature}(k) = h(L(k)) + v(k) \tag{4.19}$$

where β is an adaptive parameter, $\omega_{1,2}$ is the process noise, v is the feature noise, and $h(t)$ is the "fault-to-feature" mapping. Condition indicators are extracted in the frequency domain, and a neural network based classifier maps the selected/extracted condition indicators into a fault dimension.

Prognosis is achieved by implementing the particle-filtering framework that was described in section 4.7.1. Long-term predictions are used to estimate the probability of failure in a system given a hazard zone that is defined via a probability density function with lower and upper bounds for the domain of the random variable, denoted as H_{lb} and H_{up}, respectively. The probability of failure at any future time instant is estimated by combining both the weights $w^{(i)}{}_{t+k}$ of predicted trajectories and specifications for the hazard zone through the application of the Law of Total Probabilities; see Figure 4.13.

The resulting RUL PDF, in which t_{RUL} refers to RUL, provides the basis for the generation of confidence intervals and expectations for prognosis:

$$\hat{p}_{t_{RUL}} = \sum_{i=i}^{n} p\left(\text{Failure} \middle| X = \hat{x}^{(i)}_{t_{RUL}}, H_{lb}, H_{up}\right) \cdot w^{(i)}_{RUL} \tag{4.20}$$

The final step in the motor winding case study involved the development and implementation of a systematic methodology aimed to predict accurately the RUL of the

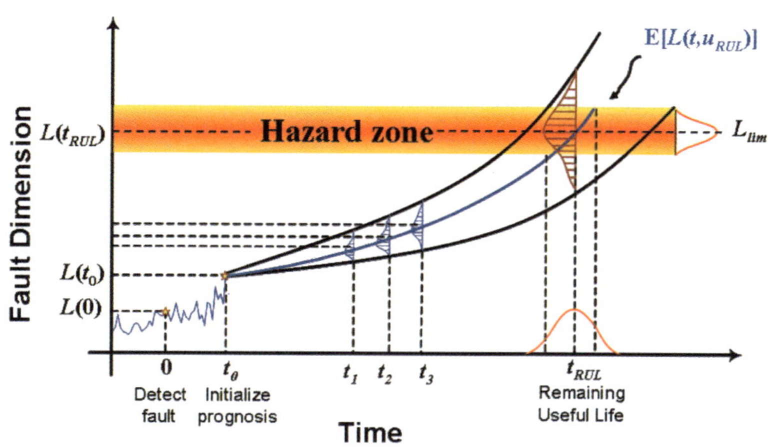

Figure 4.13 Typical prediction profiles and EOL estimate.

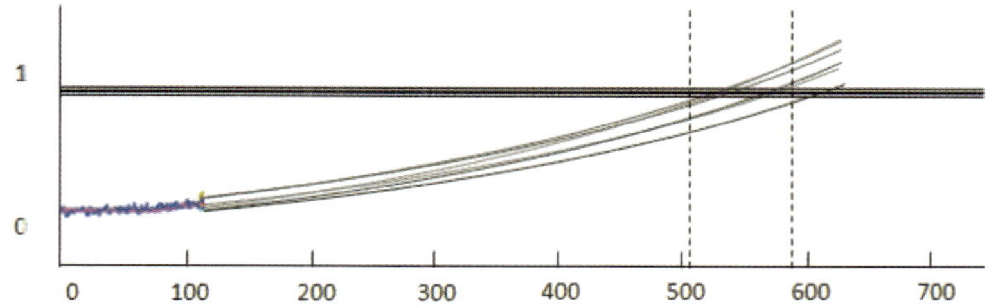

Figure 4.14 The implementation of the predictive algorithm.

failing motor insulation. The approach takes advantage of a Bayesian estimation technique using an mth-order hidden Markov model, an Adaptive Neuro-Fuzzy Inference System (ANFIS) predictor with the process noise, as the fault growth model, integrated with a high-order particle filter. Of course, other techniques may be called upon to address the prediction problem. Figure 4.14 is an illustration of the implemented prediction algorithm showing results of the estimated time to failure at $t = 120$ min.

4.9.8 Experimental Evaluation

A scaled version of the EMA was experimentally evaluated in a "proof-of-concept" test setup to evaluate the prognostic methodology. The actuator test setup consists of two commercial off-the-shelf motors; one motor simulates the actuator and the other motor provides dynamic loading. The actuator and dynamic loader are tied directly back-to-back using metal bellow couplings with a high-precision torque sensor. Both motors are controlled using a digital-space-vector Pulse Width Modulation (PWM) sinusoidal drive. This setup allows the motor under investigation to operate under dynamic loading conditions for both opposing and aiding loads, as shown in Figure 4.15.

Experimental results showed that the proposed failure prognosis architecture enabled early detection of the fault condition under analysis by anticipating failure of the system with a prognostic horizon of more than seven hours, thus providing sufficient time for corrective actions.

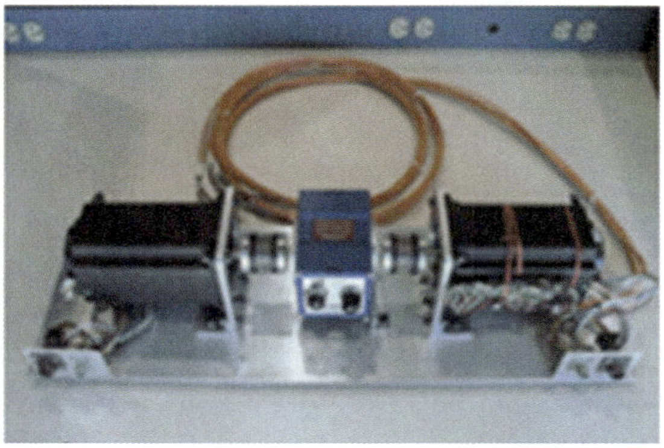

Figure 4.15 Experimental setup.

Conceptually, the performance would be tracked as shown in Figure 4.16. Here, the first prediction point is shown as a full circle

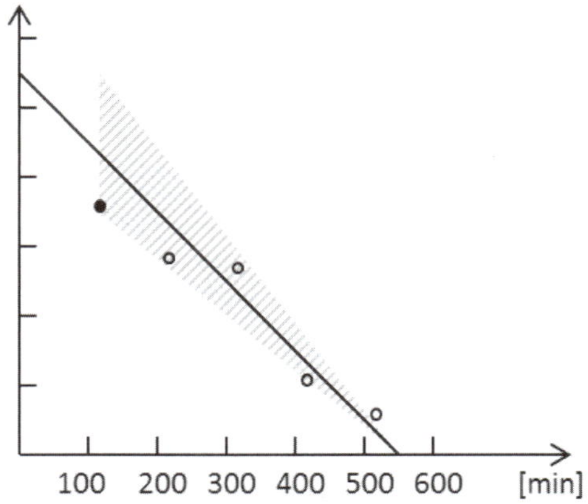

Figure 4.16 Performance evaluation with $\alpha = 0.2$.

at t = 120 min. As can be seen, the prediction is inside the α-cone of α = 0.2. Further predictions would be displayed as represented by the empty circles.

4.10 Conclusions and Recommendations

As the complexity of modern aerospace and industrial systems increases, the need for improved system reliability and autonomy follows an increased trend. Technologists attempting to develop and introduce IVHM methods for such complex systems must deal with the issues exposed in this chapter through new and novel system engineering concepts that integrate facets of modeling, testing, analysis, and algorithm development.

Prognosis is one of the more challenging aspects of the modern IVHM system. It also has the potential to be the most beneficial in terms of reduced Operational and Support (O&S) cost, and life-cycle a Total Ownership Cost (TOC) of many types of critical systems. It also ensures safe operation because the state of health of future (and, as a by-product, current) operation is continuously estimated. The evolution of diagnostic monitoring systems has led to the recognition that predictive prognosis is both desired and technically possible. A coherent and rigorous approach to develop critical health management technologies with emphasis on prognostics was introduced in this chapter, and the application of these innovative technologies was demonstrated in a typical case study. The process starts with the identification of critical components, followed by the physics-based modeling and simulation that were validated using real data, continued with the diagnostic and prognostic algorithm development, and ended with testing and validation using accepted performance metrics.

The case study introduced here exemplifies a formal and theoretically rigorous approach to the development and implementation of IVHM concepts to complex systems. A novel physical modeling and data-driven methodology via experimental testing is required to provide explanations for device behaviors and facilitate the design of fault monitoring instrumentation as well as the development of suitable diagnostic and prognostic algorithms. It is the synergy of model-based and data-driven methodologies based on an understanding of the physics of failure mechanisms for critical assets that will produce accurate, precise, and robust results that may assist the operator/maintainer to implement IVHM practices. It is worth noting the importance of methods and tools that possess the capability to address uncertainty—how to represent it and how to manage it, the Achilles' heel of IVHM.

4.11 References

Aha, D. 1997. Special Issue on Lazy Learning. *Artificial Intelligence Review, vol. 11*(1–5), 1–6.

Arulampalam, S., L. S. Maskel, N. Gordon, and T. Clapp. 2002. "A Tutorial on Particle Filters for On-line Non-linear/Non-Gaussian Bayesian Tracking," *IEEE Trans on Signal Processing 50*(2), 174–188.

Brown, D., G. Georgoulas, H. Bae, R. Chen, Y. Ho, and G. Tannenbaum et al. 2009. "Particle filter based anomaly detection for aircraft actuatoar systems," *IEEE Aerospace.*

Cheng, B., and D. M. Titterington. 1994. "Neural Networks: A Review from a Statistical Perspective," *Statist. Sci. Vol. 9,* No. 1, 2–30.

Engel, S., B. Gilmartin, K. Bongort, and A. Hess. 2000. "Prognostics, the Real Issues Involved with Predicting Life Remaining," *IEEE Aerospace*, pp. 457–469, Big Sky, MT.

Frelicot, C. 1996. "A fuzzy-based prognostic adaptive system," *RAIRO-APII-JESA, Journal Europeen des Systemes Automatises*, Vol. 30, No.2–3, 281–99.

Groer, P. 2000. "Analysis of Time to Failure with a Weibull Mode," *Maintenance and Reliability Conference, MARCON 2000.*

Jardim-Gonçalves, R., M. Martins-Barata, J. Assis-Lopes, and A. Steiger-Garcao. 1996. "Application of stochastic modelling to support predictive maintenance for industrial environments," *IEEE International Conference on Systems, Man, and Cybernetics*, pp. 117–122.

Kumar, S., N. Vichare, E. Dolev, and M. Pecht. 2012. "A Health Indicator Method for Degradation Detection of Electronic Products," *Microelectronics Reliability, Vol. 52*(2), 439–445.

Lee, S., K. Younsi, and G. Kliman. 2005. "An Online Technique for Monitoring the Insulation Condition of AC Machine Stator Windings Energy Conversion," *IEEE Transactions on Energy Conversion, Volume 20*(4), 737–745.

Lewis, F. 1986. *Optimal Estimation: With an Introduction to Stochastic Control Theory.*

Ljung, L. 1999. *System Identification: Theory for the User*, 2nd ed., New Jersey: Prentice-Hall.

Orchard, M., L. Tang, B. Saha, K. Goebel, and G. Vachtsevanos. 2010. "Risk-Sensitive Particle-Filtering-based Prognosis Framework for Estimation of Remaining Useful Life in Energy Storage Devices," *Studies in Informatics and Control*, Vol. 19, Issue 3, pp. 209–218.

Orchard, M., and G. Vachtsevanos. 2007. "A Particle Filtering Approach for On-Line Failure Prognosis in a Planetary Carrier Plate," *International Journal of Fuzzy Logic and Intelligent Systems, Vol. 7*, No. 4, 221–227.

Orchard, M., and G. Vachtsevanos. 2009. "A particle filtering approach for on-line fault diagnosis and failure prognosis," *Transactions of the Institute of Measurement and Control, Vol. 31*, No. 3–4, 221–246.

Pham, H., and B. Yang. 2010. "Estimation and forecasting of machine health condition using ARMA/GARCH model," *Mechanical Systems and Signal Processing*, 546–558.

Saxena, A., J. Celaya, B. Saha, S. Saha, and K. Goebel. 2009. "Evaluating Algorithm Performance Metrics Tailored for Prognostics," *IEEE Aerospace Conference*, Big Sky, MT.

Schömig, A., and O. Rose. 2003. "On the Suitability of the Weibull Distribution for the Approximation of Machine Failures," *Proceedings of the 2003 Industrial Engineering Research Conference*, Portland, OR.

Sharda, R. 1994. "Neural networks for the MS/OR analyst: An application bibliography." *Interfaces 24*(2), 116–130.

Studer, L., and F. Masulli. 1996. "On the structure of a neuro-fuzzy system to forecast chaotic time series," *International Symposium on Neuro-Fuzzy Systems*, pp. 103–110.

Tangirala, R. 1996. "A nonlinear stochastic model of fatigue crack length for on-line damage sensing," *Decision and Control Conference.*

Weigend, A. S., and N. A. Gershenfeld. 1993. "Time Series Prediction: Forecasting the Future and Understanding the Past." Addison Wesley, Reading, MA.

Werbos, P. J. 1988. "Generalization of backpropagation with application to a recurrent gas market model," *Neural Networks 1*, 339–356.

Zhang, B., C. Sconyers, C. Byington, R. Patrick, M. Orchard, and G. Vachtsevanos. 2008. "Anomaly Detection: A Robust Approach to Detection of Unanticipated Faults." *IEEE Conference on Prognostics and Health Management*, Denver, CO.

CHAPTER 5 IVHM Assessment Metrics

Abhinav Saxena, *SGT Inc., NASA Ames Research Center*
Michael J. Roemer, *Impact Technologies , A Sikorsky Innovations Company*

> "What's measured, improves."
>
> —Peter F. Drucker

5.1 Measuring IVHM Effectiveness

Metrics let us quantify the effectiveness of a proposed solution in a standardized manner to aid in decision making and/or process improvement. Several stakeholders of IVHM (Integrated Vehicle Health Management) have different user objectives, and IVHM performance must be measured to assess effectiveness toward fulfillment of their respective functional requirements such as identified by Jennions [ed. 2011]. Naturally, the development and use of metrics is largely guided by various user objectives of these stakeholders.

System-level requirements can be generally grouped under *safety*, *cost*, and *performance* categories. The goal of an IVHM system is to be able to maintain system performance by ensuring safety and minimizing additional costs through improvements in reliability and availability. Therefore, various IVHM scenarios are largely divided into two top-level goals from stakeholder perspectives: (1) *maintenance management*, and (2) *contingency management*, which, otherwise, are also sometimes viewed as relating to (1) *cost management*, and (2) *safety management*. Therefore, all stakeholders of IVHM fall under either Cost or Safety categories (see Figure 5.1). As pointed

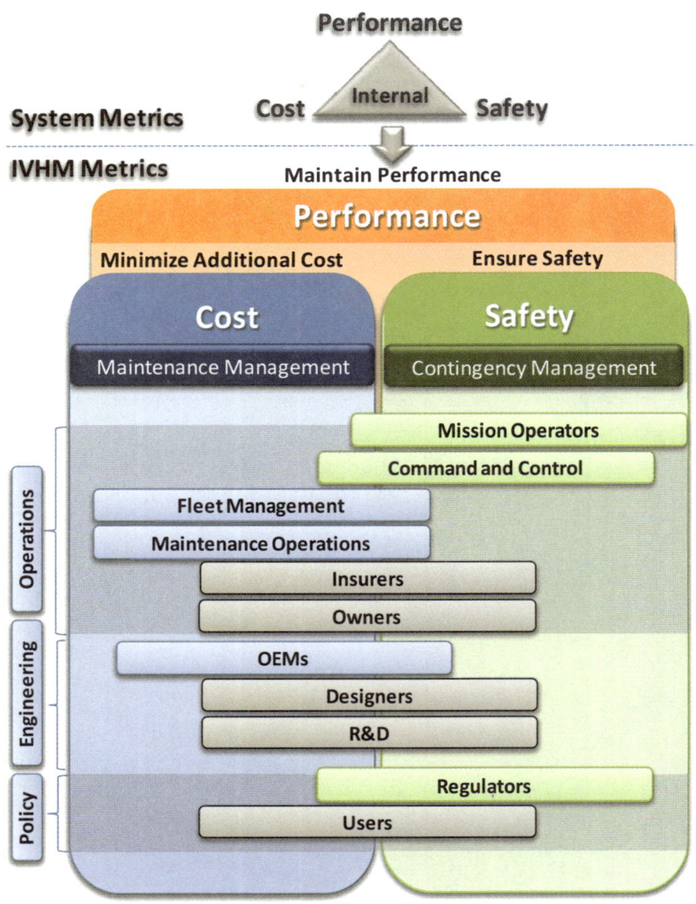

Figure 5.1 Various IVHM stakeholders and their user objectives mapped on to high-level goals of cost and safety.

out by Wheeler et al. [2009], depending upon the role of various stakeholders in the IVHM life cycle, these stakeholders and their objectives are further categorized into three broad categories in: (1) Operations (program manager; plant/fleet manager; mission operator; command and control; and maintainer, owner, insurer), (2) Regulatory (policy makers), and (3) Engineering (designer, researcher, OEMs). Several metrics have been designed that cater to these different user groups. However, the discussion here will be limited to metrics relevant to engineering development and operations of IVHM.

5.2 IVHM Metrics

5.2.1 Introduction

As identified in the previous section, metrics quantify the attributes expected from the implemented IVHM system and, therefore, vary for different stakeholders. In this section we discuss these different types of metrics and list some that are commonly found to be useful in the IVHM community. First, some of the high-level metrics that evaluate the overall performance of an IVHM system are presented [Dibsdale 2012]. These metrics, as presented in Table 5.1, directly reflect the interests of various stakeholders of IVHM.

	Metric	Description/Usage
Effectiveness	Overall Value Added	It is a general metric that defines IVHM value based on a general set of individual metrics measuring effectiveness of key factors for a stakeholder; e.g., recovery time saved (gain in availability) by using IVHM and hence the avoidance of cost due to downtime. Value can also be established by quantifying improvements in traditional metrics used for system reliability as discussed in section 5.2.2.
Design	Failure Coverage	It is specified as the ratio of failures covered by IVHM to total number of significant failures that are candidates for IVHM.
	System Coverage	It is specified as the proportion of significant components/subsystems covered by IVHM. It requires a system to be partitioned into several independent modules with significant criticality.
Cost	IVHM Implementation & Operational Costs	This measure quantifies the time and cost of installing an IVHM system on a vehicle. Installation costs include labor costs, costs for commissioning and training, licensing costs, cost of hardware and software etc. This in turn also includes the costs of computational complexity as reflected in hardware costs and power requirements (see section 5.2.3). Furthermore, these measures can be weighed against potential benefits of IVHM as discussed in section 5.2.4.
IVHM Function	Warning Times	This is the class of measures that quantifies the efficacy in terms of timeliness of IVHM output toward avoiding dire consequences. Warning times should long enough to accommodate: • Amount of time necessary to avoid secondary damage from a primary failure mode • Amount of time necessary to plan and act upon remedial actions • Amount of time necessary to carry out planned outages and minimize spare holdings
	Diagnostic & Prognostic Performance	The overall performance of IVHM heavily depends on the performance (accuracy, precision, timeliness, etc.) of underlying diagnostic and prognostic algorithms as discussed in section 5.2.5.

Table 5.1 High-Level IVHM Effectiveness Metrics

5.2.2 Reliability Analysis Metrics

5.2.2.1 Introduction

Reliability analysis stems from statistical aggregates from historical data or carefully designed experiments to collect statistics about failure rates of certain components, which can then be used to assess overall system reliability, design tolerances, and indirectly predict system life. While not a focus of this chapter, it is important to mention that reliability analysis is a very mature field, and much progress has been made in its application to the optimization of aftermarket operations. Reliability-Centered Maintenance (RCM) is the basis for much of how assets are maintained today, and it is important to draw a contrast between RCM and IVHM in the context of performance assessment.

In RCM, component failure is analyzed in great detail on the basis of experimental and field data to determine inspection, repair, and replacement intervals for all categories of components. The techniques of RCM form the basis for all aerospace maintenance practices, as embodied in MSG-3 guidelines, which are the way FAA recommends operators maintain their transport aircraft [FAA 1997]. However, much of the reliability analysis is based on past data (historical usage or lab testing), which is where IVHM differentiates itself by being focused on condition-based health management. Use of condition-based techniques calls for different metrics to assess IVHM effectiveness; these metrics are the focus of this chapter. However, it is noteworthy that reliability metrics can still be used to assess the additional value in terms of cost-benefit or the risks of using an IVHM system over the baseline system performance without IVHM because an effective IVHM system must improve these metrics as well. Therefore, a brief introduction to these metrics is provided here with some representative examples. Such metrics are generally useful for *operators, maintainers, designers,* and *regulatory stakeholders*. More discussion on reliability-based metrics can be found in [IEEE 2002]. Broadly speaking, there are two types of reliability metrics, as described next [Wood 2001].

5.2.2.2 Constant Rate Reliability Metrics

Constant Rate Reliability metrics are commonly referred to as Mean Life metrics. They represent a good approximation of the flat region of the reliability bathtub curve (see Figure 5.2) and are generally expressed as MTB*xx* (Mean Time Between *xx*), where *xx* is the domain (industry) specific quantity of interest such as number of failures.

MTBF (Mean Time Between Failures) usually assumes an exponential distribution, which

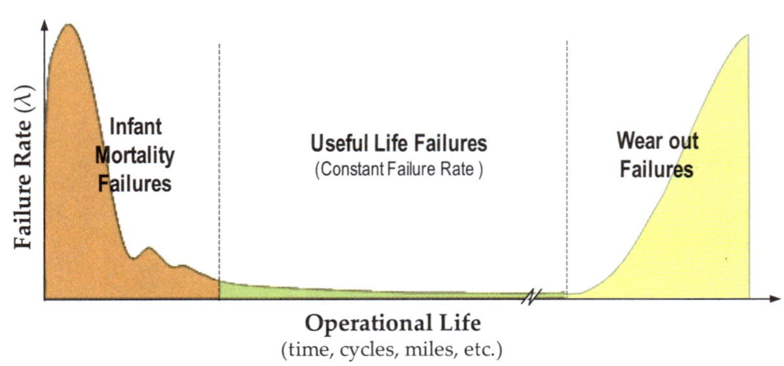

Figure 5.2 Reliability bathtub curve.

makes them an equivalent constant rate (λ) metric where the rates are either measured from field data or computed from aging experiments. It expresses observed failure rate computed as *total failures divided by total population operating time*. MTB*xx* metrics are also sometimes customized by the reference scales involved in any application such as number of cycles (MCBF), or distance in miles (MMBF) for automobiles. Some other variations of these metrics are MTBSC (Mean Time Between Service Calls) and MTBMA (Mean Time Between Maintenance Actions) that express perception of failure rates, or MTBWC (Mean Time Between Warranty Claims) that can be used for pricing warranties or setting warranty reserves. These metrics can be expressed as failures per hour, failures per year, or Annualized Failure Rate (AFR). Another variation of these metrics is Part Return/Repair Rate, expressed as total parts replaced divided by total population operating time. It is an indirect measure of failure rate when failure analysis is not available. Likewise, other metrics are used, such as MTBF/MTBUR ratio, which measures the ratio between how long a component is expected to last based on reliability estimates and how long it is actually used before replacing it. IVHM should enable the reduction of this ratio by allowing components to be used longer [Teal and Larsen 2003]. Therefore, although these metrics have been traditionally used in the field of reliability, they can still be used to track improvements due to the IVHM system by quantifying the before and after difference.

5.2.2.3 Probability of Success Metrics

Expressing a probability that a system performs a required function under stated condition for a stated period of time is useful when systems do not show constant failure rates and hence specifying mean times is not sufficient. Some examples of such metrics include Probability of Mission Success/Failure, Probability of Loss of Control, etc. Traditional methods normally finesse this by adding margins of safety, but this is not cost effective in the long run. Alternatively, measuring the percentage of population that survives a specific duration is also a useful measure of performance in some cases. These metrics are usually duration dependent (i.e., they depend on the length of the mission). These are often specified in terms of percentiles of the distributions. For instance, *Lx*-Life in mechanical systems specifies the number of hours after which at least *x*% of the population would have failed. Some metrics in the automotive industry are specified along with confidence levels (e.g., R96C90 meaning 96% Reliability with 90% Confidence). These metrics can be used to assess the effectiveness of an IVHM system by quantifying improvement in both reliability and confidence about failure-free operation during planning and decision making based on health information.

5.2.3 Computational Performance Metrics

With advancements and sophistication of IVHM approaches to vehicle systems, reasoners and models are growing in complexity, and therefore demanding high computational efficiency and performance as key engineering requirements. There is often a competing relationship between time and computational resources available vs. the desired fidelity and performance expected from an IVHM system. Computational performance metrics help specify the hardware requirements or otherwise specify constraints within which the software must work, yet still satisfy algorithmic performance requirements, hence influencing the engineering design and implementation costs. Note that some of the following metrics (Table 5.2) depend on the hardware chosen to implement the IVHM function.

Metric	Description	Unit
Computational Complexity	Generally described using "Big O" notation ($O(n)$,$O(n^2)$, $O(n^3)$, $O(n^m)$, etc.), it describes the amount of time needed for an algorithm to run (independent of a software and hardware implementation), as a function of the size of the input.	ms, s
CPU Time	Measures the time a CPU spends executing software to specify combined performance of an algorithm, its software implementation, and the hardware on which it is run.	ms, s
Elapsed Time	Measures total CPU time including wait times for input/output (I/O).	ms, s
Memory Size	Specifies memory (RAM, Flash, and/or disk space) requirements. It is a key metric for embedded applications, such as for flight computer in aerospace that usually have very limited space available.	Bits, Bytes
Data Rate	Measures how many samples per time unit the software can handle for a given hardware configuration. This is particularly important in assessing software capability to handle real-time application.	Bits/sec (bps)

Table 5.2 Computational Performance Metrics

5.2.4 Cost Benefit Metrics

A common use of IVHM performance data is to perform a Cost-Benefit Analysis (CBA) of implementing an IVHM system. Clearly, the cost-benefit equation is significantly affected by how effective the IVHM system will be. Therefore, the cost-benefit metrics try to quantify the performance of IVHM such that they can be easily converted into monetary terms to be used in a cost-benefit analysis. Some examples of such metrics are provided in Table 5.3.

Metric	Description
Return On Investment (ROI) [Sandborn 2012]	A return on investment on health management is generally realized as cost avoidance through timely maintenance and possibly prevention of downtime or lost hardware over the life of the system. $ROI = \frac{C_u - C_{HM}}{I_{HM}}$, where C_u is life-cycle cost of the system when managed using unscheduled maintenance C_{HM} is life-cycle cost of the system when managed using HM approach I_{HM} is investment in implementing and managing HM, which is further obtained as the sum of C_{NRE} (non-recurring HM costs), C_{REC} (recurring HM costs), and C_{INF} (HM infrastructure cost).

Table 5.3 Cost-Benefit Assessment Metrics

5.2.5 Algorithm Performance Metrics

5.2.5.1 Introduction

Successful implementation of IVHM technologies relies heavily on intelligent algorithms that interpret health-monitoring data and reason about the health of the vehicle systems. Therefore, much effort has been put into evaluating and refining these algorithms in the recent past. However, the notion of using fairly sophisticated algorithms to implement Prognostics and

Metric	Description
Net Present Value (NPV) [Roemer 2012]	NPV is used to project the costs for a specified time period in the future. Present value of the costs expected to be incurred in future are estimated by incorporating the discount rates. In the IVHM context it is calculated as $NPV = C_{F\mid\neg HM} - C_{F\mid HM} - C_{HM}$, where $C_{F\mid\neg HM}$ is present value of the total expected consequential cost of a failure without doing maintenance $C_{F\mid HM}$ is the present value of total expected consequential cost of a failure after doing maintenance C_{HM} is present value of the cost of maintenance.
Payback Period	Payback period is the time for which a system owner would have to wait until the benefits realized from an IVHM system break even with the value adjusted costs. It is only after the payback period that net potential benefits may be realized. A short payback period is preferred.
Technical Value [Dzakowic and Valentine 2006; Vachtsevanos et al. 2006]	Costs and benefits associated with the low-level algorithmic performance can be aggregated to be used in a cost-benefit analysis. The benefits achieved through accurate detection, fault isolation, and prediction of critical failure modes are weighed against the costs associated with false alarms, inaccurate diagnoses/ prognoses, and resource requirements of implementing and operating specific techniques. $TV = P_f(D \cdot \alpha + I \cdot \beta) - (1 - P_f)(P_D \cdot \varphi + P_i \cdot \theta)$ P_f: probability of a failure mode D: overall detection confidence metric α: savings realized by fault detection in advance I: overall isolation confidence metric β: savings realized by isolating a fault in advance P_D: false positive detection metric ϕ: cost of false positive detection P_i: false positive isolation metric θ: cost of false positive isolation
Total Value	Describes the value of a PHM technology in a particular application as the summation of the benefits it provides over all the failure modes that it can diagnose or prognose less the implementation cost, operation and maintenance cost, and consequential cost of incorrect assessments. $V_{TOTAL} = \sum_{i=1}^{FM} TV_i - A - O - (1 - P_c) \cdot \delta$ A: acquisition and implementation costs O: life-cycle operational and maintenance cost P_c: computer resource requirements δ: cost of computer systems

Table 5.3 Cost-Benefit Assessment Metrics *(Cont.)*

Health Management (PHM) concepts have been around in civil aviation since at least the late seventies when engine and aircraft data became available via Aircraft Communications Addressing and Reporting System (ACARS) transmission to the ground. A number of metrics often customized for specific applications are available in the literature, a sampling of which is provided in this section. This section is further divided into discussions specific to performance evaluation of diagnosis and prognosis, which are the core algorithmic components at the heart of IVHM systems.

5.2.5.2 Diagnostic Performance Metrics

Diagnostics is one of the key elements in implementing an IVHM function; it strives to accomplish an early fault detection, fault isolation (pinpointing the root cause), and fault identification (estimating the severity of the isolated fault). There are several flavors to diagnostic activities such as on-board vs. off-board assessments, real-time vs. long-term trending, component-level vs. system-level reasoning, etc. However, the key point to note here is that despite the granularity and approach of diagnostic methods used, the performance assessment methods share common attributes and, therefore, can be largely divided into four categories of metrics for diagnosis:

1. *Diagnostic Performance Statistics:* Performance statistics of system-level diagnostic assessments collected over time, in which multiple fault modes may be present on a variety of subsystems or components.
2. *Accuracy and Precision:* Accuracy and precision in detecting faults and estimating severity that determine the confidence levels in diagnostic assessment due to presence of noise, variability, and other uncertainties in the system.
3. *Robustness of Detection:* Robustness or sensitivity to various factors (environmental, operational, design, etc.) that could change diagnostic outcomes.
4. *Response Time:* Timeliness in assessments while maintaining accuracy and precision to facilitate responsive and useful post-diagnostic reasoning actions.

A list of representative metrics for each of these four types of diagnostic metrics is provided in Table 5.4 with a brief description and a list of original sources where these metrics have been reported.

5.2.5.3 Prognostics Metrics

Prognostics is the process of predicting a system's remaining useful life by estimating

<table>
<tr><th>Metric</th><th>Explanation and Usage</th></tr>
<tr><td colspan="2">Diagnostic Performance Statistics (Rate) Metrics</td></tr>
<tr><td>Confusion Matrix</td><td>
<table>
<tr><td rowspan="4">Predicted System State</td><td></td><td colspan="2">Actual System State</td></tr>
<tr><td></td><td>Non-faulty</td><td>Faulty</td></tr>
<tr><td>Non-faulty</td><td>a</td><td>b</td></tr>
<tr><td>Faulty</td><td>c</td><td>d</td></tr>
</table>
A confusion matrix is often constructed to document outcomes (correct or incorrect) and subsequently analyze performance of a diagnostic system. Once populated, this matrix is then used to derive a variety of performance indicators commonly used in the industry. Performance is often specified as absolute correct or incorrect outcomes, or weighted to express as rates over a period of time. Some examples are given below.</td></tr>
<tr><td>Detection Accuracy</td><td>The ratio of correctly classified cases (i.e., fault detected when the system was actually faulty, and no fault detected when the system was non-faulty) to the total number of scenarios to assess overall accuracy of the system.
$$\mathrm{DA}=\frac{a+d}{a+b+c+d}$$</td></tr>
</table>

Table 5.4 Diagnostic Performance Metrics

Metric	Explanation and Usage
Fault Detection Rate	The ratio of cases in which a fault is detected out of the total number of scenarios when the system was actually faulty. $\text{FDR} = \frac{c+d}{b+d}$
False Negative Rate	The ratio of the number of times a fault is missed (goes undetected) to the number of scenarios when the system was actually faulty. Low false negatives are preferred to avoid downtime due to missed failures. $\text{FNR} = \frac{b}{b+d}$
False Positive Rate	The ratio of the number of times a fault is detected to the number of scenarios when the system was actually non-faulty. Low false positive rates are desired to avoid unnecessary maintenance actions. $\text{FPR} = \frac{c}{a+c}$
False Positive Confidence Metric [Roemer et al. 2005]	Measures the frequency and upper confidence limit associated with false anomaly detection by a diagnostic tool. Calculation of the false confidence metric is based on the false positive function that is stated in the following equation: $F(c) = n(c)/N$ where $n(c)$ = number of false positive detection events with confidence $> c$, N = number of opportunities to detect a normal operating condition
Receiver Operating Characteristic (ROC)	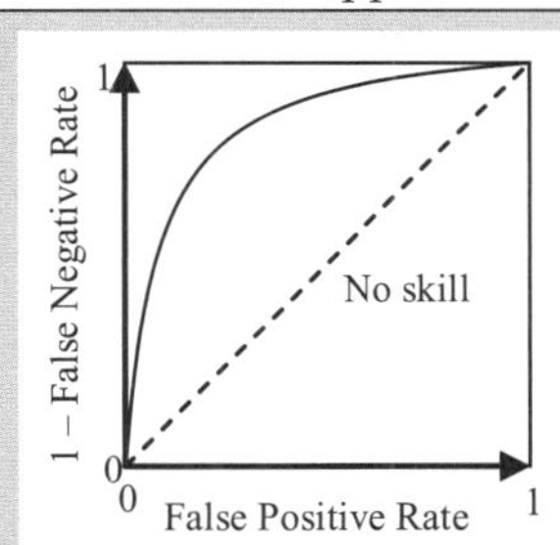 ROC gives a comprehensive overview of the tradeoff between false positives (FP) and false negatives (FN). The ideal curve would have zero false positives and zero false negatives. Such a curve cannot realistically be achieved for real-world problems and, therefore, a balance must be achieved based on cost and safety priorities in a given application.
Power of Test	Power of test estimates the overall bias of a diagnoser toward FPs or FNs over the entire range of the ROC curve. It is computed by calculating the area under the ROC curve, which represents the probability that a diagnoser favors false positives over false negatives in general.
Isolation Performance Matrix	Similar to confusion matrix above a higher fidelity confusion matrix may be constructed for a system that not only detects faults but also isolates various fault modes. Absolute or rate metrics can be constructed for isolation performance as well. In the matrix shown below all elements are weighted by total number of cases, which can also be seen as probabilities if large number of cases is observed.

		Actual State			
Predicted State	Component	$Mode_1$	$Mode_2$	…	$Mode_n$
	$Mode_1$	p_{11}	p_{21}	…	p_{n1}
	$Mode_2$	p_{12}	p_{22}	…	p_{n2}
	…	…	…	…	…
	$Mode_n$	p_{1n}	p_{2n}	…	p_{nn}

Table 5.4 Diagnostic Performance Metrics *(Cont.)* *(continues)*

Metric	Explanation and Usage
Isolation Classification Rate	The rate of correct classification. This metric equals the sum of elements along the diagonal. $\text{ICR} = \sum_{i=1}^{n} p_{ii}$
Isolation Misclassification Rate	The rate of misclassification computed as the sum of all off-diagonal elements. $\text{IMR} = \sum p_{ij} \ \forall i \neq j$
Size of Isolation Set	Often, a diagnostic system does not converge on a single isolated fault but a number of possible faults with corresponding probabilities. In such cases the size of the isolation set becomes a metric. A smaller size implies less ambiguity and hence is preferable.
Accuracy and Precision Measures	
Error	Error defines the basic notion of deviation from desired output. Applications where estimates of fault dimension are computed, such as crack length or bearing spall size, typically use estimation error as a metric of diagnostic performance. Δ = Measured Ground Truth – Diagnostic Estimate Most accuracy-based metrics are derived directly or indirectly from error as defined below.
Error-Based Metrics [Hyndman and Koehler 2006]	Several metrics are defined to aggregate errors (Δ) into absolute or weighted measures. Aggregates can be computed over multiple systems (l) (e.g., in a fleet scenario) or over multiple test cases (i) for a single system over time. Some representative examples are given here, but many more metrics exist that can be useful in different cases. Mean Squared Error (MSE) $\text{MSE}(i) = \frac{1}{L}\sum_{l=1}^{L} {}^{l}(i)^2$ Mean Absolute Error (MAE) $\text{MAE}(i) = \frac{1}{L}\sum_{l=1}^{L} \left\| {}^{l}(i) \right\|$ Root Mean Squared Percentage Error (RMSPE) $\text{RMSPE}(i) = \sqrt{\frac{1}{L}\sum_{l=1}^{L} \left\| \frac{100\ {}^{l}(i)}{r_*^l(i)} \right\|^2}$ Mean Absolute Percentage Error (MAPE) $\text{MAPE}(i) = \frac{1}{L}\sum_{l=1}^{L} \left\| \frac{100\ {}^{l}(i)}{r_*^l(i)} \right\|$
Precision-Based Metrics [Hoaglin et al. 1983]	Precision in estimation is estimated by computing spread of the error around the true value. Once again, there are several variants of dispersion measures, some of which are mentioned here to exemplify. Users pick their favorite ones based on desired robustness and acceptable computational complexity in computing these measures.
Sample Standard Deviation	Sample standard deviation measures the dispersion/spread of the error with respect to the sample mean of the error. Technically this metric is restricted to the assumption of normal distribution of the error, which may not always be a correct assumption. $S(i) = \sqrt{\frac{\sum_{l=1}^{n}\left({}^{l}(i) - M\right)^2}{n-1}}$ where M is the sample mean of the error.

Table 5.4 Diagnostic Performance Metrics *(Cont.)*

Metric	Explanation and Usage
Mean Absolute Deviation (MAD) from the sample median	A more robust estimator of the dispersion/spread of the estimation error is MAD: $MAD(i)=\frac{1}{n}\sum_{l=1}^{n}\left\|\ ^{l}(i)-M\right\|$ where $M=\underset{l}{\text{median}}(\Delta^{l}(i))$ and median is the $\frac{n+1}{2}$th order statistic.
Robustness Metrics	
In presence of noise and other uncertainties, diagnostic algorithms can be tuned to adjust detection thresholds and confidence levels per user requirements. Therefore, confidence levels and detection thresholds can be seen as tuning parameters in specifying risk acceptance.	
Detection Thresholds [Orsagh et al. 2002; Roemer et al. 2005]	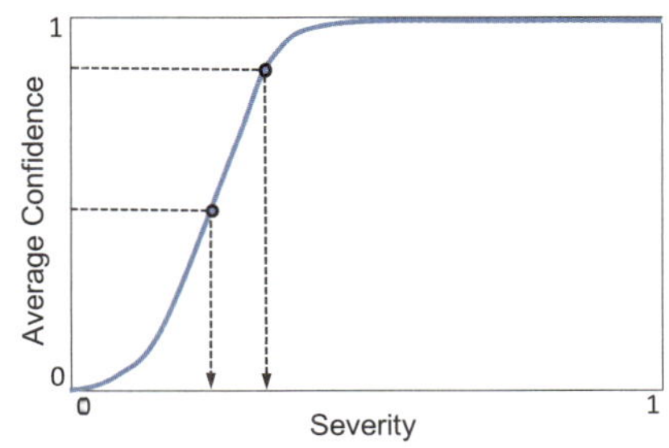 This metric quantifies the ability of a diagnostic algorithm in terms of the minimum ground truth severity an algorithm can detect with a designated confidence level and is represented on the detection success function as shown in the figure. Confidence of 67% and 95% (corresponding to one and two standard deviations of a normal distribution) are used to calculate the detection threshold metric score as: *Detection Threshold* = 1 – $S(c)$ where $S(c)$ = ground truth severity at confidence = c.
Detection Sensitivity [Bartys et al. 2006; Orsagh et al. 2002; Roemer et al. 2005]	A diagnostic system should detect anomalies over the full range of operating conditions such as loads, speeds, noise, duty cycles, etc. This metric measures the difference between the success functions of a diagnostic tool under two levels of conditions. where $\text{Condition Sensitivity}=1-\sqrt{\int_0^1 (C_1(s)-C_2(s))^2\,ds}$ $C_1(s)$ = success function at condition 1 $C_2(s)$ = success function at condition 2 s = severity
Detection Stability [Orsagh et al. 2002; Roemer et al. 2005]	This metric quantifies the fluctuation in confidence values that occurs over the fault transition by integrating the peak to peak difference at each point on the success function as stated here. Basically, a confidence level that fluctuates wildly is difficult to interpret and therefore undesirable. $\text{Stability}=1-\sqrt{\int_0^1 (C_H(s)-C_L(s))^2\,ds}$ where $C_H(s)$ = maximum value of the success function at severity s $C_L(s)$ = minimum value of the success function at severity s s = severity

Table 5.4 Diagnostic Performance Metrics *(Cont.)* *(continues)*

Metric	Explanation and Usage
Detection Stability Factor [Kurtoglu et al. 2008]	In presence of noise, a diagnostic output can fluctuate, especially when fault is less severe. In such cases decisions can be made more robust by quantifying the level of stability of the detection signal, specified as duration for which the outcome doesn't change as a percentage of total time elapsed since the first detection signal. Thresholds may be set for detection stability factors for a corrective action to be initiated.
Response Time Metrics	
It is critical in many applications that diagnoses are quick enough to enable an actionable decision in a timely manner. In such cases once can measure time taken to carry out a particular diagnostic function [Bartys et al. 2006; Kurtoglu et al. 2009; Poll et al. 2011]. Such metrics are generally useful during design and testing while tuning the diagnostic system, and require the knowledge about true fault appearance time, which is never available in real applications. Faults can be injected to a system during testing, and diagnostic response times may be measured.	
Time to Detect	The period of time from the beginning of fault injection to the moment when fault was detected for the first time.
Time to Isolate	The period of time from the beginning of the fault injection to when the isolation convergence set becomes available, i.e., final diagnostic output of an algorithm is obtained.

Table 5.4 Diagnostic Performance Metrics *(Cont.)*

the time at which the system will no longer perform its intended function within desired specifications. It is one of the advanced functions that can significantly improve IVHM performance by warning the system operators of impending failures to avoid downtimes, avoid secondary damages, and allow sufficient time to plan and execute required Maintenance, Repair, and Overhaul (MRO). However, predicting the future with limited information is not a trivial task and leads to uncertainties in the predictions. Furthermore, assessing performance of prognostics algorithms is an acausal problem, as it would require knowledge of the true End of Life (EoL) time, which is clearly not possible because this will occur sometime in the future. Therefore, most prognostics performance metrics have been developed for offline performance assessment, and relatively few exist for online applications.

To use metrics suggested in Table 5.5, some characteristics in prognostics implementation are assumed. A prognostic routine may predict system/component health from the very beginning or may be triggered by an event such as detection or isolation of a critical fault. Upon trigger, the prognostic algorithm starts predicting the Remaining Useful Life (RUL) according to a predefined EoL criteria or failure threshold, such as maximum allowable crack length, stiffness loss in a structure, etc. Furthermore, prognostic estimates should be updated continuously as more information about the system's health becomes available through time. Nominally, a prognostic algorithm is expected to produce improved predictions as time progresses, and this improvement is an important metric to track as well. Last but not least, predicting into the future inherently involves uncertainties that must be accounted for. Several approaches have used probability distributions of RUL to represent such uncertainties, and metrics should take these distributions into consideration wherever needed. These metrics can be simplified as needed to evaluate point predictions instead of distributions if that happens to be the preferred

Metric	Description and Usage
Offline Prognostic Performance Metrics	
Prognostic Horizon	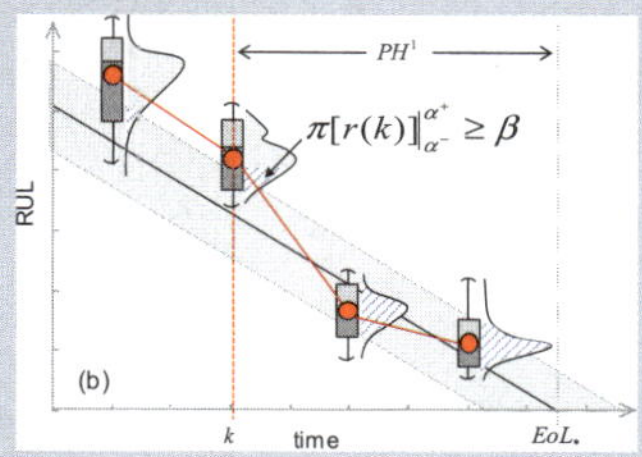 Determines the maximum advance warning an algorithm can provide with desired confidence. It is defined as the maximum predicted RUL (or the first prediction) that falls within desired accuracy (α) bounds. An algorithm with larger PH is preferred. $PH = t_{EoL} - t_{prediction} \mid \text{first prediction} \in \text{desired accuracy } (\alpha)$
α-λ Performance	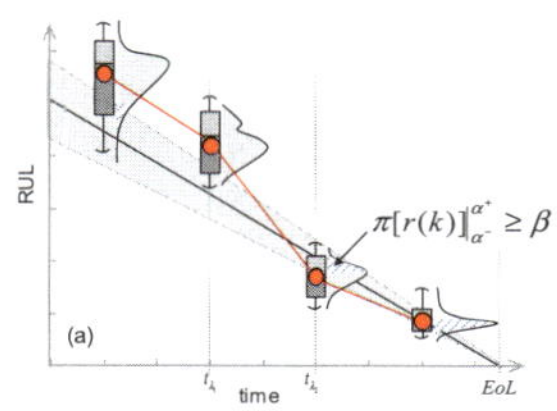 Characterizes how well an algorithm's performance (accuracy and precision) improves as EoL approaches. Improvement is measured against desired accuracy level (α) expressed as a percentage of RUL at any specified time (λ). $\alpha - \lambda \text{ Accuracy} = \begin{cases} 1 & \text{if prediction} \in \alpha \text{ accuracy cone} \\ 0 & \text{otherwise} \end{cases}$
Prognostic False Alarm Rate	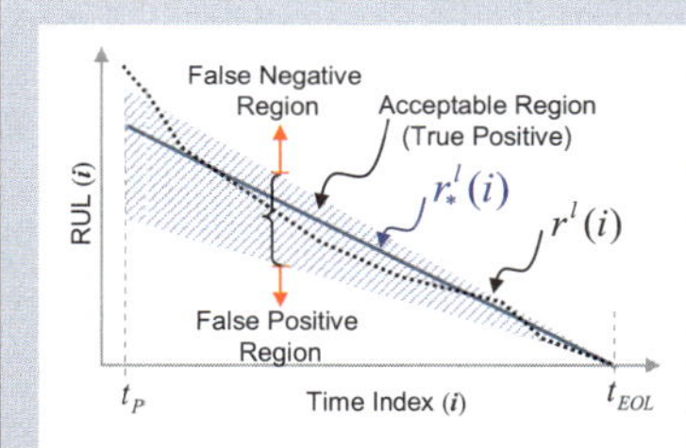 This metric is an extension from classical false positive and false negative metrics used for diagnostics performance where an early prediction is classified as false positive and a late prediction as false negative. Therefore, the α accuracy bounds mentioned above define FP and FN regions on an RUL vs. time graph. [Goebel and Bonissone 2005; Leao et al. 2008; Luna 2009; Saxena et al. 2008]
Relative Accuracy	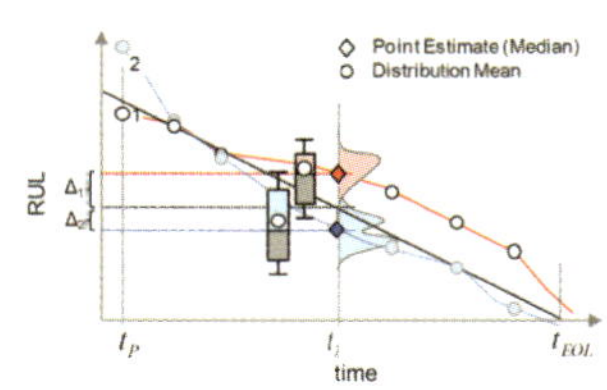 Quantifies how accurate an algorithm is at any given time (λ) relative to the EoL. It is computed as RUL error normalized by actual RUL for any given time. It captures the notion that accuracy of prediction becomes more critical as EoL approaches. Further, a cumulative relative accuracy may be computed by aggregating the RAs at different time instants to specify the algorithm's accuracy in general.
Convergence Rate	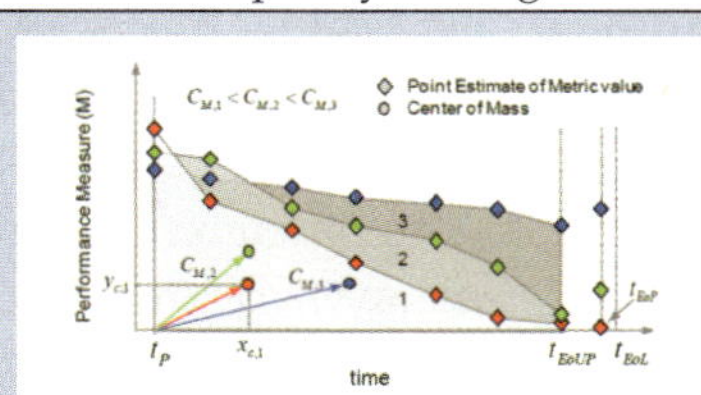 Quantifies the rate of convergence of prognostic performance improvement as predictions are updated with time. Convergence can be computed for any metric of interest such as accuracy, precision, etc. This metric quantifies the convergence rate by computing the center of mass of the area under the predicted trajectories for respective algorithms. A faster convergence rate indicates faster performance improvement and hence more time available for making actionable decisions for IVHM.

Table 5.5 Prognostic Performance Metrics *(continues)*

Metric	Description and Usage
Sensitivity [Vachtsevanos et al. 2006]	Sensitivity is a measure of how sensitive a prognostic algorithm is to input changes or external disturbances. It can be assessed against any performance metric of interest. ΔM is the distance measure between two successive outputs for metric M's value, and Δ_{input} is a distance measure between two successive inputs (e.g., failure threshold, noise level, available sensor set, sampling rate, etc.). Sensitivity may be aggregated over similar systems (units under test (l)). $S(i)=\frac{1}{L}\sum_{l=1}^{L}\left\{\frac{\Delta M^{l}(i)}{\Delta_{input}}\right\}$
Online Prognostic Performance Metrics	
RUL Online Precision Index (RUL-OPI) [Orchard et al. 2009]	RUL-OPI quantifies and tracks the precision of predicted RUL distributions by quantifying the length of 95%confidence bounds (CI(i)) normalized by the predicted RUL (r(i)) at any given time instant. An algorithm with a high index (close to 1) is preferred, which indicates high precision or narrow confidence bounds. $I(i)=e^{-\left(\frac{\sup\{CI(i)\}-\inf\{CI(i)\}}{r(i)}\right)}$
Dynamic Standard Deviation (DStd) [Olivares et al. 2013]	DStd quantifies the stability of predictions within a time window (Δ). Variance between individual predictions made within the time window is computed. The metric is normalized to a range [0,1] using the logistic function (φ) for easy comparisons. $\mathrm{DStd}=\varphi\left(\mathrm{Var}\left(E\{\mathrm{EOL}\mid y_{1:j}\}\right)_{j\in\Delta}\right)$
Critical-α Performance Measure [Olivares et al. 2013]	Looking from the perspective of actionable decision making, this measure computes the critical percentile (α) of an RUL distribution that would define a Just-In-Time-Point (JITP) for that application. JITP must always occur before actual failure, and hence the value of this metric lies in interval (0,0.5] and should be maximized to avoid unnecessary conservatism in decision making. $\alpha_{crit}=\arg\max_{\alpha}\{\mathrm{JITP}_{\alpha\%}(k_{pred})\le \mathrm{EOL}\};\quad \forall k_{pred}\in[1,\mathrm{EOL}]$

Table 5.5 Prognostic Performance Metrics *(Cont.)*

prognostic output for various reasons; however, a more general form of the metrics is presented here.

It is noteworthy that most of these metrics combine the intent of measuring accuracy and precision together with time tracking of performance by plotting the predicted RUL vs. operational time. In an ideal case, accurate predictions must fall on the −45° straight line representing ground truth, but that is rarely the case. Any deviation from the ground truth can be quantified or bounded by acceptable error bounds (α) around the ground truth line. RUL distributions can be represented as error bars showing mean, median, or percentiles as needed. Further, for better estimates one could estimate the total probability mass of the distributions that fall within desired accuracy bands and compare with a threshold specified by β to classify a prediction into good or bad categories. More details on these metrics and use cases are addressed by Saxena et al. [2010].

5.3 The Challenge of Data Availability

Performance of an IVHM system can be evaluated using various metrics described in the previous section, however, it must be noted that the usefulness of the output from these metrics depends on the quality of data on which these metrics are applied. To perform Verification and Validation (V&V) of specific PHM algorithms or technologies, various types of data sources with different levels of fidelity can be used. From the perspective of a PHM algorithm designer, the most desirable data sets would come from either seeded fault testing, accelerated mission testing, or actual field data from the operational system where detailed loading profiles are known and selected parameters were monitored that are correlated with the progression of the failure. However, these types of data sets are typically expensive to obtain and hence represent less than 10% of the data that is used for performing a comprehensive V&V program. To supplement these more comprehensive data sets, subsystem or component rig tests can be used that are more cost effective but limited in applicability and realism. Also, dynamic simulation models with different levels of fidelity can be used to simulate fault signatures and provide data at any desired sampling rate. There are obviously many benefits to the simulation-based approach but many drawbacks as well. Table 5.6 details these pros and cons of these various data sources and their applicability to PHM V&V and metrics evaluation.

Hence it is argued that performance evaluations should consider the effects of data

Method	Description	Benefit	Drawback
Simulation Models and Fault Generation	Signals, noise, and fault signatures simulated mathematically	Ideal for generalizing (varying) signals and noise; partially overcomes small sample problem	Data may not always be realistic; model may be expensive to develop to reliably simulate
Component and LRU Fault and Failure Tests	Use data collected on component or subscale test rigs	Read seeded fault, natural fault, and fault progression data available at little cost to program	Limited applicability for actual system fault observability
Subsystem/System Fault Characterization Tests	Tests on stands using actual subsystem hardware	Somewhat realistic data; usually well controlled and documented with ground truth information	Seeded faults may not be entirely realistic of natural fault
Limited Vehicle Data Evaluation	Dedicated missions for normal and off-normal test and evaluation	Realistic data of normal and some performance faults	Expensive; doesn't cover full range of potential conditions; no critical faults or progression
Technology Maturation Field Program	Data collection and tests during vehicle service	Covers full range of vehicle; mission, operational, and environmental conditions	Limited to opportunistic fault occurrence. Delays use and implementation in field

Table 5.6 Data Source Applicability to IVHM Metrics

quality while interpreting the results and their applicability toward actual stakeholders' objectives. Metrics should be chosen carefully, and an adequate thought must be given to what their outputs really represent and what is the extent of their validity toward V&V activities.

5.4 Conclusions

This chapter discusses various performance measures that are used to assess the effectiveness of an IVHM system. It is argued that metrics are developed to measure effectiveness against the user objectives of various stakeholders; therefore, whatever matters is what should be measured. Examples of overall IVHM effectiveness measures point toward broad categories of value and cost, which are further broken down into specific metrics. A special focus is given to algorithmic performance metrics because diagnostic and prognostic performance lie at the core of how effective an IVHM system can be. Considering the wide variety of metrics available for assessing effectiveness, it is suggested that this list be used as a representative set of basic metrics that can be further augmented with more customized versions that cater to a specific stakeholder's needs. Last but not least, metrics convey a message based on the data on which they are applied. Therefore, it is not sufficient to just choose the right metrics but also to decide the right set of measurements and parameters that must be tracked. In the absence of real and/or adequate field data, clever approaches are necessary that cover the gaps generated by unavailability of such data, some of which can be implemented by choosing the right metrics.

5.5 References

Bartys, M., R. Patton, M. Syfert, S.d.l. Heras, and J. Quevedo. 2006. "Introduction to the DAMADICS actuator FDI Benchmark Study," *Control Engineering Practice, 14*, 577–596.

Dibsdale, C. 2012. "Asset Management and its Implications to IVHM," In I. K. Jennions (Ed.), *Integrated Vehicle Health Management Business Case Theory and Practice* (Vol. 2), SAE International, Warrendale, PA.

Dzakowic, J.E., and G.S. Valentine. 2006. "Advanced Techniques For The Verification And Validation Of Prognostics & Health Management Capabilities," Paper presented at the Machinery Failure Prevention Technologies Conference (MFPT 60), Virginia Beach, VA.

FAA. 1997. *Maintenance Review Board Procedures AC 121-22A*, U.S. Department of Transportation.

Goebel, K., and P. Bonissone. 2005. "Prognostic Information Fusion for Constant Load Systems," Paper presented at the 7th Annual Conference on Information Fusion.

Hoaglin, D.C., F. Mosteller, and J.W. Tukey, eds. 1983. *Understanding Robust and Exploratory Data Analysis*: John Wiley & Sons.

Hyndman, R.J., and A.B. Koehler. 2006. "Another Look at Measures of Forecast Accuracy," *International Journal of Forecasting, 22*(4), 679–688.

IEEE. 2002. IEEE Guide for Selecting and Using Reliability Predictions Based on IEEE 1413™ *1413.1 TM* (p. 97): IEEE Standards Coordinating Committee 37.

Jennions, I. K., ed. 2011. *Integrated Vehicle Health Management: Perspectives on an Emerging Field*, SAE International, Warrendale, PA.

Kurtoglu, T., O.J. Mengshoel, and S. Poll. 2008. "A framework for Systematic Benchmarking of Monitoring and Diagnostic Systems," Paper presented at the International Conference on Prognostics and Health Management, Denver, CO.

Kurtoglu, T., S. Narasimhan, S. Poll, D. Garcia, L. Kuhn, J. de Kleer, and A. Feldman. 2009. "International Diagnosis Competition—DXC'09" Paper presented at the 20th International Workshop on Principles of Diagnosis.

Leao, B.P., T. Yoneyama, G.C. Rocha, and K.T. Fitzgibbon. 2008. "Prognostics Performance Metrics and Their Relation to Requirements, Design, Verification and Cost-Benefit," Paper presented at the International Conference on Prognostics and Health Management (PHM08), Denver, CO.

Luna, J.J. 2009. "Metrics, models and scenarios for evaluating PHM effects on logistics support," Paper presented at the Annual Conference of the Prognostics and Health Management Society, San Diego, CA.

Olivares, B.E., M.A.C. Muñoz, and M.E. Orchard. 2013. Particle-Filtering-Based Prognosis Framework for Energy Storage Devices With a Statistical Characterization of State-of-Health Regeneration Phenomena. *IEEE Transactions on Instrumentation and Measurement, 62*(2), 13.

Orchard, M.E., L. Tang, K. Goebel, and G. Vachtsevanos. 2009. "A Novel RSPF Approach to Prediction of High-Risk, Low-Probability Failure Events," Paper presented at the Annual Conference of the Prognostics and Health Management Society (PHM09), San Diego, CA.

Orsagh, R.F., M.J. Roemer, C.J. Savage, and M. Lebold. 2002. "Development of Performance and Effectiveness Metrics for Gas Turbine Diagnostic Techniques," Paper presented at the IEEE Aerospace Conference, Bigsky, MT.

Poll, S., J. de Kleer, R. Abreu, M. Daigle, A. Feldman, D. Garcia, and A. Sweet. 2011. "Third International Diagnosis Competition—DXC'11, Paper presented at the 22nd International Workshop on Principles of Diagnosis, 2011.

Roemer, M.J. 2012. "Maintenance Cost Analysis—Leveraging Prognostic Information," In I. K. Jennions, ed., *Integrated Vehicle Health Management Business Case Theory and Practice* (Vol. 2), SAE International, Warrendale, PA.

Roemer, M.J., J. Dzakowic, R.F. Orsagh, C.S. Byington, and G. Vachtsevanos. 2005. "Validation and Verification of Prognostic and Health Management Technologies," Paper 1344 presented at the IEEE Conference Aerospace, Bigsky, MT.

Sandborn, P.A. 2012. "Making Business Case for Health Management—Return on Investment," In I. K. Jennions (Ed.), *Integrated Vehicle Health Management Business Case Theory and Practice* (Vol. 2), SAE International, Warrendale, PA.

Saxena, A., J. Celaya, E. Balaban, K. Goebel, B. Saha, S. Saha, and M. Schwabacher. 2008. "Metrics for Evaluating Performance of Prognostics Techniques," Paper presented at the 1st International Conference on Prognostics and Health Management (PHM08), Denver, CO.

Saxena, A., J. Celaya, B. Saha, S. Saha, and K. Goebel. 2010. "Metrics for Offline Evaluation of Prognostic Performance," *International Journal of Prognostics and Health Management, 1*(1), 21.

Teal, C., and B. Larsen. 2003. "Technology Update II: Wire Systems Diagnostics & Prognostics," Paper presented at the Digital Avionics Systems Conference.

Vachtsevanos, G., F. L. Lewis, M. Roemer, A. Hess, and B. Wu. 2006. *Intelligent Fault Diagnosis and Prognosis for Engineering Systems* (1st ed.), Hoboken, New Jersey: John Wiley & Sons, Inc.

Wheeler, K.R., T. Kurtoglu, and S. Poll. 2009. "A Survey of Health Management User Objectives Related to Diagnostic and Prognostic Metrics," Paper presented at the

ASME 2009 International Design Engineering Technical Conferences and Computers and Information in Engineering Conference (IDETC/CIE), San Diego, CA.

Wood, A.P. 2001. "Reliability-Metric Varieties and their Relationships," Paper presented at the Reliability and Maintainability Symposium, Jan 22–25, 2001, Philadelphia, PA.

CHAPTER 6 IVHM Architecture

Kirby Keller, *Boeing Research & Technology*
Tarapong Sreenuch, *IVHM Centre, Cranfield University, UK*
Kevin Swearingen, *Boeing Research & Technology*
Ravi Rajamani, *Meggitt PLC*

> "They can do without architecture who have no olives nor wines in the cellar."
>
> —Henry David Thoreau

And for those less disposed to alcohol and who seek to avoid the calories:

> "Architecture is the reaching out for the truth."
>
> —Louis Kahn, architect

6.1 Introduction

One of the leading barriers to realizing the benefit of advanced Integrated Vehicle Health Management (IVHM) is access to the control and sensor data required to determine and project the condition or health of vehicle components. Acquiring this data involves the integration of additional sensing, data acquisition, processing, and communications capabilities into the vehicle—in the aerospace domain, the addition of elements is very sensitive to such cost factors as size, weight, power, and flight certification. This chapter describes some emerging technologies and architectures that can be cost effective in implementing hardware and software elements on board an air vehicle. The end goal of such implementation is to operate and support a fleet of vehicles more efficiently.

Put in terms of the sense, acquire, analyze, and act cycle; the sensing, data acquisition and analysis is put on board for two purposes: (1) to support the crew in flight, and (2) to support the ground process as shown in Figure 6.1.

The use of diagnostic or prognostic output in flight has been referred to as taking "direct action" as opposed to "deferred action," in which case the results are used to support maintenance, logistics, or operational planning [Gorinevsky and Mah 2010]. Of interest here are primarily the onboard elements. While offboard technology is obviously, a key portion of the end-to-end system, implementation of technology on board is generally the greater challenge and source of value.

Direct-action IVHM can be viewed as an extension of current built-in test systems that support redundancy management and crew alerting. The elements of IVHM that support the deferred actions are "along for the ride" in that they do not directly aid in performing

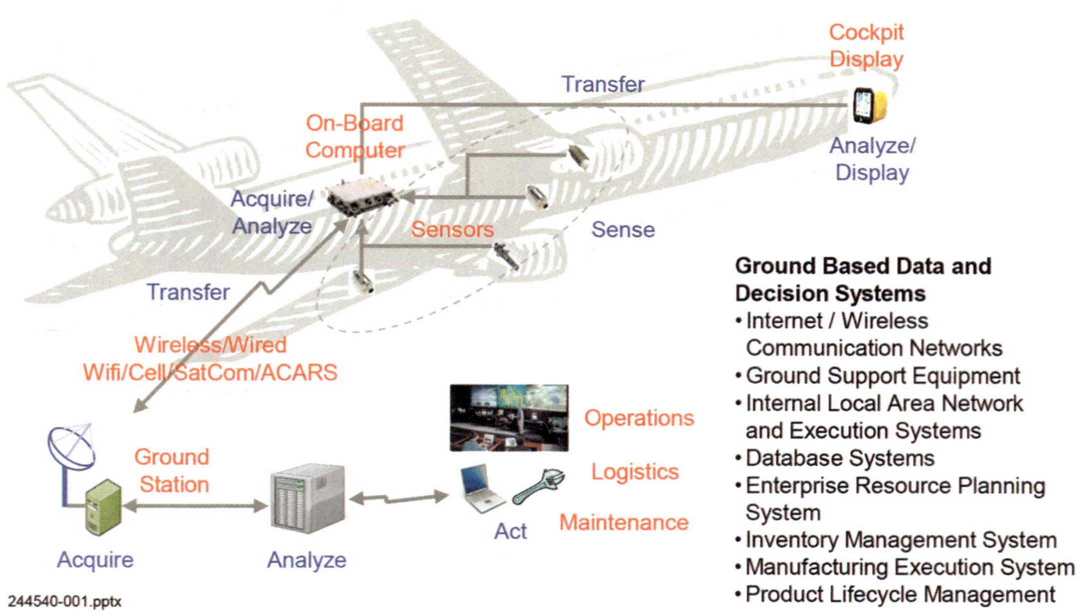

Figure 6.1 Onboard and offboard elements of IVHM.

the current flight. Because the major objective of this onboard capability is Condition-Based Maintenance (CBM), which is based on prognostic or usage-based life prediction, and because we are interested in the onboard element of CBM, we will refer to this onboard element as OCBM.

The processing continues with an offboard-based network that completes the end-to-end system shown in Figure 6.1. Because of the limitations of this chapter, the primary focus is on the onboard systems and OCBM. However, because the ground-based elements are so important to the functioning of an IVHM system and to the enterprise fleet management, we include a section describing some of the major offboard systems and functions that are involved. We also acknowledge that data may be downloaded, processed, and the results uploaded to the crew, but this is an aspect we chose not to pursue here.

Thus, the chapter provides the following:

- A background on the evolution of avionic systems and IVHM
- Guidelines on implementing cost-effective OCBM
- An IVHM functional architecture
- An advanced hardware architecture that includes IVHM and OCBM
- Hardware technologies to enable cost-effective inclusion of OCBM
- A software architecture to enable cost effective implementation of OCBM
- A description of the offboard support architecture

IVHM and OCBM are evolving and competitive, and they vary significantly with the specific vehicle application. The intent of the chapter is to provide insights and rationale behind some of the methods and technologies that have been pursued and are in the public domain. It is not an endorsement of these technologies and architectures but a description of their functionality and intended use.

6.2 Background on IVHM and Avionics Systems

In the early years of flight, the crew inspections confirmed that the aircraft was ready for flight. As electronic flight instrumentation and control features were added, it was necessary to test the capabilities both to clear the systems for flight and during flight to alert the crew to fault conditions so the crew would not rely on the instruments and/or to support reconfiguration of the system to a backup or redundant mode. The increasing complexity of the instrumentation and controls, in turn, drove the complexity of the Built-In-Test (BIT) and redundancy management functions with the goal of providing the crew with the best capability to control the aircraft.

Early avionics systems were implemented using a federated architecture. Federated vehicle electronics consist of specialized electronics or collections of electronics that perform a specific function [Ramsey 2007; Watkins and Walter 2007; Rahul 2008]. In federated architectures, each system is implemented independently, with no planned commonality that might support sharing of resources such as processing or data acquisition. Health Management (HM) in such systems amounts to a BIT capability that supports redundancy management and the generation of alerts to the crew based on the faults detected in that box or system of boxes. BIT, true to the federated philosophy, was implemented as a standalone capability for a specific system or Line Replaceable Unit (LRU), with little or no system-level integration. Its purpose is primarily to first ensure successful operation of a specific function as it contributes to usage success, and second to provide maintenance personnel with an indication that some replaceable unit associated with the federated system has failed. Troubleshooting or fault isolation was a ground-based maintenance function.

As federated avionic systems became more complex, it became increasingly difficult to detect and isolate the failed component. Faults in one component or LRU could trigger faults in other components/LRUs. The more complex systems exhibited more complex fault modes, including intermittent faults that lead to an increase in unwarranted removals and No Faults Found (NFF), especially for subsystems with complex controllers such as the engine or flight control.

In-flight data collection became more prevalent with the creation of Health and Usage Monitoring Systems (HUMS) to avoid in-flight failures of rotorcraft engines and gearboxes. HUMS units collected data that were downloaded and processed to monitor selected flight-critical systems. Analysis of the data was a manual process supported by analytical tools. As analytical capabilities progressed, it became feasible to automatically detect certain component degradations prior to failure, which in turn drove interest in collection and downloading of more-detailed data related to subsystem performance and condition.

As support costs of ever-more-complex air vehicles rose, the aerospace industry began to realize that economic benefit could be realized from acquiring and processing digital data on flight and subsystem performance, not just to help ensure safe flight but as a business to support efficient fleet operations and lower operations and support costs.

Flight recorders placed on board to support investigations of anomalous flight events and accident investigations became the focus for additional data collection and downloading for the purposes of improving aircraft maintenance. However, the cost and technical limitations of the legacy avionics systems severely limited the data that were accessible. The federated avionics architectures of military and commercial air vehicles built in the 1960s through the 1980s was developed upon a bus structure that interfaced to specialized LRUs to allow the LRUs to share data as needed. A simplified federated architecture is shown in Figure 6.2. The data shared between LRUs, and thus available on the bus to be recorded, consisted of system-level control and status messages. Bus capacity limitations encouraged designers to be frugal with bandwidth. Key systems such as flight control or propulsion execute control loops in the neighborhood of 50 to 100 Hz. An internal bus structure allowed sharing of data

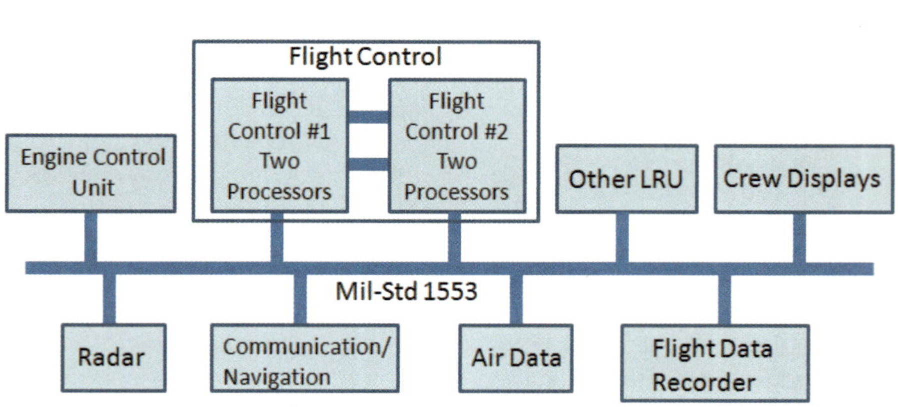

Figure 6.2 Representative federated military avionics architecture.

between redundant processors, as shown in the flight control node in Figure 6.2. However, bus throughput limitations typically dictated that only a selected subset of such control parameters could be made available on the avionics buses. Typically, such parameters were down sampled to 5 Hz or less. Thus, data that might have supported more-capable diagnostic or prognostic algorithms for such key systems are "trapped" in safety-critical controllers that are costly to modify.

Integrated modular architectures pioneered by the Boeing 777 and USAF F-22 provided common processing and interface capabilities to host avionics functions, with additional bus capacity to provide access to data [Ramsey 2007]—even to the point of enabling dynamic selection of captured parameters. However, access to control parameters was typically limited to rates less than 25–30 Hz.

Newer generation aircraft such as the Airbus A-380, Boeing 787, and Joint Strike Fighter provide even greater access to data. However, the appetite for data has increased faster than the ability to access it.

The new-generation aircraft do not readily support emerging sensor technology such as surface acoustic wave, acoustic, and high-frequency accelerometers. In addition, more electric aircraft have spawned a new generation of controllers (and failure modes) that operate at higher frequencies. For example, control of brushless DC motors used in electro-mechanical actuators may have inner loop control rates on the order of 500 Hz or greater.

The IVHM implementation "valley of death" is captured in Figure 6.3. To date, avionics system designs have not supported IVHM requirements, especially in regard to cost-effective implementation of OCBM. In addition to lack of support for accessing and processing high-fidelity performance data, the development and test cycle for aircraft systems does not support the post-flight-test

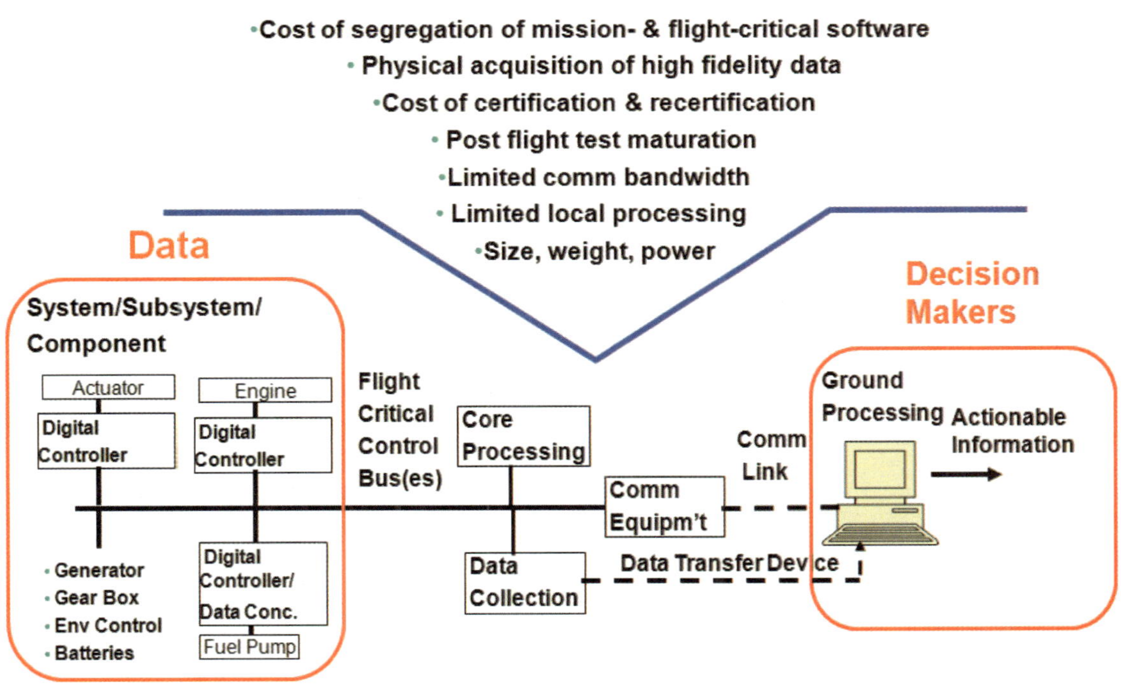

Figure 6.3 OCBM valley of death.

maturation of OCBM capabilities. Flight test data acquisition that might support maturation of OCBM algorithms is removed, and the operational data acquisition system remaining provides inadequate support.

The nature of vehicle control functions vs. OCBM heightens this situation. Control functions are developed and put through a rigorous testing and certification process. Once the control system has been certified, the hardware and software configuration is frozen and carefully controlled. In the same time frame of the control certification, any included OCBM functionality is likely to be based on laboratory seeded or run-to-failure data and/or simulation models. OCBM algorithms that detect subtle changes in health state typically require tuning and adjustment with experience. Laboratory conditions are not adequate to generate accurate real-world wear conditions, so the health management algorithms that drive CBM performance require maturation driven by vehicle use. This is contrary to most vehicle system development processes. Vehicle systems undergo intense verification and validation testing, after which changes are strongly discouraged by the significant cost of requalification.

Given the low probability of faults/degradations due to wear occurring during flight test, this maturation will likely occur after flight test when flight test instrumentation has been removed. Thus, OCBM brings with it a maturation requirement to complete the characterization of various degradations and to update the CBM software, and perhaps hardware.

On the positive side, the OCBM element, much of which is support-critical, provides a testbed to mature these technologies and architectures that may be a stepping stone to flight-critical capabilities that impact vehicle control and system safety. It also provides economic motivation to pursue such development. Also, newer vehicles do generally include greater flexibility and lower criticality frameworks to better support OCBM.

The bottom line is that attempts to retrofit OCBM onto legacy, and even new-generation aircraft, are comparatively expensive and inefficient. This situation begs for research into technologies that reduce the cost of OCBM by addressing the described issues related to the acquisition of high-fidelity control data, the addition of high-bandwidth sensors, support for post-flight-test maturation, and partitioning of support-critical OCBM elements to reduce certification costs.

There is a natural tendency to say that OCBM implementation can wait for new aircraft designs that will better support the required data acquisition and communications. Given the longevity and number of both commercial and military legacy systems for which OCBM is sorely needed, cost-effective implementation in legacy systems opens up a wealth of opportunity.

6.3 Design Guidelines for OCBM

An OCBM application may be safety critical (typically DO-178C Level A or B) if the results are used for maintenance credit to reduce or eliminate inspections, or may be support critical (typically DO-178C Levels C or D) if the results are used only as input to support operations, logistics, and maintenance decisions.

Thus, cost effectiveness is the only driver for implementing support-critical OCBM because it is not required for safe flight. The decision to include any hardware and software on board to achieve such benefit is an economic decision. The motivation for including OCBM arose within the last ten years as it became evident that OCBM is a necessary ingredient in the business of reducing sustainment costs. The magnitude of the task of including OCBM is driven by the desire for the higher-benefit prognostic applications that requires access to high-fidelity control and state data. The onboard hardware and software appropriate for specific IVHM/OCBM applications involves complex trades that include the level of performance, the timeline of the decision process, and the cost of implementing onboard hardware and software on the host vehicle. This is further complicated by the relationship of the stakeholders. The operator may be leasing the aircraft and have a "power by the hour" maintenance contact. So the actual beneficiary of IVHM/OCBM capability may vary.

The cost factors associated with the implementation of any capability on board air vehicles (see Figure 6.4) are a starting point for creating guidelines for OCBM designs [Keller et al. 2012].

The list in Figure 6.4 can be summarized by saying that the OCBM system footprint must be minimized to reduce the cost of implementation and thus maximize return. For OCBM applications to be cost effective, the design of such systems must exploit the following:

- Leverage existing platform resources such as processing, communications, power, and sensors, where possible.
- Exploit planned upgrades to the host system to reduce the cost of OCBM implementation.
- Seek opportunities to implement a new capability that is part of a multifunctional upgrade, gaining benefit from more than just OCBM.
- Mitigate the need to change the current design documentation.
- Provide a communications infrastructure and isolate design changes to the LRU or component.
- Reduce software and hardware certification costs by partitioning the support-critical from flight-critical elements.
- Leverage offboard data collection and ground-based processing, which are less expensive than doing these on board or in flight, and which should be the first option.

Cost Factors	
Minimize addition of communications wiring	Minimize thermal load
Minimize addition of power wiring	Minimize hardware volume
Minimize addition of circuit breakers	Minimize additional dedicated sensors
Minimize weight of added hardware, excluded wiring	Minimize reliability impact
Minimize uncertified components/system	Minimize maintainability impact
Minimize new hardware qualification	Minimize impact to existing interfaces (buses, protocols, etc)
Minimize the number of additional part numbers	Minimize processing requirements
Utilize high Technology Readiness Level (TRL)/proven technologies	Minimize support equipment
Maximize multi-use components	Minimize training requirements
Minimize power consumption	Minimize drawing changes

Figure 6.4 Onboard implementation cost factors.

- Seek an efficient environment for development and upgrading by using an open architecture and standards that support distributed development and contributions from multiple suppliers.
- Reduce implementation, maintenance, and configuration costs through a modular, plug-and-play approach.
- Replacement and repair of the OCBM capability must be an opportunistic task that does not impact fleet operations. This can be achieved by a mix of inherent reliability, designing CBM to be separately maintainable, and opportunistic maintenance on the OCBM system within the host's prediction horizon, if applicable.
- Configure the CBM so the host LRU/subsystem/component can be offered with and without the CBM capability.
- Support incremental OCBM upgrades.
- Maintain the flexibility to move functionality on or off board as the need arises.

6.4 IVHM Functional Architecture

As vehicle systems (including vehicle management or control functions) move from federated systems toward integrated modular designs, lower-level control migrates to digital controllers associated with the various subsystems. BIT and IVHM followed this trend. Factors such as the high rate of false alarms and NFFs in federated systems, the need to support redundancy management at the subsystem controller level, and the desire for more computationally complex OCBM functions that require fusion of data from contributing systems fuel the evolution toward hierarchical, distributed IVHM architectures.

The accepted functional organization of an IVHM system for all purposes has matured to the layered, hierarchical model shown in Figure 6.5 [Hess and Fila 2002]. This model consists of three nominal layers: member systems, area managers, and a vehicle-level manager. The member systems provide lower-level inputs that are integrated across subsystem domains at the area and vehicle level. Member system functions include data concentration, sensor interface, and signal processing. Area manager functions include system/subsystem control, communications, and system health management. Finally, the vehicle-level manager incorporates vehicle management functionality, health management, and operational management.

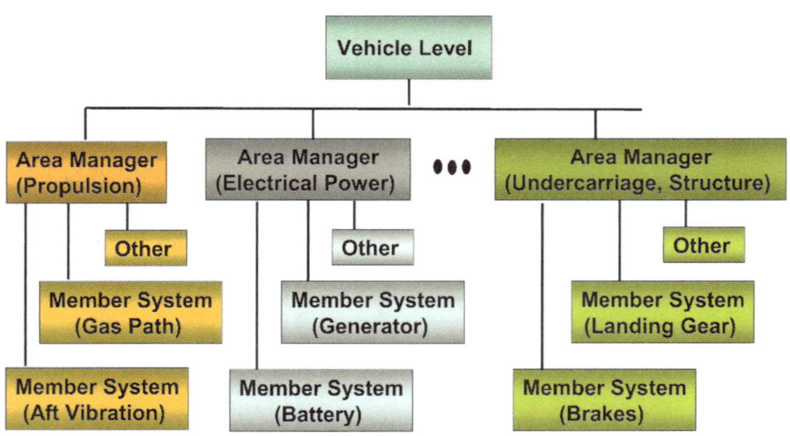

Figure 6.5 IVHM functional architecture.

6.5 The Avionics Architecture That Includes IVHM

Advanced integrated, modular avionics architectures are a good framework for enabling OCBM. The movement toward a hierarchical IVHM architecture is consistent with the greater use of digital control in subsystems in avionics systems. Lower-level, finer-grain control enhances overall aircraft capabilities and provides an improved level of subsystem and vehicle performance by enabling responses to degraded or failed systems through reconfiguration or redundancy management at any level in the hierarchy. The hardware architecture to support OCBM and vehicle control can be distributed over several hardware layers, as shown in Figure 6.6.

A distributed, hierarchical architecture requires efficient communications between nodes to achieve high-performance, real-time response. If the communications function is added to the control and IVHM function, a generic architecture node shown in Figure 6.6 results [Keller et al. 2010].

This on-vehicle hierarchical distributed architecture implements control and IVHM. It allows for the detection of faults and a local response to be generated quickly. It supports OCBM data collection and processing. It supports the handling of events that require additional processing and fusion of results up the hierarchy before the appropriate response can be generated. Compared to a centralized approach, the distributed approach can leverage parallel processing to allow better recognition of events and generate the appropriate response in the time frame required. In the non-real-time context, a data download capability and the extent of onboard processing can improve the effectiveness and timeliness of the response. Including more processing onboard generally reduces the required bandwidth of the download. Time and space partitioning of the support-critical OCBM processing and communications provides additional freedom for maturation. Such partitioning would impact the data input (sensors), processing, and communication, as shown in Figure 6.6.

Regardless of the extent of the onboard processing, the data downloaded must support accurate fault isolation or estimation of future health, which is necessary for the ground-based decision processes to quickly return the vehicle to service. This trade-off between the cost and complexity of onboard processing vs. the bandwidth required to

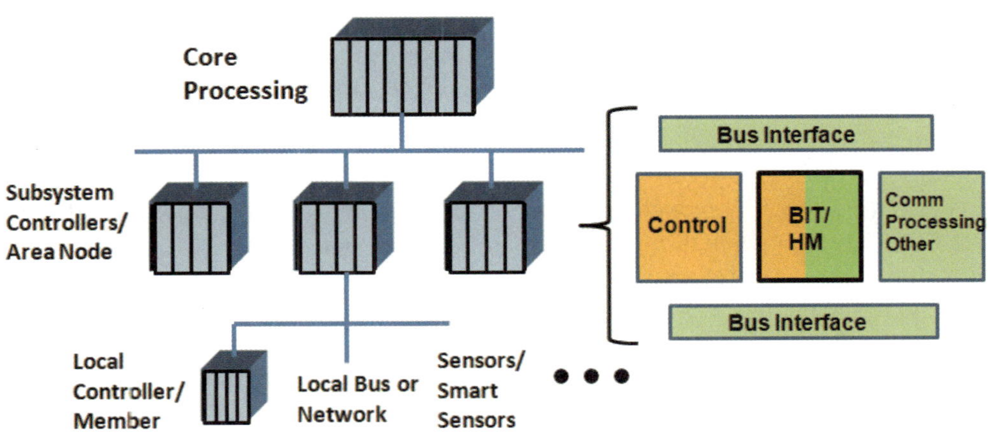

Figure 6.6 Hardware avionics architecture to support IVHM/OCBM.

communicate raw data off-vehicle frequently results in a compromise that raw data are partially processed on board to reduce communication bandwidth. This approach has a drawback in that the original source data may not be available for additional processing or system maturation. Further, this approach requires vehicle systems that incorporate computer control. On long-lived legacy vehicles such as aircraft, this is frequently not the case, which makes integration difficult, if not impossible.

The architecture must also support the control and IVHM performance requirements of the host system. Computer system performance can be divided by immediacy into hard-real-time, near-real-time, and non-real-time regimes. The hard-real-time regime typically includes responses that are necessary for critical control loops, such as flight control surface commands or electrical load monitoring for arcs or shorts. This regime is typically of the highest criticality; contains the most expensive software to host, qualify, and modify; and includes the most rigorous computing demands. The near-real-time regime includes responses that may require operator interaction, inter-system interaction, or involve off-vehicle communications. This regime is typically of lower criticality (i.e., support critical) because the information it includes may not be necessary for proper operation of the vehicle. Finally, the non-real-time regime incorporates information recorded or transmitted off board for post-flight use. Data to feed algorithms that measure long-term component wear fit in this category, as do operator evaluation, energy efficiency, and other lower priority concerns. There is little rationale for OCBM systems to operate in the hard-real-time regime. It is usually adequate for OCBM systems to operate in a combination of near-real-time and non-real-time, depending on the immediacy of the information. As the scope of, and confidence in, OCBM results increases, applications may expand from maintenance oriented to mission and flight critical, driving more time-dependent OCBM activity.

6.6 Hardware Technologies That Enable OCBM

6.6.1 Introduction

In practice, avionics designs have only recently considered other than centralized health management architectures, which have very modest communications capability and locate health management components in federated subsystems. A number of technologies are described that have emerged to enable implementation of OCBM in legacy and new aircraft. Such technologies are especially in demand for legacy systems that require such resources to be employed.

6.6.2 Onboard Wireless Communications

Among the higher costs of OCBM implementation is the communication of OCBM data within the aircraft. Wireless communications has to be considered a key technology in enabling OCBM on air vehicles. Wireless can be used to implement an inexpensive second communications network over the depth and breadth of the CBM processing hierarchy. Please consult chapter 11, Disruptive Technologies, for a more complete description of wireless technology. In this chapter we just note its effectiveness in addressing the guidelines listed.

Wireless offers the capability to achieve a low-footprint, plug-and-play communications network for the entire OCBM hierarchy. Specifically, it can be used at the local level to acquire data from smart sensors and communicate those data up to higher-level OCBM nodes.

Tests conducted with a wireless sensor network based on the IEEE 802.15.4 (ZigBee) specification on commercial aircraft show good connectivity between various compartments of an aircraft. Again, please refer to chapter 11, Disruptive Technologies.

Communication of CBM data via satellite communications from the aircraft and ground system is in use by a number of IVHM systems [Boeing 2013]. Providing data in flight gives maintenance personnel time to prepare, thus reducing the turnaround time of the aircraft.

Downloading CBM data at the gate can be done from any level in the CBM hierarchy, but is practically limited to wired power nodes at the vehicle or area manager (data concentrator). Other advantages of wireless communications (ZigBee and Wi-Fi) are its ubiquitous use in the commercial aircraft cabin for personal services and its strong Commercial Off-The-Shelf (COTS) base.

Wireless communications for CBM does have limitations for both military and commercial, including the following:

- Global frequency allocation (commercial)
- Operational signature (military)
- Security
- Energy harvesting effectiveness
- Electromagnetic interference
- Regulatory oversight, especially for critical functions
- Wireless hardware reliability

The most critical factor in enabling CBM is the effectiveness of energy harvesting. Running wires to power wireless sensors can obviate the advantage of wireless. However, in certain circumstances there are advantages to having a "wire-less" system in which the sensor is powered through wires and signals are received wirelessly. Methods to manage power as well as to increase the power/weight ratio of energy harvesting devices are key areas that will determine the usefulness of wireless for OCBM.

6.6.3 Solid-State Power Distribution

The new generation of Electrical Power Distribution Systems (EPDS) use solid-state Electronic Circuit Breakers (ECBs). These ECBs provide arc fault detection and protection by reducing the magnitude of the arcing event. The ECB has a natural secondary function as a means to support CBM for the electrical loads [Keller et al. 2007; Potter and Ballas 2012, 2012].

The cost of this IVHM capability is a communications bus to acquire the data for IVHM/OCBM purposes, and to a lesser degree the change in reliability of the EPDS as it increases in complexity. If it is desirable to support in-flight fault reporting, an architecture such as that shown in Figure 6.7 may be used. In this architecture a separate support-critical bus and processor is employed to capture and process the current signature data. The result is communicated to a central maintenance computer via a separate data bus. The safety-critical function would be implemented to the desired level of redundancy. Such an architecture could support rapid and efficient in-flight reporting.

Another configuration might capture the current data via an onboard maintenance recorder from the safety-critical bus. This would reduce the IVHM footprint but is limited by the throughput of this bus; however, it avoids the additional complexity and certification with the power distribution system. The OCBM objectives would drive the data acquisition requirement in terms of sampling rate and duration, or window, of data that must be acquired. In either of the cases, the dual role of the ECB provides a wealth of CBM data not available with thermal breakers.

A more advanced EPDS uses a hierarchical distribution system to reduce wire runs to the bottom tier of loads. Such a power system architecture provides a challenge in supporting communication of OCBM data from these local nodes. An additional challenge is the presence of controllers between the ECB and the electrical device. Such a controller can hide key electrical signals from the ECB.

In addition to acquiring digitized load current, there is ongoing research on the use of ECB to host online, active, wire fault interrogation technologies as well as using the ECB as a means of control for simple subsystems such as lighting or blowers. This provides an additional motivation to integrate multiple functions within the ECB/EPDS.

6.6.4 Multipurpose Wire Fault Detection and Location Applications

Chafing Protection System (CHAPS) is an example of a dual-use technology to detect and locate wire chafing as well as enable the low-cost and low-footprint communications infrastructure for health management [Morris et al. 2010; Born, Stratton, and Harris 2001].

CHAPS detects the onset of chafing by means of a "sense shield" consisting of a number of optical fibers (sensors) incorporated within wire bundle overbraid, as shown in Figure 6.8. Deterioration of the fiber strains inhibits their ability to communicate an optical signal, which is an indication of chafing. The obstructed fiber can be interrogated to provide a means to locate the damaged area or chafe. The degraded area is detected and located using an Optical Time Domain Reflectometer (OTDR), which is implemented as support equipment and interfaces via a test connector or can be implemented as an onboard monitoring system by means of a miniature optical transceiver.

Following up on the goal to leverage multifunctional capabilities, CHAPS can be extended beyond a standalone wire chafing detection and location system to also provide a redundant, high-bandwidth communication network for health management data. The resulting application has been referred to as CHAPS-HM. The optical fibers are terminated by a connector in which the fiber-optic "pins" are mounted in Multi-Termination (MT) ferrules that are designed for housing inside a MIL-DTL-38999 Series III Size 11 military connector shell. Split ring connectors can be used to daisy chain the fibers to link data acquisition points to higher-level HM nodes. The split ring connector mounts on the wire bundle and provides a means to couple optical fibers. It also provides a

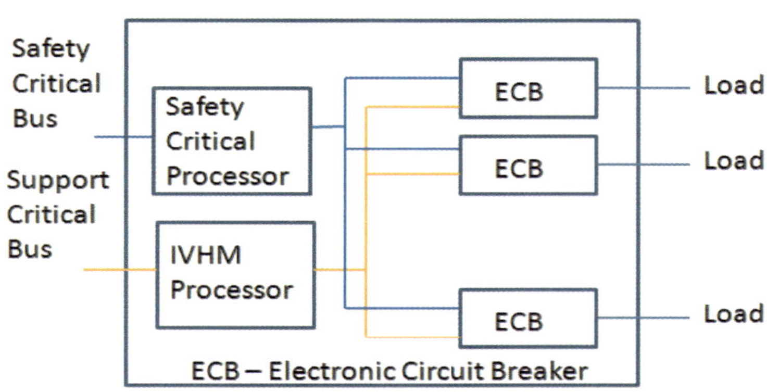

Figure 6.7 Representative architecture that leverages ECB acquisition of digital current data.

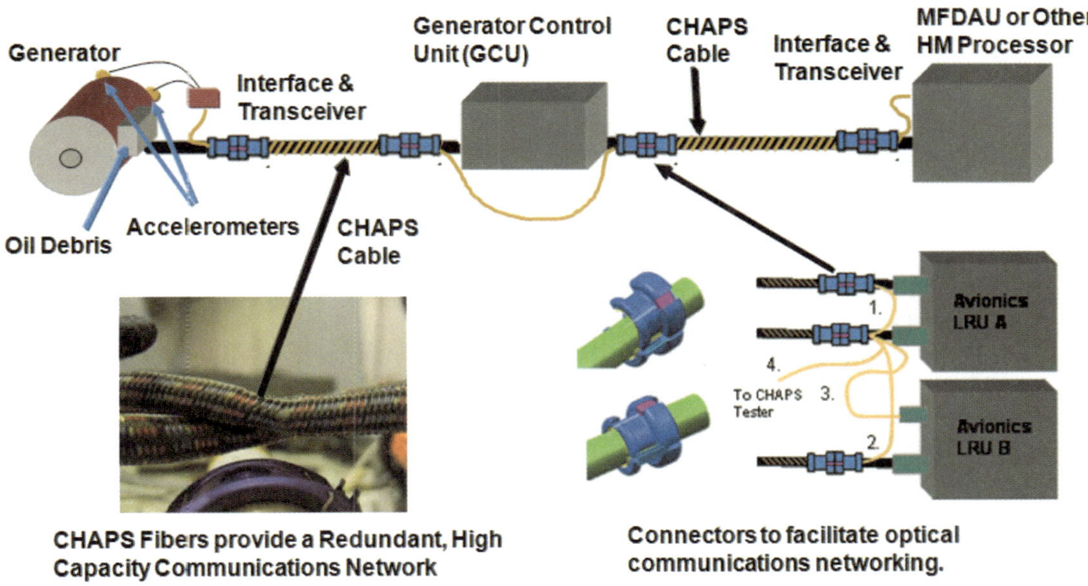

Figure 6.8 CHAPS and CHAPS-HM concepts. Picture provided by Killdeer Mountain Manufacturing.

simpler, less intrusive test connector in a CHAPS-only application.

A representative CHAPS-HM application is shown in Figure 6.8. Data from an accelerometer and oil debris monitor are processed locally in the Integrated Drive Generator (IDG) by an IVHM processing element. The results, as well as necessary raw data sets (e.g., to support maturation), are communicated to a data recorder. While Figure 6.8 shows the CHAPS-HM system bypassing the Generator Control Unit (GCU), it may be desirable to provide an interface from the GCU to provide additional electrical signal data. Because the IVHM data are support critical, loss of communication due to chafing is not a flight-critical concern. CHAPS has been flight tested on several aircraft including F/A-18, CH-53, and P-8A.

6.6.5 Other OCBM Enabling Technologies

Other technologies that have been pursued to enable OCBM include a smart connector or Embeddable Programmable Instrumentation Controller (EPIC), as shown in Figure 6.9 [Management 2013]. The small format fits in the backshell or casing of a patented electrical

Figure 6.9 Embeddable programmable instrumentation controller. Photographs provided by Management Sciences Inc.

harness connector. This unit provides passive real-time data collection, signal processing, reasoning, and optional control functions. The multitasking real-time operating system is deterministic. EPIC produces statistically accurate cognitive stochastic algorithms and models after training of baseline (healthy) and abnormal states in operating domains. This automatic programming feature dramatically reduces the need for software programming. The technology is mature, and is available for aircraft and vehicles. It is effective in acquiring and processing data from cables interfacing to subsystem controllers or from configurations using a cross-channel data link such as that shown in Figure 6.2.

The Boeing On-Line Diagnostic Reporting (BOLDR®) concept makes use of existing bus resources to communicate OCBM data. BOLDR is a scheme to make the word slots on a communication bus mode dependent on operational mode [Sudolsky 2009]. For example, the words allocated to transmit brake temperature or tire pressure may be reallocated for specific periods during the flight: see Figure 6.10. The transmitting subsystem could store and communicate the desired data when the various frame time is available on the bus.

This method does require a change to the system software, along with requalification, but does not impact system hardware.

6.7 Software Architecture That Enables OCBM

We now turn to the software aspect of low-footprint implementation of OCBM. Software costs are a major factor in the development and sustainment of any larger system. Among the desirable design features described in section 6.3 that support the implementation of OCBM are the following:

- Efficient development of a distributed system with multiple suppliers based on a well-defined architecture and industry standards
- Support for incremental OCBM upgrades
- Maintain the flexibility to move functionality on or off board as the need arises
- Reduce implementation, maintenance, and configuration costs through a modular, plug-and-play capability

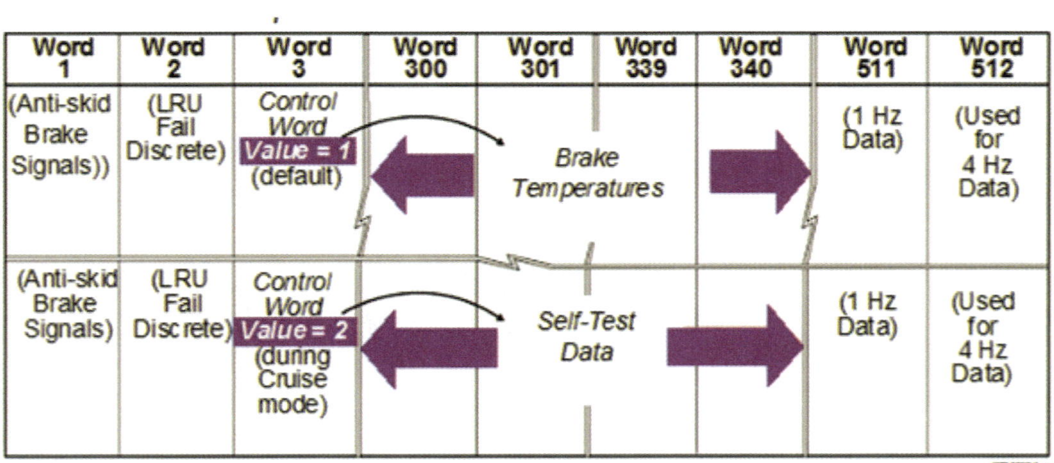

Figure 6.10 BOLDR reallocation of words in bus transmission frame.

- Configure the CBM so the host LRU/subsystem/component can be offered with and without the CBM capability
- The capability to make the best use of scarce communications, processing, and storage resources

A means to address these requirements is through an open, flexible software architecture. The architecture is based on industry and government standards and widely accepted data models. Its goal is to allow interoperability between multiple vendors' IVHM components through the use of a standard interface and data model [DoD 2008] as well as achieve the flexibility necessary for low-footprint implementation.

As discussed in the introduction, an IVHM system is part of an end-to-end support system. IVHM data are required to be shared across maintenance organizational boundaries [MIMOSA 2006]. Hence, in addition to internal data compatibility among IVHM components, the IVHM data model is also required to be externally consistent with the rest of the organization's data (i.e., reliability, logistics, training, documentation, inventory, and maintenance domains).

The Open System Architecture for Condition Based Maintenance (OSA-CBM) is an open standard that addresses interoperability and enterprise systems integration of IVHM applications [MIMOSA 2007; MIMOSA 2010; ISO 2007]. OSA-CBM is divided into the interface definition and data model. These specifications are defined using the Unified Modeling Language (UML) and are intended to be technology neutral. They can be mapped into various programming languages and middleware technologies.

The OSA-CBM data model is based on the concept of metadata. The aim is to have data that are always identifiable and traceable. Moreover, metadata enables IVHM data to be shared across maintenance-related applications. The OSA-CBM DataEvent is the dynamic data related to IVHM events generated by an IVHM application such as measurement, manipulated, or processed data (see Figure 6.11). Its metadata are used to identify and inform status of the data event. The DataEvent inheritance hierarchy is associated with six abstract IVHM data processing functionalities. The functional definitions can be viewed as higher and higher abstraction levels of data into information.

The DA data type addresses sensor data acquisition that is formatted into a consistent form. The DA data are then transformed into one or more meaningful features, which are in the DM-type data format. The features are compared against expected values, and the resulting enumerated condition indicators are stored in the state detection data event (SDDataEvent) class. The data in the HA, PA, and AG formats are the data related to current health of the machine, predicted future failures, and recommended action steps, respectively.

The interface specification defines a minimal set of actions for data events. It can also be seen as defining services offered by the IVHM applications in the Service-Oriented Architecture (SOA) sense. There are four primary types of interface to accommodate different purposes and technological capabilities: Synchronous, Asynchronous, Data Service, and DataEvent Server. The Synchronous interface returns data with the call. It models stateless service functions as founded in SOA. The Asynchronous interface allows any number of IVHM applications to establish and maintain a two-way connection for the duration that the applications need. The data are returned either on request, on alert, or by push all. The latter pushes data to the connected IVHM components every time it collects data, without the need for a request beforehand. The Data Service interface is for a one-way data input device. Two possible uses would be data storage utility and maintenance advisory receiver service. The DataEvent Server interface is similar to the return on alert or the push all modes of communication in the Asynchronous interface (see Figure 6.12). Runtime data (un-) subscriptions are enabled in the Asynchronous and DataEvent Server interfaces.

OSA-CBM does not specify the software component types or building blocks used to design and implement application systems. The internal software structure is left to the OSA-CBM implementer.

6.7.1 Configurable IVHM Software

A representative implementation of the OSA-CBM standard, such as the one developed at the Boeing-Cranfield IVHM Centre, offers a generic component framework for prototyping OSA-CBM compliant IVHM applications [Sreenuch, Tsourdos, and Jennions 2013]. The framework is aimed to simplify development of OSA-CBM applications. To allow reuse, configuration, and extension, several key features are identified and built into the component model, as shown in Figure 6.13. The framework is lightweight, hierarchical, modular, and extensible. In this framework, the subscription interface is selected for simplicity in implementation and also to naturally capture the event-driven characteristics of IVHM applications. The main components are described in the following sections.

6.7.1.1 Entry Points

The OSA-CBM interface specification defines three interfaces for an OSA-CBM module (see Figure 6.12): DataEventServer, DataEventReceiver, and DataEventObserver.

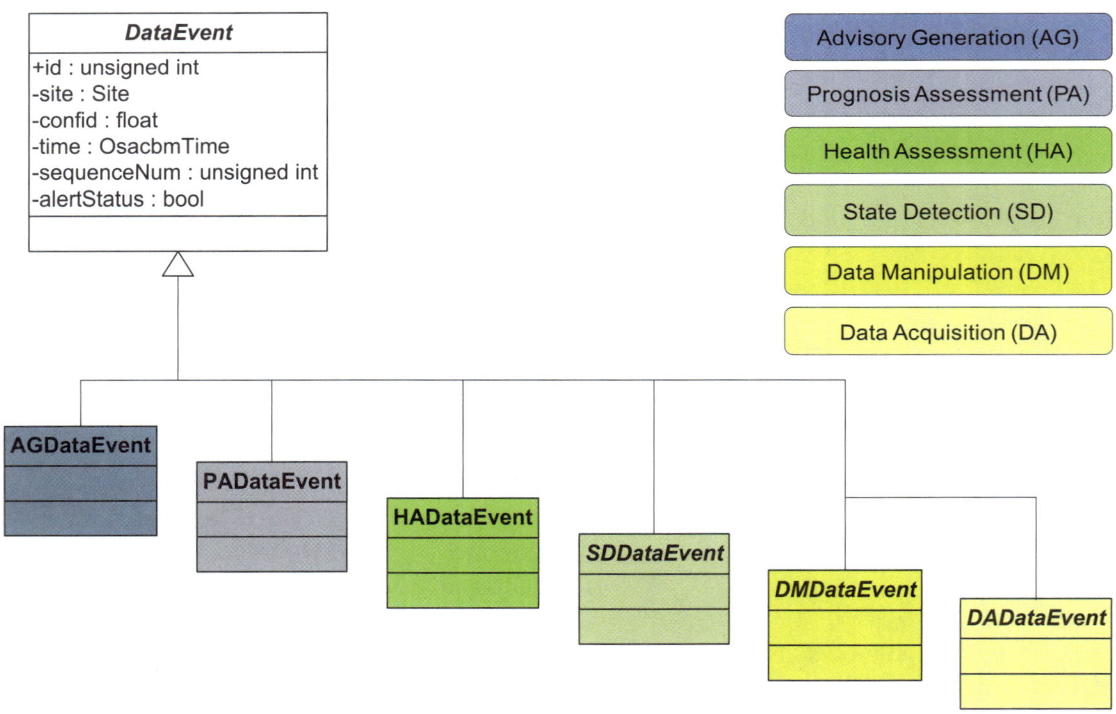

Figure 6.11 OSA-CBM's six abstract dynamic DataEvent models.

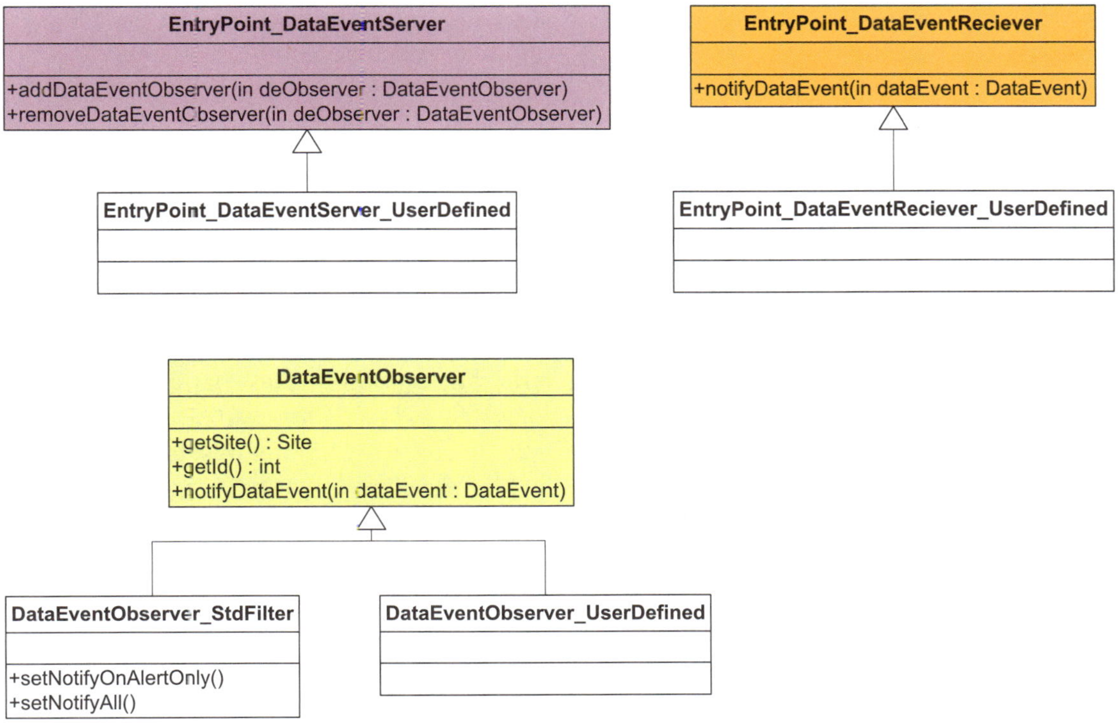

Figure 6.12 OSA-CBM DataEvent Server interface.

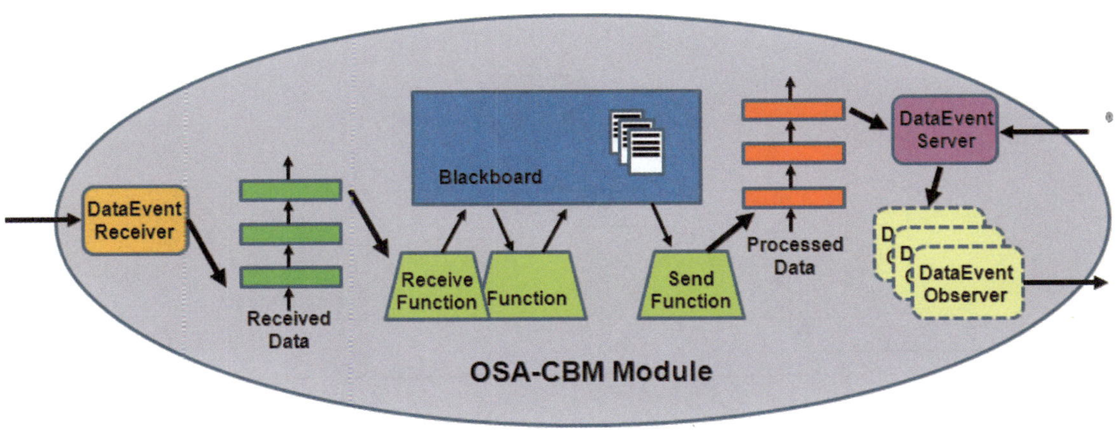

Figure 6.13 OSA-CBM component model.

DataEventServer is the interface provided by an OSA-CBM module to other OSA-CBM modules that have an interest in getting data from it. When a module wants data from a server module, the requesting/client module uses the server module's DataEventServer interface to make a request. During the requesting process, the client module will provide a DataEventObserver to the server module. The DataEventObserver contains a reference to the DataEventReceiver of the client and contains the notification method to be called by the server module when new data are ready.

6.7.1.2 Message Queues

Two message queues are defined by the specification: ReceivedData and ProcessedData. These queues provide asynchronous communication between OSA-CBM modules. A function call to update data returns immediately after the data event has been inserted in the ReceivedData message queue. The ProcessedData message queue acts as a data buffer to store the data before they are sent to the subscribed OSA-CBM modules. In the example framework shown here, queues are used to solve the call blocking performance issue. Hence, the CPU can be fully utilized for other data processing tasks.

6.7.1.3 Function

A Function is self-contained wrapper software for the data processing algorithm so that it can be used within an OSA-CBM module. Functions communicate only with the OSA-CBM module's internal components, reacting to Blackboard data events and publishing results to the Blackboard. All Functions are implemented by extending/inheriting from one or more Base Functions (i.e., Specialized Functions TransducerFunction and DataProcessingFunction in Figure 6.14 are extensions of the Base Functions: Filter, TSA, FFT, etc.). In this way, Functions are reusable and consistent and bring functionality to an OSA-CBM module; together they are the essential computing engine of each OSA-CBM module.

6.7.1.4 Blackboard

The Blackboard provides the Functions with the ability to specify and interact with DataEvents by acting as the module's shared data space. The model of communication between the Blackboard and Functions is data-driven via a publish-subscribe mechanism.

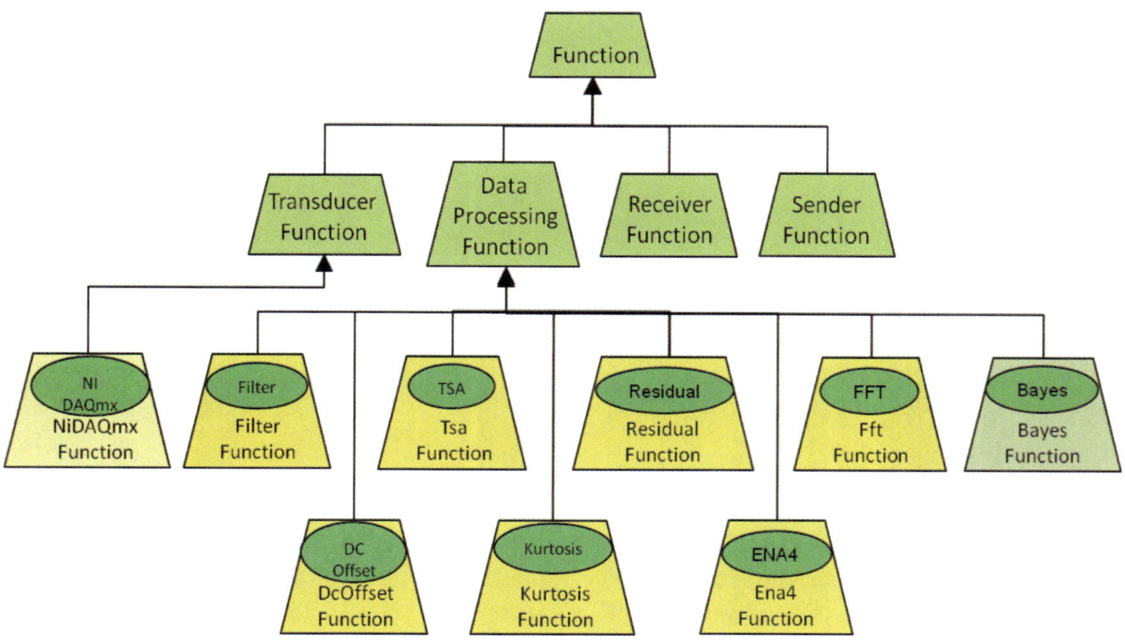

Figure 6.14 Example of base and specialized functions.

The Blackboard tracks all changes in the DataEvent and distributes the updates to all interested/subscribed Functions.

These few generic software components (i.e., Entry Points, Message Queues, Functions, and Blackboard) act as building blocks for all OSA-CBM modules. In this example, the components are dynamically loaded and connected together at application start time using a given configuration file. The component interaction is through abstract interfaces: for example, publish-subscribe API calls are used to pass data between Blackboard and Functions. Very complex functionality is possible by extending and combining very simple generic Functions. This flexibility allows the developer to easily reuse pre-developed Functions in OSA-CBM module construction or customize an application to meet a required IVHM/OCBM functionality.

6.7.2 Gearbox Health Monitoring Example

6.7.2.1 Introduction

A gearbox vibration monitoring example is chosen to demonstrate an application of the distributed software framework. The purpose of the system is to monitor the vibration signals produced by the gearbox and determine the level of damage (or fault) of its pinion gear. Figure 6.15 shows the experimental platform and the related data processing chain. In this setup, an accelerometer is mounted on top of the gearbox to measure the vibration in the axial dimension, and an optical sensor is used to provide

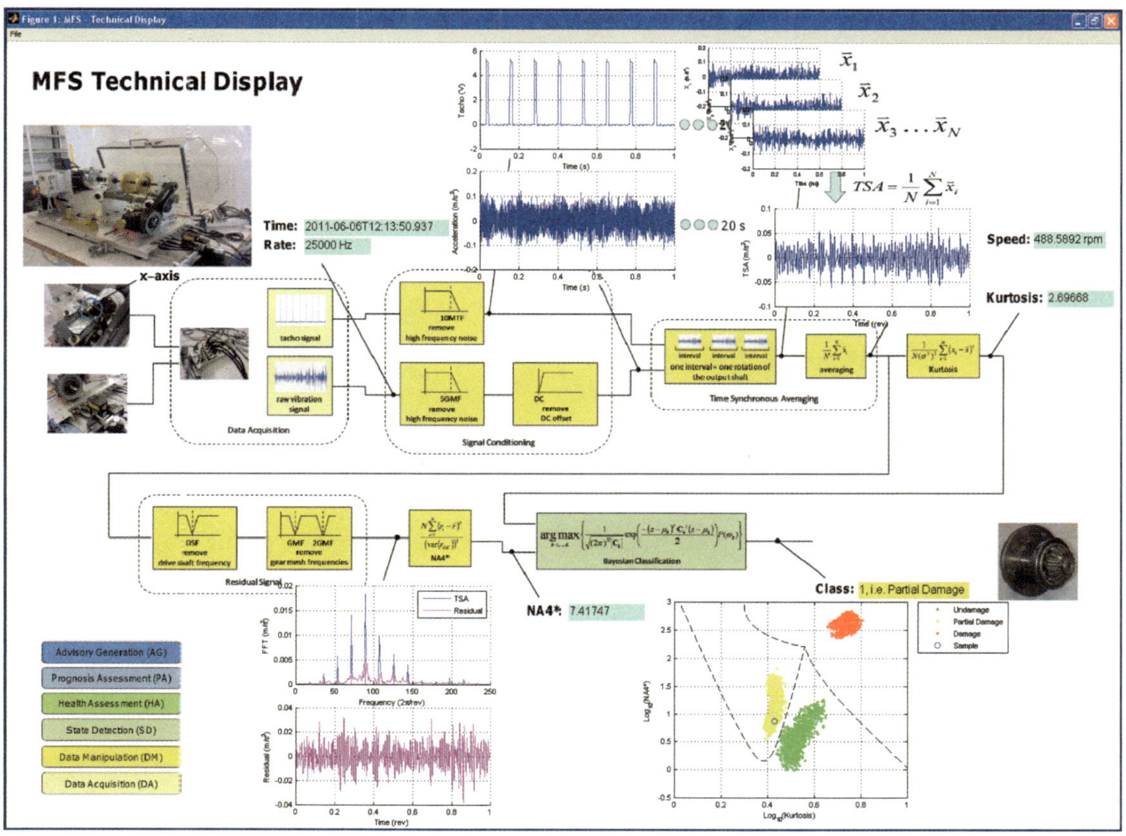

Figure 6.15 Gearbox health monitoring example.

a one-pulse-per-revolution signal used for measuring the speed of the gear shaft. A data acquisition board is used to collect batch data continuously for the vibration analysis at the rate of 25 kHz.

The data processing algorithms used in this example range from data manipulation to state detection. Based on collected data and an extensive analysis, a combination of Kurtosis and NA4* (or ENA4) of the axial vibration is sufficient for accurate fault detections in this particular system. Note that we only outline one possible example of data processing flow in this section; an extensive discussion of the gearbox vibration analysis is presented by Lebold et al. [2000]; and Vecer, Kreidl, and Smid [2005]. To determine the condition of the gearbox, the raw tachometer and acceleration signals are first conditioned by passing through the low-pass filters and the DC offset remover. The random vibrations and external disturbances are further removed by averaging multiple data segments, with each segment equal to one revolution of the gear shaft. This technique is known as Time Synchronous Averaging (TSA). In addition, the shaft rotation speed can be obtained during the TSA computation. The Kurtosis and NA4* features are then extracted from the TSA and residual signals, respectively. The residual is computed by removing the shaft frequency and gear mesh harmonics from the TSA signal. The aim is to have an analytic signal (i.e., the residual signal) that is less dependent on the speed and load of the gearbox. The Bayesian classifier processes the Kurtosis and NA4* input features and determines the gearbox condition based on the statistical properties of the training data.

6.7.2.2 Practical Application Development Process

The first step of an implementation is to categorize the underlying algorithms according to the available Base Functions, as shown in Figure 6.15. In this example, these generic vibration algorithms are then wrapped to form an OSA-CBM algorithms library by extending/inheriting either the TransducerFunction or DataProcessingFunction. Note that the implemented Specialized Functions are generic and configurable. The configuration parameters can be the number of inputs and outputs, expected input data type, and algorithm parameters.

The network setup of the distributed IVHM system is shown in Figure 6.16. This system features partitioned data processing and distributed algorithms that form a clearly defined hierarchy. In this representative setup, OSA-CBM module #1 hosts comm and some gearbox Functions. It runs on an Intel Atom D510 single-board computer attached to an NI data acquisition board and has Windows XP

Embedded, Java SE 6, and NI DAQmx C library installed. This OSA-CBM sensor node is for acquiring the tachometer and vibration signals and to perform low-level data processing: filtering and TSA. In this phase of data processing, the vibration information is compressed from 500,000 samples each of tachometer and acceleration signals to a TSA signal of 4096 data points.

OSA-CBM module #2 hosts higher-level analysis and interpretation Functions. It runs in the health reasoner, which is a Freescale MPC8641D flight-worthy single-board computer and has a precompiled Linux file system from the Embedded Linux Development Kit (ELDK) version 4.2 and OpenJDK 6 installed. The health reasoner extracts the Kurtosis and NA4* features from the TSA data generated by the other module and determines the current health state of the gearbox.

At each OSA-CBM module start time, the module specific binary configuration file is loaded, and the module is created by means

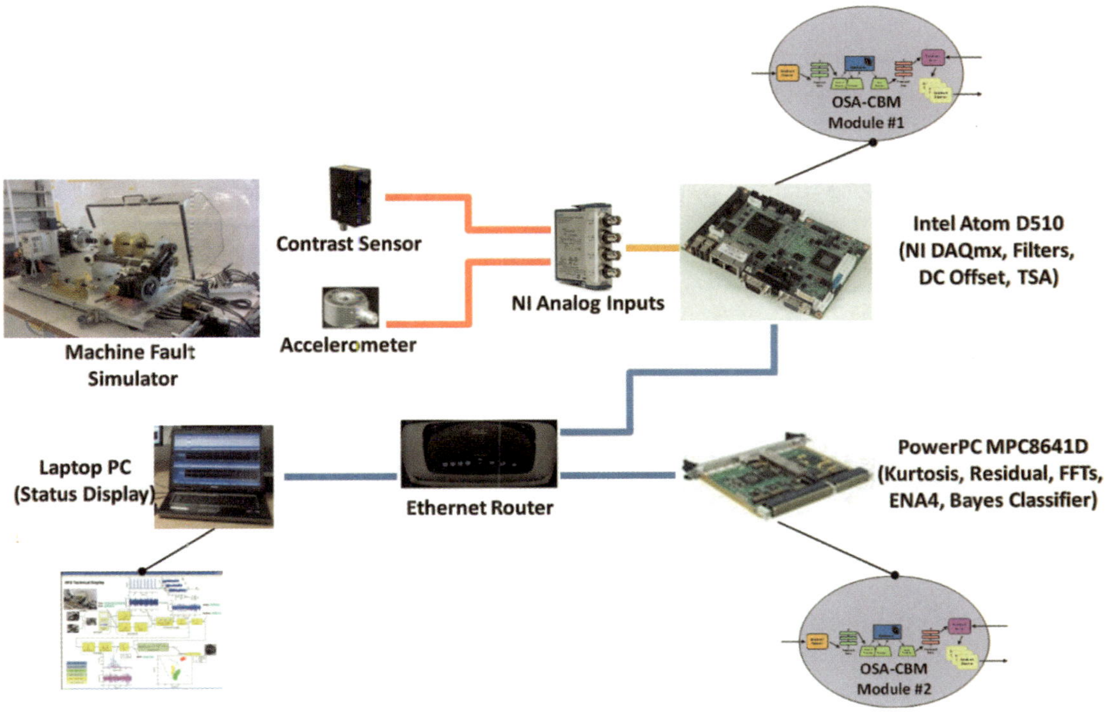

Figure 6.16 Distributed gearbox health monitoring system.

of runtime instantiation and configuration of the precompiled Functions. The configuration file contains the module description, a description of algorithms (including parameters) used for data processing, and lists of individual algorithms' inputs and outputs.

6.8 On-Ground Network Elements

To complete the end-to-end picture for CBM as laid out in the introduction, we will describe some of the key elements needed on the ground to make the entire system work. Data from the assets are received by a number of means on the ground. The very first offboard element that plays a significant role in the architecture of an IVHM system is the ground station or other ground support equipment that can download data from the asset. For aerospace systems, much of the data today come down via the Aircraft Communications Addressing and Reporting System (ACARS), but these data are limited by transmission capacity and cost, and all the processing for the data must be done on board. In the future we will have the means to stream data to the ground via SatCom or cell modem.

Ground Support Equipment (GSE) is typically used on the line to service the asset. Guided troubleshooting algorithms are used to get to the root cause quickly so that the precise LRU can be replaced and the asset turned around quickly. Availability of the asset is of paramount importance both for military and commercial operators.

Once the data are on the ground, they can be routed to a centralized server or to any other system via the internet. This is what is done today with ACARS data. Clearly, security and encryption is a key ingredient of this step. Archiving the data for further analysis, future reference, and even incident investigations is an essential element of IVHM. For this very reason, archival databases must be mirrored and secured so that the chance of data loss is minimized. Clearly, configuration management will be important in this context, as will be data security, for which only authorized personnel can gain access to the data. It is likely that the same database will hold data from different customers, especially in the commercial world—further increasing the need for access management. In recent years, cloud computing and storage have increasingly become popular for storing data, but not for safety-critical applications. However, this is changing as companies have started to market cloud-based data systems that are not on the public Internet but within a "private" cloud.

With the data in a secure place, many other elements of the offboard system come into play to actually use these data to make intelligent decisions about maintaining the assets. We briefly list these systems here and, instead of going into the details, we point the reader to another source for more information [SAE 2013].

To be able to achieve all the benefits (such as reducing cost, increasing asset availability, and enhancing safety), this maintenance decision information has to be used by the Maintenance, Repair, and Overhaul (MRO) supply chain to carry out the prescribed action in an efficient manner. To be most efficient, the logistics and business processes within the MRO supply chain all must be coordinated. This supply chain consists of processes including parts and supplies manufacturing and ordering, inventory management, repair and overhaul scheduling, resource management, and financial management.

Because these processes are spread across multiple suppliers and entities, this desire for coordination seems daunting. However, many large organizations (both Original Equipment Manufacturers (OEM) and asset owners/operators) that are deploying IVHM systems have already started developing links between large IT systems such as Enterprise Resource Planning (ERP), Product Lifecycle Management (PLM), Manufacturing Execution System (MES), Inventory Management Systems, Database Management Systems (DMS), etc. These systems are fast automating the process of managing the entire life cycle of any manufactured or sourced and inventoried part. The IVHM system is an additional input into this enterprise value chain that provides the necessary predictive capability to allow the system to be further optimized.

Typically, the parts OEMs and the service providers will make gross predictions of future repair and overhaul needs based on parameters such as fleet sizes, expected usage, expected part lives, and failure statistics. Since the IVHM system is constantly monitoring the assets, and analyzing the usage against nominal systems models, it is able to continuously provide better and better estimates of future maintenance needs for each part. Because of the nature of these statistical tools, these predictions become more and more precise as the parts reach the ends of their lives, but even approximate, not-as-precise information, fed to the MRO supply chain over a period of time, can allow the system to optimize its production, inventories, as well as shop capacities. This optimization can work at the line and intermediate maintenance levels as well as in the overhaul and repair shops. One of the key aspects of making such an optimized MRO chain a reality is to allow different enterprise-level IT systems to be interconnected so that the data they are operating on are the most recent, and hence the most precise. The need for near-real-time delivery of information to various business processes will be defined by the asset type and the business needs for the particular service business. For more details, the reader is referred to [SAE 2013]. A simulation study of a fleet of aircraft operating with and without IVHM capabilities showed improvements in several notable metrics such as availability, inventory levels, and shop turn-around times [Rajamani, Wang, and Jeong 2004]. This simulation took into consideration many aspects of the MRO supply chain and used the IVHM information to generate key maintenance decisions.

6.9 Summary and Conclusions

The aerospace vehicle design community seeks ever-higher levels of design fidelity and integration to achieve greater vehicle performance and more-economical operation. The Air Force Research Laboratory and NASA notion of a digital twin [Glaessgen and Stargel 2012] embodies a paradigm shift that "integrates ultra-high-fidelity simulation with the vehicle's onboard integrated vehicle health management system, maintenance history, and all available historical and fleet data to mirror the life of its flying twin and enable unprecedented levels of safety and reliability." To realize this goal, the components and systems of the vehicle will be monitored with increasing fidelity, bringing on new challenges in regard to acquiring, communicating, and processing data. These data will feed a correspondingly high-fidelity enterprise ground-based network used to store and retrieve data, analyze the data, and help make intelligent decisions over the whole spectrum of enterprise fleet management from vehicle upgrades and future designs to operations, maintenance, logistics, and business planning. In such an environment, it seems inevitable that IVHM/OCBM designs will become an agile, flexible capability that is adapted throughout the life cycle to support flight test as well as changing in-service requirements and design upgrades.

6.10 References

Boeing. 2013. http://www.boeing.com/boeing/commercial/aviationservices/information-services/real-time-operations/ahm.page, Accessed June 10, 2013.

Born, Frank H., Roy F. Stratton, and Lamar R. Harris. 2001. "Apparatus and Method for Detecting Conduit Chafing," U.S. Patent No. 6,265,881 Issued July 24, 2001.

Deputy Chief of Staff. 2007. *G-4, U.S. Army CBM+ Roadmap*, Department of the Army, Washington, D.C.

DoD. 2008. Deputy Under Secretary of Defense for Logistics and Materiel Readiness, *Condition Based Maintenance Plus DoD Guidebook*, Department of Defense, Washington, D.C.

Glaessgen, E. H., and D. S. Stargel. 2012. Digital Twin Paradigm for Future NASA and U.S. Air Vehicles, 53 Annual Structures, Structural Dynamics and Materials Conference, Special Session on Digital Twin.

Gorinevsky, Dimitry, and Robert Mah et al. 2010. "An Open Architecture for Integrated Vehicle Health Management," AIAA InfoTech Conference, April 21, 2010.

Hess, Andrew, and Leo Fila. 2002. "The Joint Strike Fighter (JSF) PHM Concept: Potential Impact on Aging Aircraft Problems," IEEEAC paper #403, IEEE Big Sky Conference.

IEEE. 2007. "IEEE Standard for a Smart Transducer Interface for Sensors and Actuators," IEEE Standards Association. doi:10.1109/IEEESTD.2007.4346346, http://ieeexplore.ieee.org/xpl/mostRecentIssue.jsp?punumber=4346344.

ISO. 2007. *13374-2 Condition Monitoring and Diagnostics of Machines—Data Processing,*

Communication and Presentation—Part 2: Data Processing. Geneva: International Organization for Standardization.

Keller, Kirby J., Jeanne Maggiore, Robab Safa Bakhsh, Michael Walz, and William Rhoden. 2012. "Sensory Prognostics and Management System (SPMS)," SAE Paper No. 2012-01-2095, SAE International, Warrendale, PA.

Keller, Kirby J., Jeanne Maggiore, Robab Safa Bakhsh, Michael Walz, and William Rhoden. "Sensory Prognostics and Management System (SPMS)," FAA report DOT/FAA/AR-DTFACT-09-A-80001.

Keller, Kirby J., Kevin L. Swearingen, Dan D. Gilbertson, and Jim Peck. 2010. "Architectures for Affordable Health Management," AIAA InfoTech Conference, April 21, 2010.

Keller, Kirby J., Kevin Swearingen et al. 2007. "Aircraft Electrical Power Systems Prognostics and Health Management," IEEEAC Paper #1380, Big Sky IEEE Conference 2007.

Lebold, M., K. McClintic, R. Campbell, C. Byington, and E. Song. 2000. "Review of vibration analysis methods for gearbox diagnostics and prognostics," in *the 54th Meeting of the Society for Machinery Failure Prevention Technology*, Virginia Beach, VA, pp. 623–634.

Management Sciences, Inc. 2013. www.mgtsciences.com.

MIMOSA. 2006. *Open Systems Architecture for Condition Based Maintenance (OSA-CBM) Primer*, Tuscaloosa, AL: Machine Information Management Open Systems Alliance.

MIMOSA. 2007. *MIMOSA Techology Update*, Machinery Information Management Open Standards Alliance.

MIMOSA. 2010. *OSA-CBM UML Specification 3.3.1 Release*, Tuscaloosa, AL: Machine Information Management Open Systems Alliance.

Morris, Charles E., Kirby K. Keller, Kevin L. Swearingen, and Dan Hedger. 2010. "A Dual Use Fiber Optic Technology for Enabling Health Management."

Potter, Fred, and Michael Ballas. 2012. "Advanced Aircraft Electrical Power Management—Maximizing Utility Function Integration with Solid State Power Control," SAE Paper No. 2012-01-2183, Power Systems Conference, Mesa AZ Nov. 1, 2012, SAE International, Warrendale, PA.

Potter, Fred, and Michael Ballas. 2012. "Solid State Power Control as a Network Backbone for Aircraft System Health Management," SAE Paper No. 2012-01-2233, Power Systems Conference, Mesa AZ, Nov. 1, 2012, SAE International, Warrendale, PA.

Rahul, G. 2008. "Avionics Industry Moving Towards Open Systems," NASSCOM, http://www.nasscom.in/Nasscom/templates/NormalPage.aspx?id=55495, December 2008.

Rajamani, R., J. Wang, and Jeong K-Y. 2004. "Condition-based maintenance for aircraft engines," *Proc. ASME TurboExpo 2004*, Vienna, Austria, June 2004.

Ramsey, J. 2007. "Integrated Modular Avionics: Less is More," *Aviation Today*,

http://www.aviationtoday.com/av/commercial/Integrated-Modular-Avionics-Less-is-More_8420.html#.UXmMZErbWrA, February 2007.

SAE International. 2013. SAE Aerospace Recommended Practice, HM-1 Cornerstone Document, ARP6803, under preparation, Aug. 2013.

Sreenuch, T., A. Tsourdos, and I. K. Jennions. 2013. "Distributed Embedded Condition Monitoring Systems based on OSA-CBM Standard," *Computer Standards & Interfaces*, Vol. 34, No. 3, pp. 123–123.

Sreenuch, T., A. Tsourdos, and I. K. Jennions. 2013. "A software framework for prototyping embedded IVHM applications," Accepted to publish in *AIAA Journal of Aerospace Computing, Information and Communications.*

Sudolsky, Michael. 2009. ARINC 573/717, 767 and 647A: The Logical Choice for Maintenance Recording And IVHM Interface Control or Frame Updates, Annual Conference of the Prognostics and Health Management Society.

Vecer, P., M. Kreidl, and R. Smid. 2005. "Condition Indicators for Gearbox Condition Monitoring Systems," *Acta Polytechnica*, Vol. 45, No. 65, pp. 35–43.

Watkins, C. B., and R. Walter. 2007. "Transitioning from Federated Avionics Architectures to Integrated Modular Avionics," 10.1109/DASC.2007.4391842, December 2007.

CHAPTER 7 Data Management

Karl Reichard, *Pennsylvania State University*

"As a rule, he or she who has the most information will have the greatest success in life."

—Disraeli

7.1 Introduction

There was a time when the collection of data about the performance or health of a piece of machinery required a maintainer, technician, or engineer to make a measurement with a dedicated, specialized piece of measurement equipment and record the reading in a notebook. This paper trail record of the machinery condition could then be used to determine if the machinery was operating within specifications, and the measurement history could be used to identify trends in machinery health and performance. Over time, electronic systems (first analog, then digital) were developed, which automated the collection and storage of a diverse range of sensor data about the current operating point of a machine (temperature, vibration, flow rates, etc.). Thanks to increases in the capability of digital electronic devices (such as processing speed and memory capacity), commonly linked to the familiar Moore's Law[1], the ability to collect and process data from sensors on machinery has also increased. As a result of the increase in electronics capability, many consumer electronics devices such as cellular telephones now contain sophisticated data acquisition and processing electronics in compact and portable packages. All of these factors combined have reduced the per-channel cost of acquiring and managing sensor data for IVHM applications.

Reductions in size and power requirements, improvements in capability, and market-based reductions in the cost of data acquisition and processing electronics, means that on-platform IVHM systems are now capable of collecting more health-related data than can be efficiently transmitted to and used by consumers and customers of the information. Consequently, an important aspect of IVHM system design is the management of the data collected from the sensors.

A typical IVHM system consists of on- and off-system measurement, processing, and storage components, in which the system can be a machine in a manufacturing plant, a platform such as an automotive vehicle or airplane, or a network or system of systems. On-system components include sensors and electronics for collecting, processing, and storing health and performance related data. Processing may be limited by time constraints due to the need to sample sensor outputs at uniform time intervals, or by the available electrical power and space because the size and power demands of computing systems often increase with computational ability. On-system software typically implements simple filtering, transformations, and data fusion to produce real-time warnings of system condition or impending failures to operators.

On-system data storage is often limited, as is connectivity to external systems, and is analogous to the maintainer's notebook in the opening example. Consequently, at-system devices are often used to download data for transmission to consumers and customers. The volume of data collected by the on-system components determines the frequency and duration of operations required to transfer the data from the on-system to at-system components if the data are not processed by the monitoring system and loss of data is not permissible. At-system devices may also implement additional data processing, benefiting from the ability to analyze longer time histories of individual data channels and to compare data collected across multiple subsystems. The primary consumers of information generated

1. The law is named after Intel cofounder Gordon E. Moore who, in a 1965 paper [1], described the trends in the semiconductor industry and noted that the number of components in integrated circuits had doubled every year from the invention of the integrated circuit in 1958 until 1965 and predicted that the trend would continue. It has been noted since then that those capabilities in digital electronics, such as processor speed, memory size, and even the numbers of pixels in digital cameras, have continued to grow in a similar exponential manner.

by at-system components are usually maintainers and operators planning near-term operations.

Off-platform systems provide access to data for the largest community of data consumers and information customers and generally have the greatest computational ability. Off-system components also have the benefit of access to IVHM data from multiple machines or platforms across an entire plant or fleet. The bandwidth and availability of the communications link between the at-system and off-system components determines the frequency and timeliness with which the data are made available to off-system consumers and customers.

The ability to move data from the monitored system to the intended consumers and customers is a key factor in the effectiveness of the IVHM system. One potential solution to this problem is to increase the bandwidth of the transmission system—analogous to installing high-speed home Internet to enable on-demand, 24/7 access to streaming movies instead of receiving movies on DVDs through the mail or a neighborhood retail outlet. In many cases, however, the systems being monitored are in remote locations where access to high-speed or wide-bandwidth communications is impractical. Thus, a key aspect of data management is reducing the amount of space required to store IVHM data or equivalently the bandwidth required to transmit the data to the end users.

This chapter provides an overview of techniques for management of data in an IVHM system. It focuses on the movement of data from sensors to enterprise-level information systems[2], and the conversion of data into useful information. This ability depends on both on- and off-system architectures, the application of data fusion to combine sensor data on-platform, the application of data reduction and compression techniques to reduce the size and hence bandwidth requirements for data transmission, and preprocessing the data into useful forms for off-system analysis techniques such as data mining. Each of these areas is covered in the sections that follow.

7.2 System Architectures

The architecture of the system components, both on vehicle and off, has a significant effect on how data are managed. The on-platform architecture defines how data are acquired, processed, and managed within the system being monitored. At-platform systems facilitate the collection and transfer of data from sensors to off-platform systems. The off-platform architecture defines how data are managed once they are off-loaded from the platform. Traditional on-platform monitoring systems use a centralized architecture with a single, centralized data acquisition and processing system, as shown in Figure 7.1. The centralized architecture arose because the data collection system was often the most expensive part of the monitoring system in terms of cost, power, and volume. The data acquisition subsystem may simply collect and record the data or may include the ability to process the data on-platform (for example, processing an analog signal to

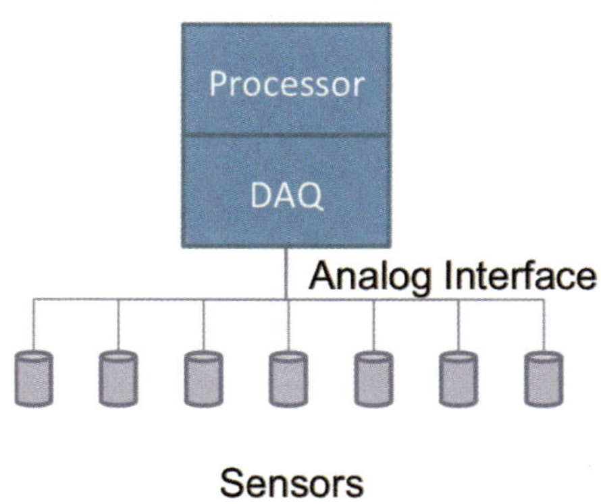

Figure 7.1 Centralized monitoring architecture.

2. Business information systems supporting the entire organization and external customers.

extract low- and high-level threshold exceedances). The data acquisition and processing hardware in the centralized architecture must be sized to handle all of the sensors and data sources on the platform.

A major disadvantage of the centralized architecture is a lack of scalability—if additional sensors are not planned for in the initial monitoring system design, then it may not be possible to add additional sensors later. Manufacturers of measurement systems have tried to address this by using expandable chassis with modular components for acquiring the data and a single management or control computer. The centralized architecture requires long sensor data cables to connect the sensors to the data acquisition system; they can be a significant overhead when weight is a key driver, for example in aircraft. The raw and/or processed data are stored for later transmission off-platform. The advantage of having an on-platform IVHM system for data processing is that the algorithms can perform preprocessing and data fusion to reduce the volume of the data and the corresponding bandwidth required to move data from the on-platform data collection system to off-platform archives and analysis systems.

Decentralized and distributed architectures employ multiple interconnected data collection and processing nodes. Distributed nodes may store data locally or be connected to a central data collection and communication node for transmission to at- or off-platform system components. Figure 7.2 shows a distributed, network-based architecture. Distributed data collection and processing nodes are all connected to a communication network. One advantage of the network-based architecture is expandability—as with a local area computer network, new devices can easily be added to the network to generate or to process data. A main processor or gateway connects the network to off-platform systems. The expandability of the network-based architectures also makes it easier to capture vast amounts of data, which must either be processed locally or transmitted off-platform, increasing communication bandwidth requirements.

Regardless of the monitoring architecture, it is desirable to move the data off-platform to provide actionable information to various users. Maintainers need access to data either on or at the platform (at the base or maintenance facility supporting the platform). Planners, logisticians, engineering support

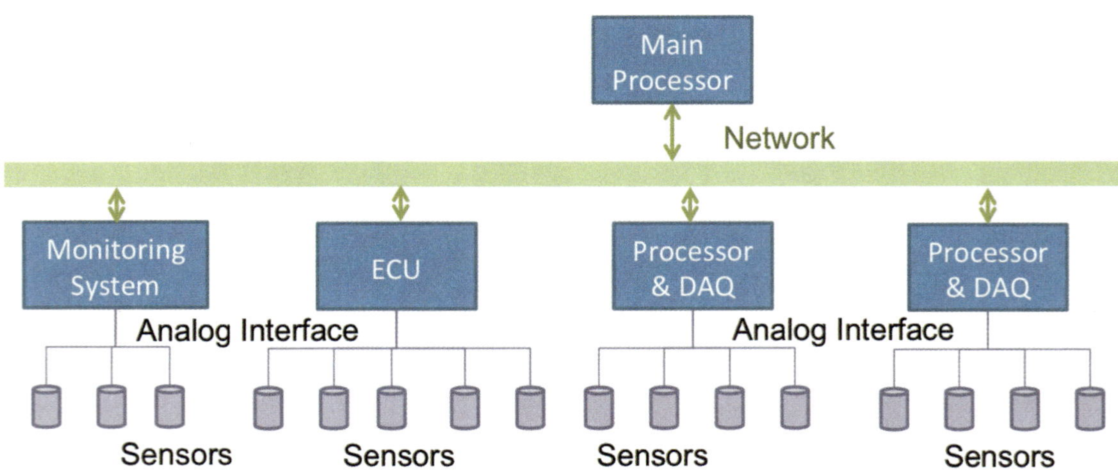

Figure 7.2 Network-based monitoring architecture.

personnel, and design engineers benefit from access to the health and performance data at remote locations away from the platform. There are two primary challenges to providing the data from on-platform monitoring systems to the remote customers for the data and information: the volume of the data collected, and the often-limited bandwidth of the connection to the platform. Many platforms operate in remote locations with limited bandwidth or limited availability of communication resources with which to connect the platform systems to external communication networks. This is typically referred to as the "dirty" mile or "last tactical" mile and is a significant challenge in implementing IVHM systems, especially for oil and gas or military applications. Therefore, there is a need to either process or compress the data on-platform to reduce the requirements on communication.

7.3 Data Processing and Fusion

In the most general sense, data processing is the representation, transformation, and manipulation of data and the information they contain. Data fusion combines data and information from multiple sensors and sources to infer information that cannot be made based on a single sensor or data source [Hall 2001]. Data fusion is a general term, which covers many levels of data and information refinement. Data- or sensor-level fusion combines data from multiple sensors or low-level data sources to produce refined or derived measurements. In the context of IVHM, sensor-level fusion is used to validate sensor data and to compute features or condition indicators (statistical measures such as means or standard deviations, power spectra, etc.) related to the health of system components. Feature-level fusion combines multiple features or condition indicators (and may incorporate new sensor data) to detect and classify faults at the component, subsystem, or system level. Decision-level fusion combines assessments of component, subsystem, or system health to produce actionable information for data and information customers. This multi-level view of how data fusion is applied at various levels in an IVHM system is captured in Figure 7.3.

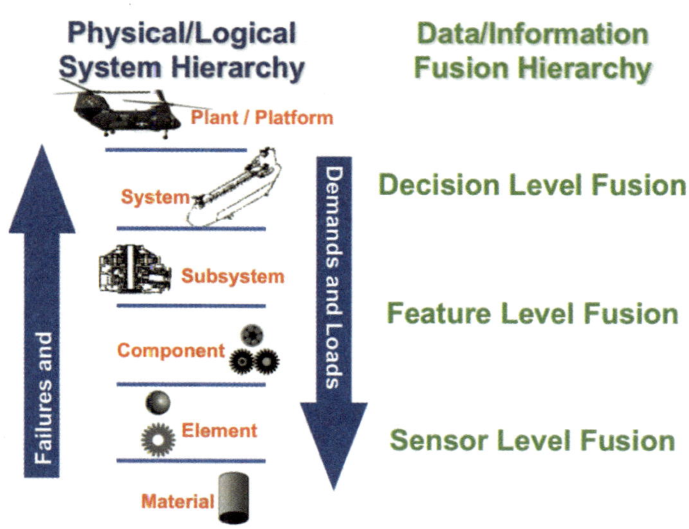

Figure 7.3 Alignment of physical/logical system hierarchy and data/information fusion hierarchy.

Another view of data fusion was developed by the Joint Directors of Laboratories (JDL) Data Fusion Subpanel and later refined by the Data Fusion Information Group (DFIG) to include additional levels of situational awareness, user refinement, and mission management [Llinas et al. 2004]. The JDL and subsequent DFIG models were developed to provide a common reference model for the development of information and data fusion systems. While the data fusion models were originally developed for defense applications such as target identification and tracking, the models have also been applied to nondefense applications such as medical diagnosis and IVHM.

The DFIG model includes six levels of data fusion:

0. Source preprocessing/subject assessment
1. Object assessment
2. Situation assessment
3. Impact assessment (threat refinement)
4. Process refinement
5. User refinement (cognitive refinement)

Each level in the DFIG model can be thought of as a building block for a larger data processing system. Depending on the implementation and the processing architecture, these can be linked serially, so Level 0 feeds Level 1 and so on, or can be distributed so functions within each level can access the outputs of other levels. Level 0 includes functions to allocate data to processing algorithms and functions for prescreening data: for example, identifying bad sensor channels or monitoring a channel to decide when sufficient data have been collected to pass them on for higher-level processing functions such as Fast Fourier Transform (FFT). Object assessment encompasses transforming data into consistent engineering units, aligning separate data channels in time or space, assigning data objects for tracking, and identifying or classifying data. Level 1 encompasses many of the signal processing functions associated with fault detection and classification, and corresponds to both signal- and feature-level processing as described in Figure 7.3. Level 2, situation assessment and refinement, includes functions to develop relationships between data elements and using relational information to determine the meaning of data. Examples in IVHM include identifying correlations between changes in different variables and identifying whether the increase in a particular sensor value is due to a change in environmental conditions, load conditions, or health. Impact assessment or threat refinement incorporates the projection of data into the future and developing alternative hypotheses for the meaning or interpretation of data. This level is the home for prognostic health assessment. Level 4, process refinement, is responsible for identifying data gaps and for the adaptation of fusion and processing algorithms to improve classification or decision making. Decision-level fusion incorporates functions from Levels 3–5. User refinement, Level 5, provides the data fusion results to information customers and to higher-level fusion systems.

7.3.1 Sensor-Level Fusion

Figure 7.4 shows a typical IVHM data processing flow from sensors to a fault diagnosis or decision. The data processing flow depicted in Figure 7.4 includes sensor-, feature-, and decision-level fusion, in addition to signal preprocessing, feature extraction, and classification. Individual sensor inputs are processed to remove noise, perform calibration, or otherwise condition the signals for feature calculation. Features indicative of the health or condition of the system are calculated from the sensor signals. In some cases, data from multiple sensors may be combined (or fused) to create a new signal from which to compute features. Features are used to detect and classify faults. Some faults

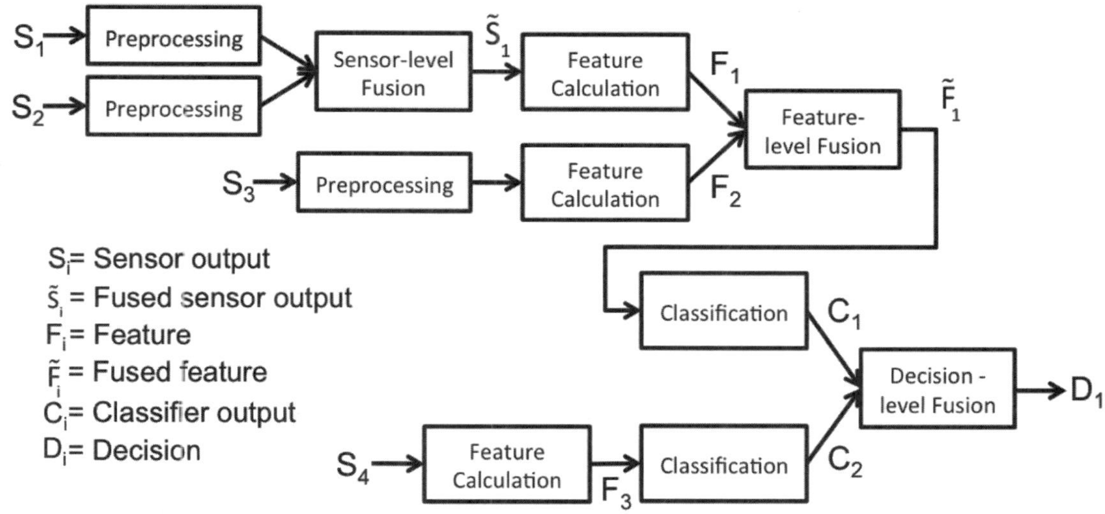

Figure 7.4 Typical IVHM data processing flow diagram.

can be classified using single features, while others require combinations of features. For example, detecting gear breakage with a vibration sensor may require monitoring both total vibrational energy and higher-level statistical features such as Kurtosis. In the same manner, the final decision about the existence of a fault or the appropriate action in the presence of a fault may depend on the output of one or more classifiers.

Many feature calculation and classification algorithms are generic and function the same regardless of whether they are operating on sensor output data or the results from the fusion of lower-level data. But many feature calculation algorithms require the fusion of multiple sensor inputs as an input, and many classification algorithms operate more efficiently when the number of input features is reduced through feature-level fusion.

The primary goal of sensor-level data fusion is to associate data from different sensors in time or space. In a missile tracking system, this may entail aligning measurements from multiple radar tracks in time and/or associating multiple radar tracks based on the spatial locations of the targets. Once measurements are aligned in time and/or space, features extracted from each data stream can be compared to perform target identification. An example of time alignment of data in IVHM applications is time alignment of vibration measurements with tachometer data to compute time-synchronous signal averages. Spatial alignment is frequently used in structural health monitoring to align data from different physical locations.

Image and video processing algorithms often require multiple frames or images to identify objects, detect changes, or compute motion. Multiple images are compared at the individual pixel level to align the images spatially for comparison. Spatial filtering algorithms may be applied to detect edges and other features within the images to assist in the pixel alignment. Pixel-level image alignment is also used when combining information from multiple imagers. Examples include multi-spectral imaging that may combine images from visible and infrared cameras, combining two-dimensional image data from traditional cameras to produce three-dimensional images, or combining two-dimensional image data with three-dimensional data from a LIDAR (LIght Detection And Ranging or Laser Imaging Detection And Ranging), which produces a 3-dimensional point cloud representing distances to each point.

When fusing successive images (or data samples) from a single sensor, there is an assumed, fixed, time relationship between images or successive samples. When combining data from multiple sensors or imagers, both spatial and temporal alignment is required to align pixels within the images. Often, data are recorded with a common time base to mark or time stamp the data. This is a relatively straightforward problem when all data are collected with a single system (recorder or computer); it is a little more challenging in a networked architecture in which different sensors are recorded by different systems, each with their own clock. Strategies for improving time alignment include providing a single synchronization "clock" for all recording systems (for example, a GPS one-pulse-per-second signal) or using network-based time synchronization protocols such as IEEE 1588. Another approach is to calibrate the time across recording systems by recording a common synchronization event on all sensor channels that is easily detectable and can be used to calculate the offset in the time alignment. A familiar example of the latter approach, for those of a certain age, is the use of a clapperboard in film production before the start of a scene. The visual cue provided by the shutting of the clapstick, and the sharp "clap" sound, allow synchronization of the visual and audio channels. Modern clapperboards even display a universal time counter to synchronize the video with other systems.

A very common example of sensor-level fusion in IVHM is the use of time synchronous averaging in the computation of features for gear fault diagnosis and classification [Campbell et al. 2001]. Many well-known gear fault features, such as NA4, FM0, FM4, NB4 and others, require the calculation of a time-synchronous averaged vibration signal. Time Synchronous Averaging (TSA), or synchronous time averaging, removes harmonic signals that are not harmonically related. The FFT improves the signal-to-noise ratio by boosting the levels of sinusoidal signals while suppressing broadband, random noise. In IVHM applications, such as gearbox or drive train health monitoring, almost all of the rotating elements contribute sinusoidal components to the measured signal. To compute vibration-based features for a specific rotating element, it is advantageous to remove the sinusoidal signal components not related to the element of interest. Classic TSA uses a tachometer signal along with the vibration signal to identify the time period of one rotation of the element of interest. The vibration signal is divided into blocks corresponding to one or more rotations of the element of interest and then averaged across multiple blocks. This suppresses the sinusoidal components not harmonically related to the frequency of the element of interest, making it easier to detect changes and compute the required diagnostic features.

As another example of sensor-level fusion, consider the use of Structural Intensity (SI) for damage detection in monitoring the health of a platform structure (i.e., a wing). When power is injected into a structure at one or more locations, the input power travels along multiple flow paths and is dissipated by material damping or sound radiation. Structural damage can change the flow paths and the power dissipation in the structure. Structural intensity is a vector quantity, which can be broken down into contributions from shear, bending, and torsional waves in the structure. Structural intensity is approximated using collocated measurement of surface velocity and strain, as shown in Figure 7.5. The photograph in the top left corner shows the sensors mounted on the gearbox; the drawing shows the relative sensor orientations. The plots on the right show several features used to detect and classify structural damage in a gearbox. The top plot on the right shows the magnitude and direction of the structural intensity computed by fusing the vibration and strain measurements [Banks 2000; Shepherd et al. 2012]. The lower plot shows statistical features, crest factor, and kurtosis, computed from the vibration measurement alone. These features act as inputs to a classifier to detect and classify the fault within the gearbox.

7.3.2 Feature-Level Fusion

At the most basic level, data fusion involves associating data to ensure that it represents a single event in space or time. This association can be applied at multiple levels of processing and refinement, as shown in Figure 7.4. Sensor-level fusion involves the direct fusion of sensor data. Feature-level fusion involves the fusion of features or feature vectors derived from the sensor data. Sensor-level fusion typically involves commensurate sensors (sensors measuring the same physical phenomena such as two or more image sensors, vibration sensors, or acoustic sensors), which can be directly combined. If the sensor data are non-commensurate, meaning that they are measuring different phenomena (e.g., blade tip timing and pressure measurements), then the data must be fused at the feature- or decision level.

Feature-level fusion involves the extraction of representative features related to the object or event whose detection and classification is the goal of the sensor processing. Features can be statistical in nature, such as the root mean square, Kurtosis, crest factor, or frequency spectrum computed from a vibration measurement. In imagery, features can include shapes (such as faces, airplanes, or terrain), colors, or identifiable patterns (the spatial Fourier transform). Humans use feature-based cognitive processes to

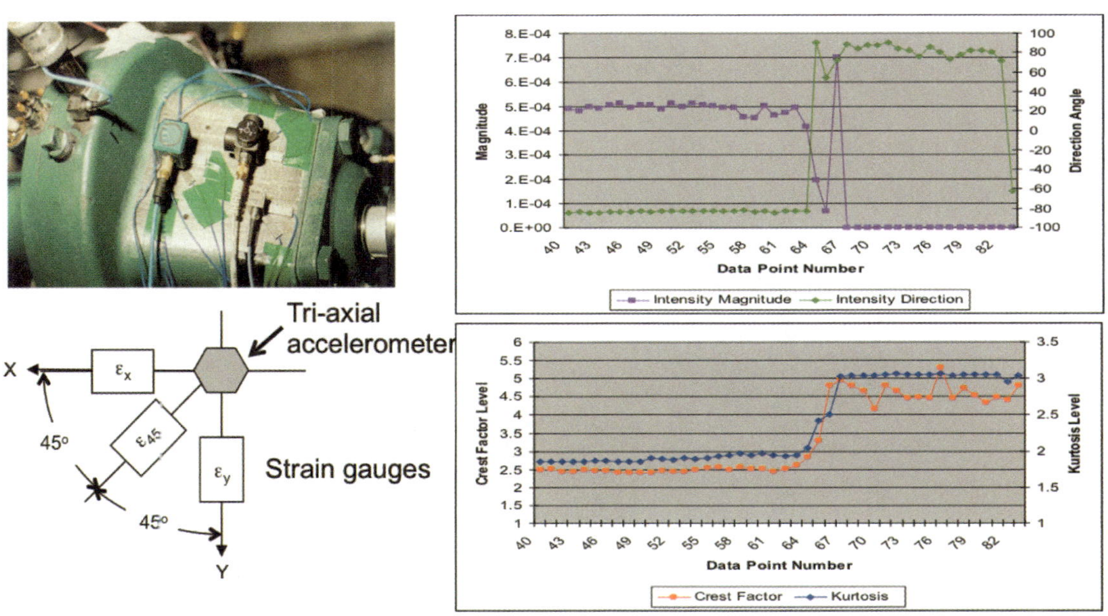

Figure 7.5 Surface structural intensity measurement for gearbox health monitoring.

recognize and distinguish between different objects such as apples, oranges, or baseballs. In multi-sensor feature-level fusion, features are extracted from multiple sensor observations and combined into a single feature vector that is used as the input to a pattern recognition algorithm.

In the TSA example described in the previous section, features computed using different average block lengths can be combined and presented to a classifier to both detect a fault and determine which rotating component is responsible for the change in the vibration signal. As in sensor-level fusion, one of the key elements is proper alignment of the data in time to ensure that the classifier is seeing changes in the features in the proper sequence.

In some cases, feature-level fusion can be used without first performing sensor-level fusion. Consider the earlier discussion of image registration. Instead of spatially aligning images from multiple sensors on a pixel basis, we could instead search each image for distinguishing features (edges, shapes, colors, etc.). As long as the imager outputs are time-aligned, the features from each image can be combined to detect and classify the object of interest. For example, a feature in a visible image may show a cylindrical object, while a feature from an infrared imager may show a hot plume-like shape. Combining these two features along with rough knowledge of their location in the respective images (vs. pixel-by-pixel alignment) may be enough for a classifier to determine that the object is a missile.

Figure 7.6 shows three features used to detect tooth failure in a gearbox: images from a borescope used to visually examine the gear for damage, fatigue particle counts from a debris sensor, and the Root Mean Squared (RMS) vibration level (indicative of the vibrational energy). The three features are time aligned to show the correlation between increases in the fatigue particle count and RMS vibration as a function of damage (indicted by the visual evidence provided by the borescope images). Fusing features from non-commensurate sensors allows the different features to be used to corroborate changes in the system health and can provide robustness in the classification against individual sensor performance variation and reduce false alarms.

Many techniques exist for combining features including fuzzy logic, Bayesian networks, and neural networks [Duda, Hart, and Stork

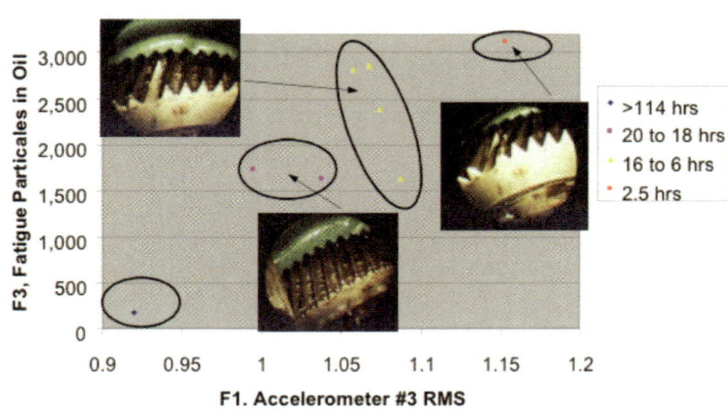

Figure 7.6 Multi-sensor fusion for gear fault detection.

2001]. Shown next is a sample fuzzy logic rule demonstrating how feature-level fusion is used for determining if a filter has become clogged based on measurements of filter differential pressure, fluid temperature, and volumetric flow. In this example, *Filter Δ P*, *Temperature*, and *Volume Flow* are all features. Historic data on system performance and operational characteristics are used to define fuzzy limits for High, Not Low, Not High, Increasing, or Changed.

If (*Filter Δ P* is High) and (*Temperature* is Not Low) and (*Volume Flow* is Not High) Then (*Filter Clog* is Yes) *CF* 0.81

A similar type of rule could be developed for the features shown in the structural intensity example:

If (*Magnitude* is Increasing) and (*Direction* is Changed) and (*Crest Factor* is High) and (*Kurtosis* is High) Then (*Gearbox Damage* is Yes) *CF* 0.90

Another example of feature-level fusion is the use of feature reduction techniques to simplify classification. Figure 7.7 shows a graphical representation of a classifier for helicopter gearbox fault classification using vibration sensors. The plot shows the separation of six different faults and one no-fault condition using two features. These two features, however, are not calculated directly from single (or even multiple) vibration measurements. The two features F1 and F2 shown in Figure 7.7 are computed from linear combinations of a larger set of features. A technique called Fisher linear discriminant analysis reduces the set of features to a minimal set giving optimal classifier performance. Linear discriminant analysis is closely related to ANalysis Of VAriants (ANOVA), regression analysis, and principal component analysis. These same techniques are used in data mining to identify relationships between variables without a priori knowledge for correlations or patterns.

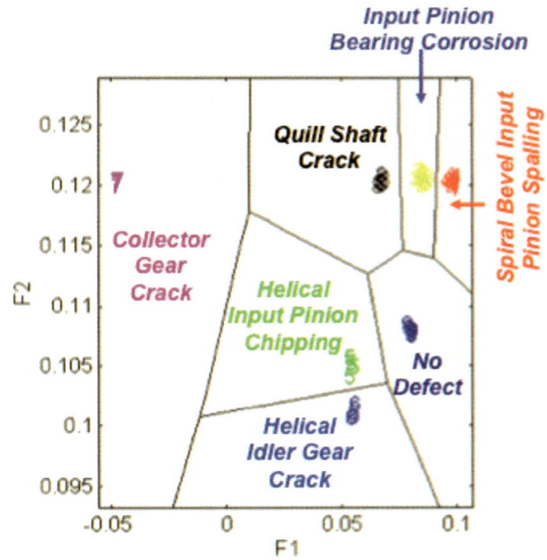

Figure 7.7 Classification of gearbox faults using vibration data.

7.3.3 Decision-Level Fusion

Decision-level fusion combines sensor or feature information after each sensor or combination of sensors has been used to make a preliminary determination of an entity's location, attributes, and identity (in IVHM, the fault, severity, and location). Examples of decision-level fusion techniques include weighted decision-making (voting), inference networks, Bayesian inference, and Dempster-Shafer's method [Hall 2001]. Whereas sensor-level fusion combines signals from multiple sensors to calculate features, and feature-level fusion combines individual features to feed pattern recognition and classification algorithms, decision-level fusion combines multiple classifier outputs to produce a refined classification or decision.

Decision-level fusion techniques fall into two primary categories: integration and selection. In an integration approach, all of the classifiers contribute to the final decision. The underlying assumption in an integration approach is that all of the classifiers are different and are competitive so that each performs differently. Integration approaches can employ familiar mathematical rules to combine classifier results: computing the mean, computing a sum (with or without weights), and computing minimums, maximums or medians.

Selection-based techniques choose the best classifier from a group and do not take advantage of combining the different classifier results. Voting schemes are often used to select the classifier output with the strongest belief (Dempster Shafer) or highest probability given the available data (Bayesian). Hybrid fusion techniques may employ a combination of integration and selection.

This section has shown how sensor data generated on board an IVHM system can be processed and fused into features, which indicate the state of the components. These features can then be used to infer the fault and the severity such that a maintainer can make an informed decision on any necessary maintenance actions.

7.4 Data Reduction and Compression

In this section, we will consider data volumes in IVHM systems, and techniques used to reduce both data volumes and the bandwidth required to transfer or transmit IVHM data from on-platform systems to off-platform systems. In most IVHM systems, sensor and operational data are collected by real-time embedded computing systems, which are configured based on the frequency characteristics of the individual sensor signals. Data are typically collected and stored with minimal processing. Processing of the sensor data is required to detect and isolate faults and failures and is performed quickly with the raw and processed data, and then stored in on-platform storage devices for later retrieval. The steady improvement in computational capability and reduction in cost of the embedded data collection and processing systems has lead to the ability to collect increasing amounts of data. As discussed earlier in this chapter, however, these data must eventually be retrieved and transmitted to local at- or remote off-platform systems for vehicle maintenance and operations planning, and to enterprise systems for fleet-wide asset management.

Connectivity from the platform to off-platform systems and network connectivity between at-platform systems and enterprise resources limit the ability to provide the on-platform information to external customers rapidly. Therefore, a critical issue in the management and utilization of IVHM data is ensuring that the data can move from the platform to enterprise systems in an efficient and timely manner.

Because many systems operate remotely, accommodating limited communication bandwidth between the platform and enterprise systems is a serious consideration. If increasing the bandwidth of the communications link between the platform and the enterprise is not an option, then the bandwidth requirements for transmitting the data must be reduced. There are several options for reducing the bandwidth requirements for moving data off-platform:

1. Eliminate sensors or channels monitored by the on-platform system
2. Reduce the volume of data saved and transmitted by using expert knowledge of a data set to remove unnecessary data
3. Compress the data by using patterns in the data to store them more efficiently

Once the monitoring requirements for the IVHM system are established, they typically drive the requirements for which sensors are needed, and the subsequent sample rates and basic data formats; therefore, option 1 typically is not feasible. Figure 7.8 illustrates

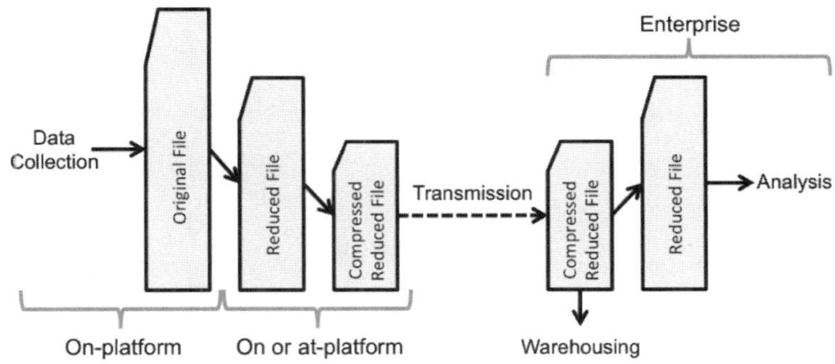

Figure 7.8 Managing data volume to facilitate transmission from platform to enterprise.

the use of data reduction and compression to reduce the storage volumes and the bandwidth required to transmit data off platform.

7.4.1 Data Volumes

The Pennsylvania State University Applied Research Lab (PSU/ARL) has conducted several studies on the impact of health-monitoring data on requirements for on-platform data storage, requirements for off-loading platform health-related data, and requirements for transmitting data from platforms to enterprise-level computing and storage resources [Walter, McCall, and Boylan 2013]. In one such study, PSU/ARL analyzed CBM data production rates for a U.S. Army helicopter. The helicopter's data management system was examined to identify typical issues encountered in the collection and movement of health-related data from a system (the helicopter) back to the organization responsible for its life-cycle management.

Most helicopters have a Health and Usage Monitoring System (HUMS) installed. Data are often stored on-aircraft in an internal storage device and on removable storage media, such as a thumb drive. In order to extract data, a maintainer may use a cable to connect the HUMS system or the removable storage device to an electronic logbook or portable Maintenance Support Device (MSD)—typically a hardened, laptop computer. A unique logbook computer is often used for each aircraft, and software on the logbook computer automatically ingests each recorded file into the aircraft logbook and converts all the data into a single file, which aggregates the data from multiple raw data files into a binary file for transmittal to a server. If a connection to the server is not available, the data can be put on a portable storage device, and loaded on the server manually (sneaker net). Local servers are typically connected to additional servers maintained by the organization responsible for life-cycle management of the helicopters.

The HUMS may generate a variety of different file types listed in Table 7.1. Raw sensor data are contained in Asynchronous Time Domain (ATD) and Asynchronous Frequency Domain (AFD) files. A basic manipulation of the data occurs to improve data quality through a process known as synchronous averaging, and the resulting data are stored in Synchronous Time Average (STA) and Synchronous Order Domain (SOD) files. The raw and synchronously averaged data are used to obtain Condition Indicators (CI) that are compared against threshold values to provide warnings and alerts. The steady-state flight regime history is stored in the RGM files, and the XM files include

File Type	Description
AFD	Asynchronous Frequency Domain (vibration data)
XM	Recordings of MIL-STD-1553 data bus messages and analog inputs (if any)
RGM	Regime (steady state flight regime information)
CI	Condition Indicators (processed IVHM health indicators)
STA	Synchronous Time Average (vibration data)
LOG	Event log of crew initiated changes to default configurations and settings
SOD	Synchronous Order Domain (vibration data)
FIF	Flight Information File
TRK	Rotor track and balance data
ATD	Asynchronous Time Domain (vibration data)

Table 7.1 HUMS Data Types Saved by Onboard Data Collection System

the MIL-STD-1553 data bus messages and analog inputs (if any are used). Log files capture crew-initiated changes to certain configuration settings on the aircraft. Flight information files contain specifics about software and hardware versions, serial numbers, configurations, etc. In addition, there are several files used for data associated with rotor track and balance (TRK, HTM, and IDX).

The raw and processed data files vary in size based on sample rate, data type, and data reduction resulting from onboard processing. Figure 7.9 shows the file size distribution for each type of file created by the aircraft's HUMS. AFD and XM data account for over 80% of the data generated on the aircraft, which then must be transmitted back to the data customers over the supporting network infrastructure.

While the distribution of data volume by file type shown in Figure 7.9 is from a helicopter HUMS system, it is representative of most IVHM systems: vibration data often have the

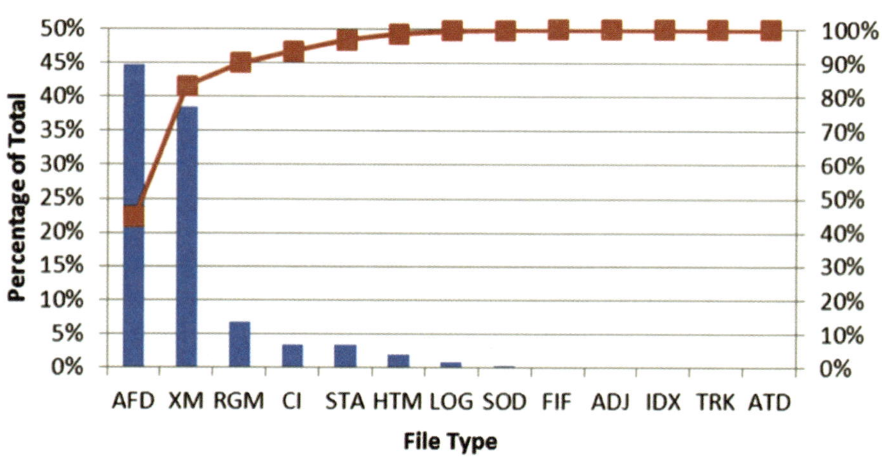

Figure 7.9 Helicopter HUMS parametric data file sizes by type.

highest data acquisition rates and therefore often account for the highest data volumes. Vibration data for rotating equipment are typically collected with accelerometers in one or more axial directions (up to three separate measurements per location). Vibration sensor sample rates vary from 10's of Hz (samples per second) for shaft vibrations to 100 kHz or greater for shock events and acoustic emissions associated with material damage. Strategies for reducing the file sizes required for vibration-related data are discussed in the next section.

7.4.2 Data Storage Standards and Formats

An important aspect of data management is storage of the data in formats that make the data accessible to a wide range of users and customers. In the early days of digital data acquisition systems, data were often stored in proprietary formats developed by the manufacturers of the data acquisition systems because they were also the primary provider of analysis tools. As additional analysis tools were developed by outside vendors, tools were developed to convert data from one proprietary format to another (the format preferred by the analysis tool vendor) or to an open format.

Formats used to store data vary depending on the location and the tools used to access the data. Most open data storage formats use some form of ACSII data representation. One of the earliest attempts to develop a common format for sharing data between computers was the Universal File Formats (UFF), a series of formats for different types of data [Phillips and Allemang 2012]. The UF formats were originally developed as 80-character (punch card image), ASCII records in specific orders depending on the UF type.

Another common file format is the Comma Separated Variable (CSV), in which data are written to a file in ASCII format with commas or other text delimiters between data values. CSV and related ASCII file formats have the advantage that they are human readable using almost any text editor, and can be opened for analysis and visualization in most spreadsheet programs. A limitation of the CSV and other ASCII file formats is the inability to accommodate different types of data or different measurements within a single file because data are typically stored as records with a fixed number of entries within a single file. To provide flexibility in file content while maintaining human readability, there has been a move from straight ASCII file formats to file formats based on the Extensible Markup Language (XML). XML preserves human readability using a text edit, but includes additional metadata to describe the measurement data in the files and can accommodate blocks of data (bulk data) in different formats.

Since the 1990s there has been a push to develop standards for the design of IVHM systems, and the exchange of information between different IVHM system components. On the data acquisition side of the data management flow described in Figure 7.8, data formats are described in numerous data bus standards such as IEEE1451, Mil-Std-1553, SAE J1939, SAE J1850, SAE J1708, and others. The International Standards Organization (ISO) and others have developed standards for the exchange of information within and beyond the on-platform IVHM system components. ISO 13374 provides a schema that describes the structure of an IVHM system, breaking it into functional layers such as data acquisition, data manipulation, state detection, health assessment, prognostic assessment, and advisory generation [ISO 2013]. The Open Systems Architecture for Condition Based Maintenance (OSA-CBM) defined interfaces between each layer in ISO 13374 and provided reference designs for implementing the messages using different message standards including

XML [OSA-CBM 2013]. The Open Systems Architecture for Enterprise Application Integration (OSA-EAI) defined schema for moving data from the on-platform IVHM system to network or enterprise levels for storage and analysis [OSA-EAI 2013]. The Machinery Information Management Open Systems Alliance (MIMOSA) now manages the OSA-CBM (see chapter 6) and OSA-EAI standards efforts.

The U.S. Army has been working to develop a common file format for the storage of IVHM and logistics data across all of its platforms. The goal is to have a single file format around which the U.S. DoD can develop the necessary architecture for transmission, analysis, and long-term storage. The Army Bulk CBM Data (ABCD) format is based upon the Common Data Format (CDF), which was developed by the National Aeronautics and Space Administration (NASA) [NASA 2013]. The CDF is a data format used for storage and manipulation of scalar and multidimensional data in a platform, in a discipline-independent fashion. The ABCD file specification integrates specific MIMOSA metadata into the CDF scalar and/or multidimensional data. The contents of a CDF related to ABCD fall into two categories. One is a series of records comprising a collection of variables consisting of scalars, vector, and n-dimensional arrays. The other is a set of attribute entries describing the CDF in global terms or specifically for a single variable.

The CDF library is a scientific data management package used to manage and manipulate scalar, vector, and multidimensional data arrays. There is completely transparent data formatting for the user and it is accessible through sets of interface routines. Version 3.3.0 of the CDF library was released on 7/16/2009 and contains a number of utilities for creating, accessing, manipulating, and managing CDF files.

There are also several file formats available in the CDF library. A variable is the generic name for an object that represents data in a CDF file, where data can be 0-dimensional (scalar data) or multidimensional (up to ten dimensions). A single-file CDF contains the control information, metadata, and the data values for each of the variables in one file. The advantage of this file format is that it minimizes the number of files one has to manage and makes it easier to transport CDF files across a network. Multi-file CDF is metadata files and data files, one for each of its variables. The restrictions of this format are three fold: compression, sparseness, and allocation. Compression is not allowed for the CDF or any of its variables. The sparseness revolves around sparse records or arrays for variables that are not allowed.

7.4.3 Data Reduction

Data reduction reduces the size of the collected data by permanently removing or transforming the content of the data. *Data compression* reduces file size by changing the file at a bit level; the meaning of the file contents is ignored. The two can be used together to reduce the size and transmission bandwidth requirements for IVHM data, as shown in Figure 7.10.

While data reduction and compression both reduce the size of data files, they differ in the way the file sizes are reduced and the knowledge required to implement the approach. Data reduction requires the expert knowledge of a data set by an engineer or analyst and assessment of the needs of data customers to determine which data are unnecessary and how they can be removed without the loss of important information. Data compression reduces file size by changing the file at a bit level; the meaning of the file contents is ignored. Parametric data compression is lossless, restoring all the data in the file to its original state when decompressed.

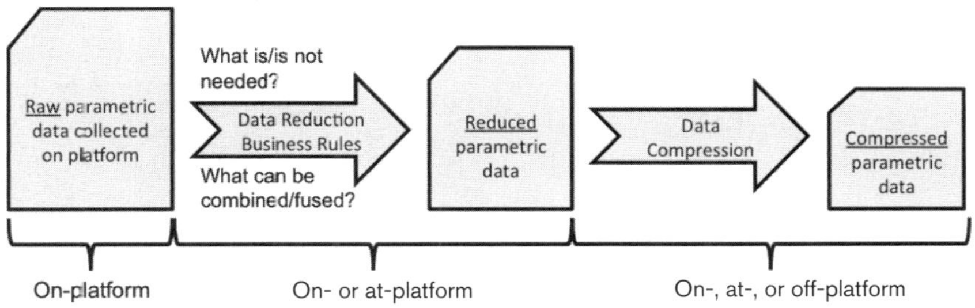

Figure 7.10 Data processing flow to reduce file size and transfer bandwidth requirements.

Information needs may change over time; hence, data reduction strategies may need to be changed over time as well to reflect the changes in data requirements. In most (but not all) cases, data reduction introduces losses, trimming unnecessary data that can never be recovered.

There will always be arguments against data reduction: if data are unnecessary and can be removed, why collect the data in the first place? On the other hand, having the data collected in combination with a user interface on-platform to allow engineering to reset what data are reduced (or preprocessed on-platform), allows updating of the IVHM system as new characteristics or conditions are discovered. In some cases a valid argument can be made for eliminating particular data channels, but periodic reevaluation of on-platform data collection and reduction requirements is recommended; nevertheless, there are cases for which data reduction makes sense.

Multiple Data Customers: Sometimes high-bandwidth data are used immediately to produce actionable information to a driver or maintainer; in those cases a reduced data set may be sufficient for archival purposes.

Changing Bandwidth Requirements: It can be useful to abstract a data collection process from the data bandwidth and storage restrictions placed upon it by network administrators.

Inflexible Data Collection: Data collection devices can be low-powered and difficult or impossible to configure. A separate data reduction tool allows greater flexibility for placement on-, at-, or off-platform, and can be independently configured and managed.

Variable Usefulness of Data: Often, data are only considered interesting if some algorithm has detected an anomalous condition at the time. An analyst may require high-bandwidth data only in a certain time frame when an alarm, fault, or event occurs. Data reduction allows for selective preservation of interesting data that a data collection device might not have the intelligence to manage.

Table 7.2 lists several common data reduction approaches. Several of these approaches are briefly described next. All of the methods listed in Table 7.2 apply to single-dimensional data. If a file contains multidimensional data (i.e., x-y position data, magnitude, and phase), the methods may have to be altered or tailored based on the relationship between dimensions. One problem with many data reduction methods is that they may miss (eliminate) anomalous or unexpected data, which may be very important for detecting unknown or unexpected faults. Another challenge is that the data reduction methods may

Reduction Approach	Parameters	Effect
Decimate	Decimation Factor (n)	Low-pass filter data to avoid aliasing, then only retain every nth data point (where n is the decimation factor). Effectively resampling the data at a lower sampling rate.
Average	# of points in average (M)	Partition data into M-points bin long and average each bin.
Suppress	Threshold	Remove data entirely which does not meet threshold requirement.
Transform	Data Type Scaling Factor Offset	Change data into smaller data type (range compression or coding). For example, scale floating-point numbers and replace with integers requiring fewer bits.
On Change	Delta	Only store data if it changes (by at least *Delta*).
Range	Value Condition	Only store data if the value meets a range and condition requirement (can also be used to remove invalid data).
Exclusion	Condition Window	Store data for some period of time before and after the value meets a condition.

Table 7.2 Summary of Data Reduction Approaches

not capture precursors or consequences of fault events. The **exclusion** method attempts to solve these problems by capturing data (without reduction) before and after the occurrence of an "event" (described by the condition). The use of exclusion is illustrated in Figure 7.11.

For many types of data, the important IVHM information is derived from the frequency domain representation resulting from a Fourier transform. In fact, several of the data types included in Table 7.1 are extracted from the FFT or power spectral density of time-series data. Figure 7.12 shows an example

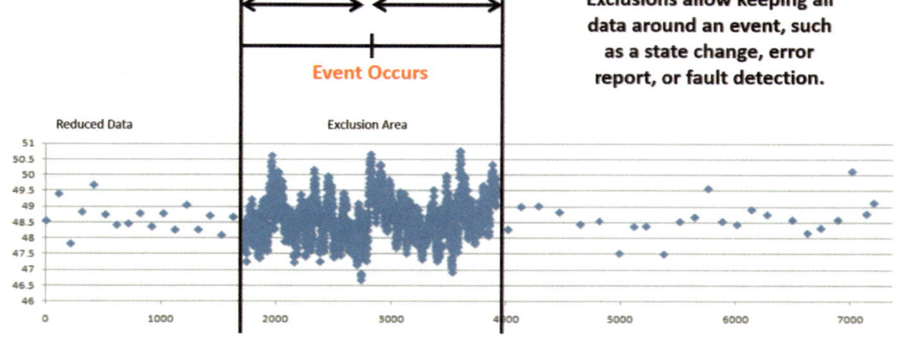

Figure 7.11 Example of exclusion approach to data reduction/retention.

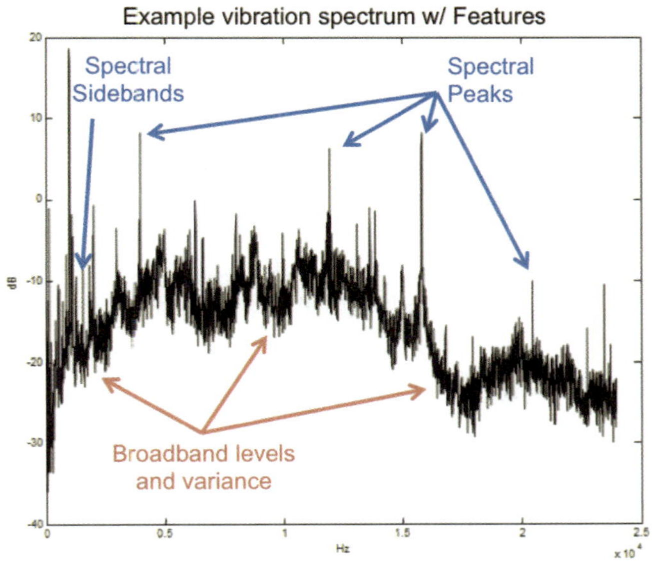

Figure 7.12 Example vibration spectrum showing important spectral features.

vibration frequency spectrum plot along with features of interest for IVHM analysis. The Fourier transform itself does not reduce the number of bits required to represent the data—the FFT of an N-point data sequence still requires at least N numbers to completely represent the data. Nevertheless, opportunities exist to apply data reduction rules to spectral representation of the data, which can reduce the number of bits required to transmit and store the spectral data. Spectral features of interest in Figure 7.12 include spectral peaks (often associated with the rotational rates of key system components and their associated faults), spectral sidebands (associated with faults or other variations in key rotational rates), and broadband vibration levels and their associated variances (associated with both nominal and faulted operating conditions). By detecting the key spectral features and their associated frequencies (or frequency ranges), the data required to represent the important information in the spectrum can be significantly reduced by only saving the feature information. The vibration spectrum can be represented in terms of its background level and variance over different frequency ranges and significant spectral peaks or features and their respective frequencies. The upper plot in Figure 7.13 shows a vibration spectrum with spectral features identified (spectral peaks, background levels, and corresponding variance); the lower plot shows the reduced spectral representation consisting of only the saved feature information. Retaining only the spectral feature information in the lower plot reduced the storage and transmission requirements of the vibration data by 16:1 (93.75%). Selection of the final processing parameters for reducing the content of the spectral data will depend on the impact of the content reduction on the computation of spectral health condition indicators.

7.4.4 Data Compression

In computer science and information theory, data compression refers to the process of encoding information using fewer bits than its original representation. Data compression is widely used for pictures, video, and audio files to reduce their storage and transmission requirements. Data compression is also used to reduce the size of electronic documents and data files. Compression can be either *lossless* or *lossy*. *Lossless compression* reduces the number of bits required to represent data by identifying and eliminating redundancy [Shannon and Weaver 1949] within the data, resulting in no information loss. *Lossy* compression reduces the number of bits required to represent the information by eliminating redundancy (as in lossless compression) and by identifying portions of the data with little information value and removing it. The process of reducing a file size is popularly referred to as data compression, but from an information theory standpoint, it falls under the category of data encoding (effectively re-encoding the data) [Lelewer and Hirschberg 1987]. For the purpose of IVHM data, we typically use lossless compression to avoid the loss of potentially critical information.

The design, or selection, of data compression schemes involves trade-offs between several factors: the degree (or ratio) of compression, distortion or loss of information (for lossy compression), and the computational resources required to compress (and decompress) the data. The last factor is more or less important depending on whether the compression/decompression must occur in real time (as in video or audio playback) or can be accomplished off line (for example in data archiving operations). While one typically compresses the entire contents of a file, in certain file formats individual variables can be compressed. Popular or familiar compressed file formats include MP3 (lossless or lossy depending on settings), JPG (lossy), and ZIP (lossless). In addition, many popular computer programs such as Microsoft Word and MATLAB now use embedded compression to reduce document and data file size. The general characteristics of several common data compression techniques are described next.

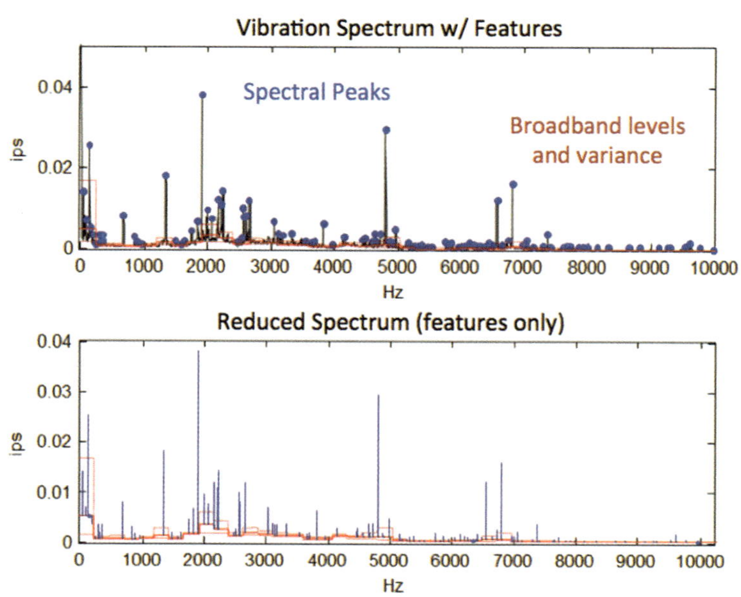

Figure 7.13 Demonstration of spectral-based data reduction.

Run-Length Compression is a lossless compression scheme that takes advantage of repeating bytes in the data. A table is generated that captures commonly occurring bit patterns, which are then replaced throughout the file with their representative code (requiring fewer bits).

Huffman is a lossless compression scheme that takes advantage of the frequency at which certain byte values occur in the data. A sequence of bytes that contain a high percentage of a limited number of byte values will compress better than if each byte value occurs with equal probability.

Adaptive Huffman takes advantage of the frequency at which certain byte values occur in the data and provides slightly better compression than Huffman.

GZIP stands for the Gnu ZIP Compression algorithm in which significant compression occurs over a wide variety of data sets. This compression has one parameter, which may be set to a level in the range from 1 (least amount of compression but fastest) to 9 (most amount of compression but slowest).

Often, no compression is done on platform prior to saving data to local storage. This is driven by the capability of the on-platform, embedded computing resources and the real-time nature of the data acquisition and analysis performed in the monitoring system. However, compression techniques can provide significant reductions in terms of file size and, therefore, communications bandwidth required to transmit the data.

The Pennsylvania State University Applied Research Laboratory applied several compression techniques to representative IVHM data files originally stored in the U.S. Army ABCD data format. The algorithms were tested to approximate the potential file size reduction that can be achieved. First, the four built-in CDF compression modes were employed: Run-Length Compression, Huffman, Adaptive Huffman, and GZIP. Next, three free desktop application compression methods were tested: RAR (via WinRAR), Zip (via 7-Zip), and 7z (via 7-Zip). Finally, PSU/ARL provided the ABCD files to a commercial company, which compressed the files using two proprietary techniques. The company noted that they expect higher compression ratios if the data in the ABCD files can be regrouped before being compressed to take advantage of redundancy across data entries instead of processing each individually.

Each of the nine methods compressed one day's worth of ABCD files (173) from a platform, which totaled 2.937 GB. The results are listed from best to worst in Table 7.3.

Compression Type	File Size (bytes)	% Compression
Original File—No Compression	*2,936,828,868*	*0.00*
Proprietary Algorithm #1 with Delta Encoding (1)	516,881,881	82.40
7z (Normal)	916,791,063	68.78
Proprietary Algorithm #2 without Delta Encoding (2)	917,734,473	68.75
RAR (Normal)	1,157,832,987	60.58
Zip (Normal)	1,166,438,437	60.28
GZip (level 9)*	1,249,463,849	57.46
Adaptive Huffman*	2,022,211,104	31.14
Huffman*	2,493,953,296	15.08
Run Length Encoding*	2,952,691,564	–0.54

* Included in the CDF specification

Table 7.3 Compression Test Results

7.5 Conclusions

IVHM systems provide information on the health and capability of the monitored platform or system. They are capable of collecting and storing large amounts of data. Customers of the data include operators, maintainers, planners, and other support personnel. The ability to move data from the monitored system to the customers is an important factor in the effectiveness of the IVHM system. Therefore, a key aspect of IVHM data management is reducing the amount of space required to store IVHM data, or equivalently, the bandwidth required to transmit the data to the end users.

This chapter provided an overview of techniques for management of data in an IVHM system. Data processing and fusion techniques convert raw sensor data to assessments of system health and can produce recommended actions for maintaining or operating the system. The data processing and fusion may reduce the volume of raw data needed to adequately represent the state of the system. Many techniques, however, actually increase the volume of data as a result of the need to save both the raw data and any data products.

Data reduction techniques reduce the size of the collected data by permanently removing or transforming the content of the data. Data compression techniques reduce file sizes by changing the file at a bit level, independent of the meaning of the file contents. The two can be used together to reduce the size and transmission bandwidth requirements for IVHM data.

A final aspect of data management is the storage of the data once they are transferred from the IVHM system to off-platform information systems. Data must be stored in formats that facilitate access and processing of the data and information by the intended customers. Common, open data formats and processing architectures have been developed to facilitate transfer, access, and additional processing of IVHM data.

7.6 References

Banks, J. 2000. "Characterization of Machinery Condition Using Structural Surface Intensity," Masters Thesis, The Pennsylvania State University, 2000.

Campbell, Robert, J. Banks, C. Begg, and C. Byington. 2001. "Diagnostic Feature Comparisons for Experimental and Theoretical Gearbox Failures," *Proceedings of the 55th Meeting of the Society for Machinery Failure and Prevention Technology*, Virginia Beach, Virginia, April 2–5, 2001.

Duda, Richard O., Peter E. Hart, and David G. Stork. 2001. *Pattern Classification*, 2nd Edition, John Wiley and Sons: New York.

Hall, David L. 2001. *Handbook of Multisensor Data Fusion*, CRC Press LLC.

ISO. 2013. "ISO 13374-1:2003, Condition Monitoring and Diagnostics of Machines—Data Processing, Communication and Presentation—Part 1," International Organization for Standardization, http://www.iso.org/iso/catalogue_detail?csnumber=21832, Accessed March 24, 2013.

Lelewer, Debra A., and Daniel S. Hirschberg. 1987. "Data Compression," *ACM Computing Surveys*, Vol. 19, pp. 261–296.

Llinas, James, Christopher Bowman, Galina Rogova, Alan Steinberg, Ed Waltz, and Frank White. 2004. *Revisiting the JDL data fusion model II*. Space and Naval Warfare Systems Command, San Diego, CA.

Moore, Gordon E. 1965. "Cramming more components onto integrated circuits," *Electronics Magazine*, Volume 38, Number 8, April 19, p. 4.

NASA. 2013. "CDF Homepage," National Aeronautics and Space Administration.

http://cdf.gsfc.nasa.gov/, Accessed March 22, 2013.

OSA-CBM 3.3.1: Open Systems Architecture for Condition-Based Maintenance, Machinery Information Management Open Systems Alliance, http://www.mimosa.org/?q=resources/specs/osa-cbm-331, Accessed March 24, 2013.

OSA-EAI 3.2.3: Open Systems Architecture for Enterprise Application Integration, Machinery Information Management Open Systems Alliance, http://www.mimosa.org/?q=resources/specs/osa-eai-323, Accessed March 24, 2013.

Phillips, Allyn W., and Randall J. Allemang. 2012. "Evaluating a Format for Viable, Long-term Dynamic Data Archiving," *Sound and Vibration Magazine*, June, pp. 14–17.

Shannon, C. E., and W. Weaver. 1949. *The Mathematical Theory of Communication*, University of Illinois Press, Urbana, IL.

Shepherd, Micah R., Stephen C. Conlon, Fabio Semperiotti, and Stephen A. Hambric. 2012. "Structural Intensity Modeling and Simulations for Damage Detection," *Journal of Vibration and Acoustics*, Volume 134, Issue 5, October 2012.

Walter, Robert L., Douglas H. McCall, and David C. Boylan. 2013. "Parametric Data Reduction Took Overview," presented at American Helicopter Society Airworthiness, CBM and HUMS Specialists' Meeting, Huntsville, AL, 11–13 February 2013.

CHAPTER 8 Vehicle-Level Reasoning Systems

A. N. Srivastava and R. W. Mah, *NASA Ames Research Center*
R. Bharadwaj and D. Mylaraswamy, *Honeywell Aerospace*

"Knowledge is power only if man knows what facts not to bother with."

—Robert Staughton Lynd

8.1 Introduction

Early in the development of complex aerospace systems, built-in testing became an important method to verify the integrity of onboard systems. The main idea is that it allows a machine to test itself and in some cases perform diagnostics to determine the source of a problem. The hope was that these systems could help improve system reliability and increase mission assurance. In many cases, the built-in test capability was implemented in hardware and helped a technician determine the state of health of the system. Early systems had little software and relied primarily on the combination of human analysis of the results of the built-in test. As a result, such systems were not designed to operate in real time or during a mission.

In contrast, modern aerospace systems are fundamentally different. These systems are comprised of a combination of hardware and software operating on embedded systems. Because of these highly complex systems, developing techniques that can analyze their state of health in real time with little to no human intervention is essential.

This chapter discusses a few of the techniques and approaches that can be employed in a Vehicle-Level Reasoning System (VLRS). A case in point is the development of a vehicle-level reasoner with the capability to detect anomalies in real time, diagnose the source of the anomaly, and in some cases predict the remaining useful life of a subsystem or component on a real-world aircraft. We begin by providing a background on real-time diagnostics and prognostics and also discuss the use of data mining to improve the diagnostic and prognostic capability of the reasoner. We give an overview of systems that are similar to the vehicle-level reasoner but that are used in other domains. After we provide the background, we give a more precise definition of what we mean by vehicle-level reasoning and we define the scope. We provide distinction between vehicle-level reasoning and traditional diagnostics and prognostics that can be found in many implementations. After that, we discuss the vehicle-level reasoner's architecture, touching on the ground and flight operations and associated communications.

Because the vehicle-level reasoner is inherently a real-time system, we discuss different diagnostic and prognostic methods and their associated standards that can be applied in real time. The next section gives an overview of anomaly detection algorithms that have been developed to detect unusual signatures in discrete and continuous signals. The motivation for vehicle-level reasoning is the potential benefit from the economic and safety standpoint. We discuss some of the benefits of vehicle-level reasoning in terms of economics and safety and the way operators can benefit from it. As is the case with many system health management approaches, the economics and safety can be the drivers for successful implementation. We discuss this in some detail and provide conclusions at the end of this chapter.

8.2 Background

Onboard vehicle health monitoring is becoming a must have for the aerospace industry. Real-time diagnostic and prognostic systems are being installed in a wide range of commercial and military vehicles from airplane, to helicopter, to spacecraft. Boeing first incorporated onboard health monitoring into the 747-400 in the late 1980s in the Central Maintenance Computer (CMC) to collect fault data from various system components and perform real-time diagnostics. Since then, a growing number of Vehicle Health Management (VHM) systems are being installed in commercial aircraft. A case in point is the Honeywell Aerospace crew information system/maintenance system that is being fitted into the Boeing 787 within the CMC [Aerospace 2011]. This overarching technology manages all member systems on board the airplane from flight control systems to in-flight entertainment. On the Boeing 787, more than 150 key systems are monitored. In addition, Boeing provides a real-time datalink with the ground to enable the airline operators to be ready with the right Line Replaceable Unit (LRU) or part when the airplane lands, thus cutting the cost of running repairs and minimizing flight delays. Airbus has a similar onboard system called the AIRcraft Maintenance ANalysis (AIRMAN) as well as a ground center called the Airbus Technical Aircraft-on-ground Centre (AIRTAC). The underlying technology is a model-based fault propagation model and a suite of knowledge management tools to define the relationships between member systems and LRUs and the defined behaviors of these systems. Over 100,000 parameters and 30,000 fault conditions can be analyzed and interpreted to provide concise crew alerts and specific maintenance actions.

Boeing's Airplane Health Management (AHM) product has a database of past trends and outcomes and uses real-time prognostics to analyze the historical data and present data coming from real-time acquisition of airplane data while in flight. Built-in decision support tools are used to predict when a component, subsystem, or system might fail so they can be replaced or repaired at the next opportunity. European Aeronautic Defence and Space Company (EADS), Europe's premiere aerospace and defense company, has a similar approach using real-time prognostics to monitor usage and the life-cycle environment of a system. Key system parameters are measured in-situ, and load profiles in conjunction with models are used to assess degradations due to cumulative load exposures. The EADS Sensor-based Aircraft Maintenance Support (SAMS) tool is installed on the Airbus A380 and Boeing 787 to monitor engine health [Pecht 2008]. It is said that this tool, together with in-service analysis of vibration signals to identify potential bearing failures, provides accurate and fast location of defects in the airplane and shortens maintenance times and lowers costs.

Data mining technology is being used by the National Aeronautics and Space Administration (NASA) to enhance real-time onboard vehicle health monitoring. A system health monitoring tool called Anomaly Monitoring Inductive Software System (AMISS) has been deployed on the International Space Station (ISS) to provide real-time health monitoring. This tool is based on the Inductive Monitoring System (IMS) [Iverson and Hillin 2012] technique to establish a baseline of normal behavior from archived telemetry for the monitored system. IMS uses the baseline to identify off-normal behavior in real-time telemetry rather than on threshold exceedance monitoring, which is used in many caution and warning systems. The tool, initially installed to monitor the ISS Attitude Determination and Control System and the External Thermal Control System, provided earlier warnings of unusual behavior in one of the control moment gyros used to control ISS attitude, and ammonia bubble growth in the thermal tank six days prior to detection via standard techniques. Results such as these motivated NASA to generalize the tool for use by any flight control discipline, and the tool is now used for real-time flight control room monitoring of 19 ISS subsystems, with additional deployments underway.

Similar systems to VLRS have been used in other domains. A case in mind is the Remote Agent deployed on a spacecraft in 1999 [Miles and Peters II 1999]. It was the first "artificial intelligence" to be given primary command over a spacecraft. The Livingstone software is a key component that monitors the health of the spacecraft and attempts to correct any problems that occur. It receives information about critical subsystems via monitors located at various places on the spacecraft. Livingstone knows what a healthy spacecraft should be, and if it detects a problem, it diagnoses the most likely cause of the problem and suggests a course of action. If the problem is one that cannot be fixed, Livingstone reports this and relies on the Remote Agent's Smart Executive and Planner Scheduler to devise an alternative plan to reach the mission's goals. The Remote Agent worked well in handling several simulated failures on board the spacecraft.

This success led to development of more advanced Livingstone 2 software that is reusable for monitoring complex systems. It was deployed on the Earth Observing One remote sensing satellite in 2004. The onboard software uses machine learning and pattern recognition, continuous planning, and robust task and goal-based execution to increase the scientific return radically by enabling intelligent downlink selection and autonomous retargeting. This achievement demonstrated the potential for space missions to use onboard decision-making to detect and diagnose instrument faults, analyze status, and respond to scientific events, and to downlink only the highest-value science data [Chien 2012]. Prior to Livingstone, unmanned spacecraft have flown blind with little or no ability to make autonomous decisions based on the content of the data they collect.

8.3 Defining VLRS

8.3.1 Scope

The primary function of a VLRS is to ensure aircraft safety, and it does this by detecting and predicting faults and failures at the aircraft level. The VLRS receives health information from individual subsystems and fuses the information to derive an overall health state for the aircraft. Similar to a conventional diagnostic algorithm, deriving a health state involves the following functional elements:

1. **Measure:** associated with sensing, measuring, or reading data from data bus messages.
2. **Inference:** associated with extracting the evidential information from the data provided, and estimating the current and future health of the asset(s) being monitored.
3. **System Model:** encodes the designer's knowledge of how the system/asset behaves under different conditions and operating modes.
4. **Interact with User:** functions for displaying health data and status (text and graphics), displaying documents, controlling indicator lights and gauges, and also receiving commands and data inputs from the user.

Figure 8.1 shows the vehicle-level health management system. In this approach, the subsystem-level health information such as the status of the built-in tests and the monitors are aggregated by the area-level reasoner and the VLRS to determine the vehicle health status. In the most rudimentary form, the VLRS diagnostics and prognostics approach is similar to that at the subsystem level. This is especially true for complex subsystems such as the propulsion and flight controls, which use complex hardware and software. However, the choice of VLRS approach used to aggregate the aircraft subsystem diagnostic and prognostic information is essentially dependent on the available knowledge of the system and the subsystem dependencies.

8.3.2 Different Approaches for VLRS

In a broad sense, reasoning at the vehicle level can be approached either from a model-based or a case-based point of view—these viewpoints arise based on how we approach the reasoning solution. In addition, we also recognize an emerging approach to VLRS

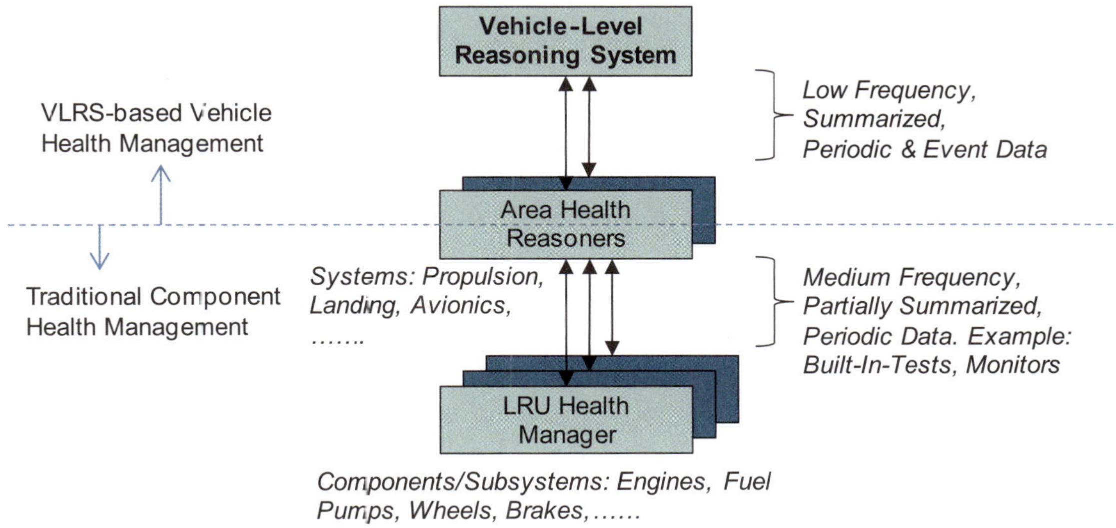

Figure 8.1 Hierarchical view of component, area/system, and vehicle-level health management system.

arising from the ubiquity of digital data being collected and archived by most modern vehicles. Figure 8.2 shows the taxonomy for describing these approaches together with exemplary methods for VLRS. The model-based approach assumes that we have a description of vehicle behavior under various conditions, some of which are faulty. In other

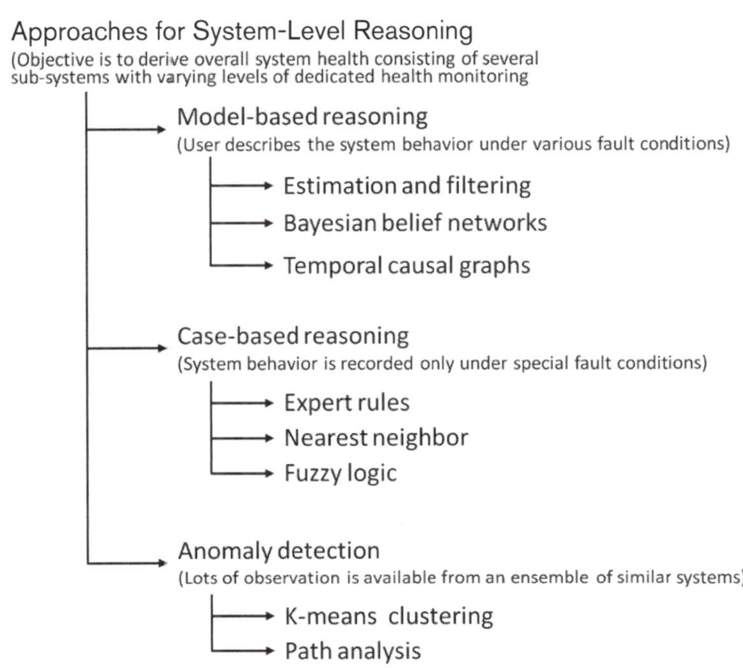

Figure 8.2 Approaches for VLRS.

words, we start with the knowledge "cause A" leads to system "behavior B" manifested as "observations C." The case-based approach takes an opposite viewpoint: when the system behavior was described using "observations C," analysis led to the conclusion that "cause A" may have started it. While this distinction is important to enumerate a good algorithmic choice, most practical systems tend to be a blend of these two approaches.

8.3.2.1 Model-Based Reasoning

Model-based reasoning approaches can be further classified based on the mathematical form of the underlying model: that is, depending on how the cause → effect relationship is expressed. If the mathematical description takes the form of differential-algebraic equations with parameters, the algorithm will try to "find" the best set of parameters to minimize the error between the observations made from the vehicle and the model predictions. In this case, VLRS takes the form of estimation and filtering methods. If the description includes probabilities and probabilistic approximations, VLRS can be achieved using variants of Bayesian methods. If the cause → effect relationship is expressed as a temporal sequence of events, VLRS is effectively achieved using temporal causal graphs such as the finite state machine.

System Model-Based VLRS

System model-based VLRS is used for cases in which the vehicle and subsystems can be expressed in closed mathematical form. Figure 8.3 shows a block diagram example of this VLRS approach. The system models can use an observer to generate the residuals. The observer/filter can be linear or nonlinear; examples include a Kalman Filter (KF), Extended-Kalman Filter (EKF), or particle filter. The subsystem residuals can be used to generate the fault symbols or signature. At runtime, the VLRS function compares the fault signatures to the temporal causal model to isolate the fault. This approach is best suited when the underlying physics are known for the system and the temporal relationship between faults can be expressed in a closed mathematical form [Manders, Narasimhan, and Biswas 2000]. This approach can be used for both the component-level diagnostics and prognostics and by the VLRS. At the component level, due to the availability of mathematical models, this is the preferred approach for diagnostics and prognostics. However, the need for mathematical models makes it difficult to integrate software monitoring. That, coupled with the federated approach to aircraft design and development, makes the use of this approach impractical for aircraft VLRS.

Probabilistic Approximation-Based VLRS

In cases where the cause-effect relationship between failure modes and evidences is known probabilistically but the complete system model is not known, the VLRS can be driven by the system reference model that

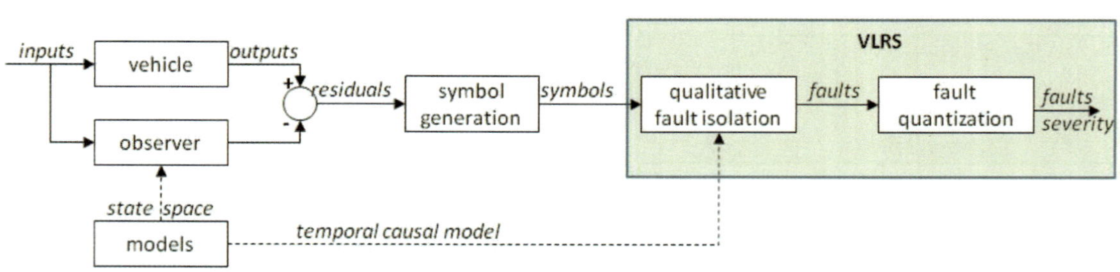

Figure 8.3 Block diagram of system model-driven VLRS. The dashed lines show the model types to observer and fault isolation modules. The solid lines show the signal flow.

captures the relationship between the *cause* and *effect* and applies some form of Bayesian reasoning. The VLRS does not need knowledge of the system in a closed mathematical form. A VLRS using a "noisy–OR" (discussed in section 8.6.2) is a special case of this type of reasoning. The conditional probability between failure modes is not required for a noisy–OR-based reasoner. In this type of VLRS, a priori probabilities are computed using historical data and are captured in the system reference model. The system reference model captures the relationship between the causes (failure modes) and the effects (symptoms/evidences). At runtime, on arrival of new evidence, the system fault probabilities are updated using some form of Bayes rule. Prognostics-enabled VLRS using this approach is covered in detail in section 8.6.

8.3.2.2 Case-Based Reasoning

Case-based reasoning approaches can be classified depending on how the effect → cause empirical observation is stored, retrieved, and used to determine how "similar" the current vehicle behavior is to those captured by various cases. The simplest form of VLRS would be expert systems, which merely express each and every effect → cause as IF-THEN rules. Variants include applying weights based on subject matter expertise and a reinforcement learning loop that counts the number of times a rule fires. Fuzzy logic and rules provide a formal mechanism to express subject matter expertise to alter the effect → cause relationships selectively based on business rules. Nearest neighbor methods define an explicit distance metric to define similarity. While Euclidian distance is the most commonly used metric, variants include Manhattan and Mahalanobis distance that provide selective weighting to one or more observations and hence bias the effect → cause relationships.

In this class of VLRS, the correlation between evidence and failure modes is known from historical cases. This is a truly data-driven method in which the prior data are used to derive a heuristics model to capture the relationship between evidences and the failure modes. The direction of the relationship as captured in the system model is the primary difference between the Heuristic and cause-effect-relationship-based VLRS. A fuzzy logic-based reasoner is a special case for this type of reasoning system. The Fuzzy Inferencing System (FIS)-based VLRS computes the failure probabilities by applying the fuzzy rules and input-output membership information and sequence information to the fault signatures [Uluyol and Hickenbottom 2010; Mendel 1995]. A number of hybrid VLRS systems use a combination of both of the approaches.

8.3.2.3 Anomaly Detection Methods

Irrespective of the method used, model-based VLRS always generates an ambiguity group of plausible "causes" that can explain observed vehicle behavior. That is, it generates a set of plausible hypotheses that can best describe the current vehicle observations with respect to the knowledge we have about the vehicle. The term "reference" is often used as a prefix to describe this knowledge because it provides the frame within which the VLRS operates. The fidelity of the method can be determined by the number of elements and ranking of these elements in the ambiguity set.

An emerging approach to VLRS is grouped under the category, anomaly detection. Often observations from multiple, homogenous (not identical) vehicles can be used to detect incipient failures that may be difficult to detect if one had observed only one vehicle. In other words, individually each vehicle may not have any failures, but collectively the ensemble exhibits behavior that deviates from the normal [Chandola, Banerjee, and Kumar 2009]. In this case, VLRS takes the form of anomaly detection—an outlier is defined as something that does not adhere to the norm; the norm is defined as something that most of the vehicles exhibit. In our case, the only way we can express this is as observations that follow the norm and outlier that do not follow the norm. K-means clustering is one of the most popular methods for establishing this norm; its popularity arises from its simplicity. In this case, VLRS takes the form of determining outlier observations with respect to established behavior clusters. In some cases, nominal behavior is established as a sequence of events, and deviations from this "path" can be labeled as outliers. In this case, VLRS takes the form of path analysis and looking for events that do not adhere to dominantly occurring sequences.

Perhaps the simplest way to perform anomaly detection, assuming that all data are continuous and appropriately normalized, is to compute the Euclidean distance from one point and all other points that are known to be nominal. Such a computation can be extremely expensive because it requires computing the distance between the query point and the entire database of nominal observations. Bay and Schwabacher [2003] introduced Orca, which improves the computation time significantly through the use of a simple pruning rule. This algorithm has become one of the key methods for anomaly detection on continuous signals.

Many distance-based anomaly detection techniques do not build an internal model. From the standpoint of a reasoning system, however, it can be more advantageous to perform anomaly detection using an internal model. In many applications, using a clustering technique that builds an internal model of nominal behavior can be advantageous. The Inductive Monitoring System (IMS) builds an internal model and then measures the similarity of a point by its distance to the nearest cluster [Iverson and Stop 2004]. This approach yields models that can be transferred to a reasoner for further diagnostics and evaluation, although the approach integrating a system such as IMS with the reasoner described here has not yet been done. These models can also be used to improve diagnostic quality because they can help identify the source of the anomaly.

Although many anomaly detection techniques exist for studying continuous signals, the analysis of discrete signals is more difficult because in many cases we are interested in identifying anomalous sequences in discrete signals. In the analysis of human-machine interactions, these types of anomalous switching sequences can be important to discover [Das et al. 2010] because in some cases they can indicate issues in the piloting commands.

8.3.3 Similarity and Differences between VLRS and Component Diagnostics and Prognostics

At the basic approach level, VLRS is very similar to the approach used in component-level diagnostics and prognostics. Both use expert knowledge and data to design and develop the models used for diagnostics and prognostics. The algorithms such as the KF, EKF, particle filters, Bayesian reasoners, and fuzzy logic can be used at both the component level and the VLRS for diagnostics.

However, VLRS differs from diagnostics and prognostics at the subsystem level as follows:

1. **Measure Function**—The major difference is the heterogeneity and volume of input data. A typical VLRS does not deal with time-series, high-frequency sensor data. It deals with summary information generated by component-level health-monitoring algorithms. The following list includes some examples that contrast summary information with raw-sensor data.

a. Pass/fail result of an over-speed condition vs. 10-Hz engine speed data
b. Number of exceedance vs. temperature signals
c. Spectral bin counting vs. 10-kHz vibration signal

Further, input to a VLRS is often "exception-based" rather than based on continuous signals. That is, VLRS may receive an input when the component-level health-monitoring algorithm decides it has something interesting to report.

2. **System Model**—The primary difference is the mathematical format for encoding knowledge needed to support the analysis. By its very nature, VLRS needs to understand how causality flows among various aircraft subsystems. This can arise due to signal flow, physical connectivity, or proximity. This information is often encoded using simple input-output transfer functions rather than differential equations. Further, these transfer functions are expressed using point-wise conditional probability values rather than analytic statistical distributions.
3. **Inference**—The main difference is the number of outputs or hypotheses that must be considered at any given time. At the vehicle level, single fault assumption cannot be assumed. Consider an aircraft consisting of five subsystems, each of which can fail in four distinct ways. At the subsystem level, the diagnostic algorithm needs to select the 1-out-of-4 health state, while at the vehicle level, this space increases to $2^5 * 4$, or 128. Here the number 2^5 arises from the fact that we may have multiple combinations of failed subsystems. Hence, VLRS needs to manage this complexity and apply some "rules" to prune this hypothesis space.
4. **Predict**—The biggest difference is the estimation of functional impact based on prevailing health state involves multiple world scenarios. Any look-ahead function must make assumptions of how the asset shall operate in the future. The most common approach is to extrapolate current health states. Using our simple example of an aircraft consisting of five subsystems, one needs to make assumptions regarding all subsystems. This gives rise to multiple future worlds, and hence the functional impact must be estimated for each of these future worlds. To manage this complexity, VLRS must make simplifying assumptions about a worst-case scenario or most-likely world to calculate the functional impact.

8.4 VLRS Architecture

The goal of a VLRS health management system is to ensure vehicle safety. There are, however, operational and business goals that need to be satisfied as well, including timely and cost-effective vehicle health management. These goals pose significant challenges in developing a comprehensive VLRS health management capability. They include working around onboard vehicle constraints such as limited monitoring, computational, and communication resources; and issues of asynchronous communications and protocols, as well as dealing with the complexity of current and future vehicle system architecture.

Boeing and Airbus are working around the constraints and complexities via a real-time datalink from the vehicle to the ground. This approach allows a simpler VLRS on board the vehicle and enables airline operators on the ground to bring resources to bear as needed. Looking ahead, standardization of a VLRS design through the use of open system architectures, modularizing health management functions, and enforcing component interchangeability and reusability will help to keep development, operational, and business costs down, improve portability, and facilitate the adoption of new technologies as they become available.

A consensus in the aerospace industry is that there is a need to establish open architectures and standards because they would facilitate interaction of aircraft primes, subsystem OEMs, aircraft operators, government regulators, and the research community in development and integration of health management capabilities from the best available subsystems and components. Several IVHM open architecture studies by the aerospace industry and NASA are applicable here [Felke et al. 2010; Gorinevsky et al. 2010] as they describe architectural frameworks for IVHM interoperability. The interoperability framework includes standardization of operations, functions, protocols, and information management models. The NASA paper reflects the view of a NASA IVHM Architecture Advisory Working Group that included representatives of aircraft prime, avionics integrators, aircraft propulsion OEMs, and DoD along with key people from the NASA IVHM Project. More detailed discussions on architecture are in chapter 6, IVHM Architecture.

8.5 Standards for VLRS

The existing standards deal with the central aircraft maintenance function and communications with the maintenance functions. There are some upcoming standards on maintenance and test reasoners that are also applicable to different aspects of the VLRS.

ARINC 624 is a standard for the Onboard Maintenance System (OMS) [ARINC 1991]. The standard includes a general philosophy, guidance, and recommendations for the design and use of OMS. It also includes the use of ARINC 429 for data transmission between subsystems/member systems. This standard, however, does not cover the aspects of reasoner design for the VLRS.

ISO 13374 covers the interface requirements for many functions that can be used to implement the VLRS [ISO]. However, it does not cover the design and inner workings of the reasoner. Many aspects of reasoner design are covered by IEEE 1232 [IEEE 2010]. The IEEE 1232 family of standards was developed to provide standard exchange formats and software services for reasoning systems used in system test and diagnosis. The development of IEEE 1232 standards has been targeted for automatic test equipment, but the standard exchange formats developed under this standard can be applied to the VLRS.

In summary, the existing standards cover different aspects of design, communications, and exchange for the VLRS function but do not provide a comprehensive coverage for a prognostics-enabled VLRS.

8.6 Example: Emerging VLRS Implementation

8.6.1 Vehicle Integrated Prognostic Reasoner

The Vehicle Integrated Prognostic Reasoner (VIPR) for VLRS research was developed through a NASA research agreement contract from the System-Wide Safety and Assurance Technologies (SSAT) project under the NASA Aviation Safety program [Mylaraswamy 2010; Hadden et al. 2011; Mylaraswamy et al. 2011; Mack et al. 2011; Srivastava et al. 2011; Bharadwaj et al. 2012]. VIPR proactively generates evidence and takes an active role in aircraft-level health assessment. VIPR established the technical feasibility and a design trade-space for this next-generation VLRS. Significant accomplishments for the VIPR program include:

1. Prognostic reasoning. Established technical feasibility of a Bayesian prognostic reasoner for onboard detection of adverse safety events. Demonstration

included working prototype software and design documentation for implementing the algorithms as expansion of the central maintenance computer functions. Results show the proactive role of VIPR in detecting three safety incident precursors from a regional airline's historic data, and its ability in detecting a potential cascading avionics fault based on hardware demonstration with an inertial reference system.

2. Semi-supervised learning loop. The system reference model within VIPR encodes aircraft-specific data such as failure mode definitions as well as evidence that can be generated and their relationships. With this member-system, supplier-provided information drives onboard diagnostics and prognostics of known problems; detection of emergent/unknown events that may impact safety is enabled by a semi-supervised learning loop. The technical feasibility of such a learning loop was established using:
   a. Tree-augmented, network-based supervised learning. This supervised learning method generates an evidence-failure mode map when trained with annotated fault cases. This map is readily incorporated in the VIPR reference model as (1) new evidence nodes, (2) updating selective arc information in the evidence-failure mode bipartite graph (see section 8.6.2).
   b. An anomaly detection function for generating outliers based on information entropy in time-series parametric sensor values. Demonstrated a Kolmogorov-complexity metric to detect these outliers and download contextual bad actors data for expert analysis. Some of these outliers presented interesting cases for offline expert analysis as precursors of incipient faults.
3. Onboard inferencing. Simple on/off threshold-crossing evidence is insufficient for prognostic reasoning. VIPR standardized three formats to express more complex and heterogeneous forms of evidence such as noisy evidence, prognostic vectors, and condition indicators. A naïve Bayesian framework was extended to handle uncertainty in evidence, causal cascades, and trend information. Standardization of the evidence enabled making these inference rules aircraft agnostic.
4. Prototype implementation of VIPR to establish the technical feasibility of these features as extensions to the Central Maintenance Computer Function (CMCF). This prototype provides a cost-effective pathway for realizing VIPR as an extension to existing aircraft hardware and software.

8.6.2 VIPR Mathematical Basis: Noisy-OR

Model-based strategies for diagnosing large, complex, real-world systems rely on domain experts to craft reference models used for monitoring and isolating faults. The complexity of the system makes it almost impossible to create complete physics-based models with reasonable resources; a more pragmatic solution is to rely on expert-generated cause-effect models. In simple terms, the reference model of the system being monitored can be represented as a bipartite graph consisting of two types of nodes: failure modes and evidence. The set F defines all distinct failure modes fm_j defined or enumerated for the system under consideration. The Noisy-OR approach for VLRS allows only binary (occurring or not-occurring) states for the failure mode: $fm_j = \{1, 0\}$. Every failure mode has an a priori probability of occurring in the system given by $P(fm_j)$.

To isolate and disambiguate the failure modes, the model also defines an entity called evidence m_i. Evidence can be present $m_i = 1$ or absent $m_i = 0$; this state can be asserted with some uncertainty $P(m_i = 1)$; this state can be asserted at current time $(t, P(m_i = 1))$ or some future time $(t + 1, P(m_i = 1))$, $(t + 2, P(m_i = 1))$, ... The diagnostic monitor m_i associated with the ith evidence can either indict or exonerate a subset of failure modes called its ambiguity group. This relationship is expressed using a detection probability d_{ij} and a false alarm probability ϵ_i. This relationship is shown in Figure 8.4.

Using the example illustrated in Figure 8.4, monitor m_i has three failure modes in its ambiguity group, and hence it has three detection probabilities. An "OR" gate is used as a simplification to describe the impact of failure modes on appearance of indicting monitoring. That is, any of the failure modes fm_1, fm_2, fm_3 can cause the appearance of an indicting monitor. An indicting monitor can also appear when none of the failure modes are present due to background noise. Hence, this algorithm is named "noisy-OR."

The a priori probability of observing an indicting monitor for our example is given by:

$$P(m_i = 1 | fm_1 = 1) = 1 - (1 - \epsilon_i)(1 - d_{i1})$$

Here h_p denote a hypothesis that at the system level any of the failure modes fm_1 or fm_2 or fm_3 may be occurring.

When one observes an indicting monitor, the Bayesian rule is applied to calculate the probability of a failure mode occurring in the system as follows:

$$P(fm_1 = 1 | m_i = 1) = \frac{P(m_i = 1 | fm_1 = 1) \times P(fm_1 = 1)}{P(m_i = 1)}$$

A given failure mode can cause more than one indicting monitor to appear; conversely, an indicting monitor may appear because of more than one failure mode being present in the system. This results in a bipartite graph. Figure 8.5 shows an example of such a graph for a propulsion subsystem.

The run-time algorithm is typically implemented as a sequence of two steps:

1. Abductive step: Whenever an indicting monitor is issued, failure modes that may have triggered them are identified. Right to left traversal of the bipartite graph is shown in Figure 8.4.
2. Deductive step: For each of the failure modes, all indicting and exonerating monitors are gathered together. Left to

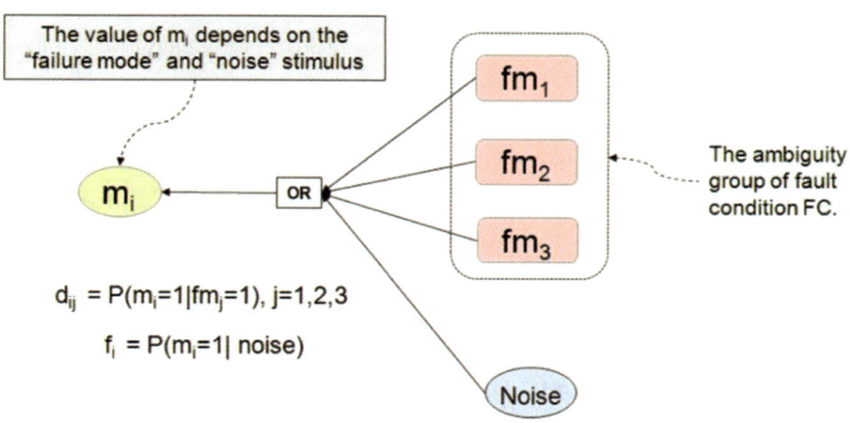

Figure 8.4 Noisy-OR model for VLRS inferencing.

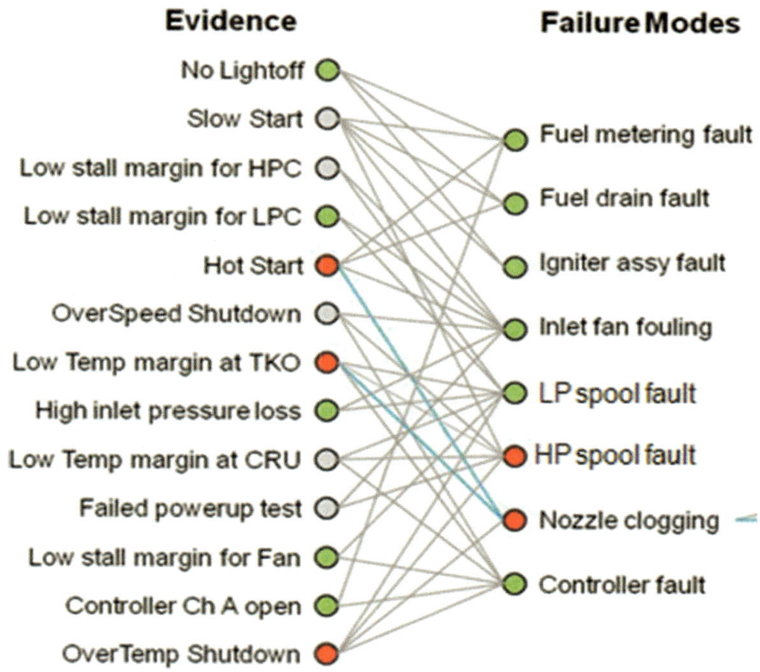

Figure 8.5 NOISY-OR works on a bipartite graph to maximize the probability of explaining all observed indicting (red circles) and exonerating (green circles) evidence. Failure modes marked in red form an ambiguity group that can explain all observed evidence. Failure modes in green are exonerated based on exonerating evidence.

right traversal of the bipartite graph is shown in Figure 8.4. The probability of this failure mode is calculated using the evidence independence assumption (noisy-OR) formula listed earlier.

As this process is repeated, the result is a set of failure mode hypotheses that can explain the current health state of the system. Complexity is often managed by applying additional heuristics or business rules that operate on this failure mode hypothesis. A few rules are listed below:

1. A lower bound on $P(fm_j = 1|\ldots)$ to eliminate weak hypotheses.
2. An upper bound on $P(fm_j = 1|\ldots)$ to report the results to the user or higher enterprise entities.
3. Cascade rule that specify $P(fm_j = 1|fm_k = 1)$ to merge two or more hypotheses.
4. Time-out rule that checks for hypotheses whose probabilities have not been updated over a predetermined time interval: that is, $\Delta P(fm_j = 1|\ldots) \le \alpha$.

The primary simplifying assumption—namely independence of evidence—makes the algorithm suited for fast computation and real-time application of VLRS to avoid safety excursions and pilot notifications. While the system model is primarily encoded in the form of a bipartite graph, the algorithm enables a domain expert to encode additional rules such as failure mode cascade and symptom cascades. Further, because the Noisy-OR algorithm is a simplified form of a standard Bayesian network, a number of machine learning techniques [Friedman, Geiger, and Goldszmidt 1997] for building Bayesian networks from data can be used to refine the model as aircrafts accumulate more operational data. This feedback learning loop can be partially automated using data mining techniques discussed later in this chapter.

8.6.3 VIPR Impact on Safety

VIPR research showed three detailed case study examples to illustrate the potential impact on a safety issue. The first example surrounds an In-Flight engine Shutdown (IFSD) incident recorded by the Federal Aviation Administration Aviation Safety Information Analysis and Sharing (ASIAS) database. An in-flight engine shutdown is a highly undesirable, adverse event that significantly impacts aviation safety. The flight crew immediately responded and, per the operating procedures, turned back and safely landed the aircraft. Investigation by the maintenance crew indicated a faulty fuel metering unit, although no exceedances had been reported for that engine in the recent past. If detected earlier, correcting such a problem would have been a routine line maintenance activity that would have prevented the IFSD and potential safety impact. This case has been previously discussed in detail [Srivastava et al. 2011].

While this single event may be statistically insufficient for machine-learning validation metrics, the clear explanation discovered by the data mining method and the change to the reference model, together with a VIPR, seemed sufficient for the engine domain expert to approve.

VIPR analyzed two additional safety incidents. The second incident also involved an in-flight engine shutdown caused by a high vibration event. In this case, VIPR correctly identified a broken blade that eventually led to high vibration and a safety engine shutdown. The third case was an engine-on fire safety incident. In this case, the data mining discovered a rather "noisy monitor." However, VIPR was able to handle this noisy monitor and, through active query and cascade analysis, conclude that the problem was a leaking fuel manifold, which led to an engine-fire incident. A leaking manifold affects engines #2 and #3 simultaneously, while failing fuel metering affects only one engine. These three cases not only showed the prognostic accuracy of VIPR but also revealed its impact on avoiding safety incidents. The prediction window produced by the VIPR approach for all three cases is summarized in Figure 8.6.

8.7 Economic and Safety Consequences

VLRS is an intelligent sensing and reasoning system. For mission operations, it improves safety and mission success through accurate and detailed assessment of health data to provide early identification of failure onset, and replacement of critical parts before they can cause accidents. It can also improve mission survivability through mitigations such as mission reconfiguration and adaptive control to compensate for failures or degradation in the performance of a subsystem.

For maintenance operations, VLRS can reduce the need for frequent inspection, provide advanced notification of maintenance needs, reduce fault ambiguity, and reduce time for repairs. This leads to reduced flight delays, reduced operational and business cost through Condition-Based Maintenance, and better fleet management. Likewise,

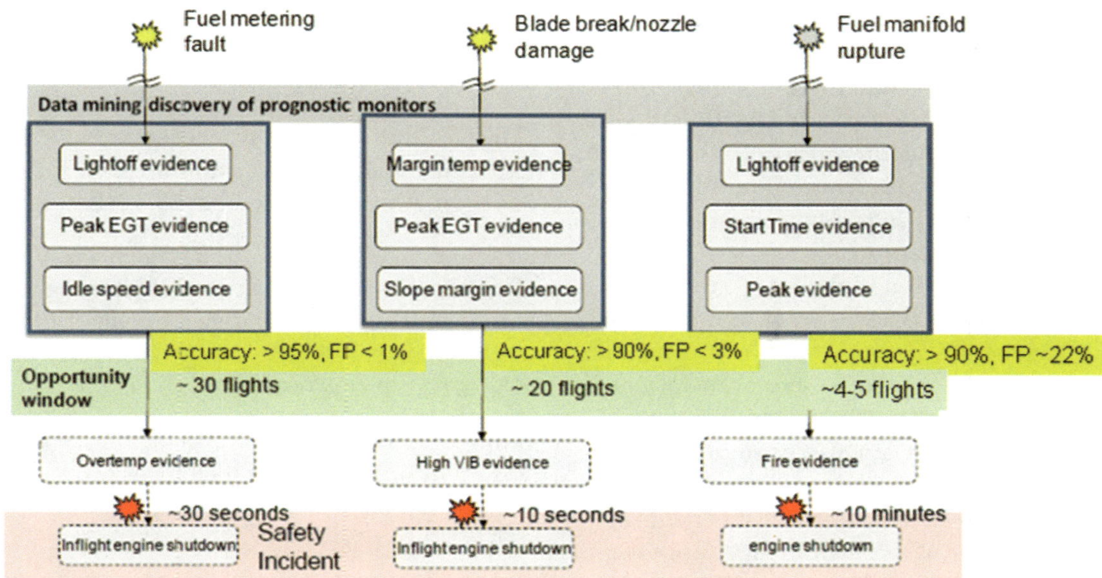

Figure 8.6 VIPR results for three safety incidents.

automating the monitoring and analysis functions minimizes the need for human input, hence resulting in enhanced reliability and maintainability, and reduced manpower costs. Commercial aerospace experience has shown that nearly 95% of aircraft life-cycle costs are attributable to maintenance activities.

The scope of a VLRS is not an all-or-nothing approach. It can vary and include different functions as well as different subsystems, components, and elements. VLRS technologies will bring the greatest benefits when they are selectively applied to areas that have the most critical impact on vehicle performance and support costs.

Because of the cost of the hardware and software, added onboard weight and processing/communication resource requirements, and potential false alarms, accurate safety and cost-benefit analyses are the key drivers for a successful VLRS implementation [Benedettini et al. 2009]. The level of implementation that is most appropriate for a vehicle would depend on an accurate assessment of the costs, risks, and rewards. Moreover, early assessment of the appropriate level of implementation would lead to a wider adoption. In any case, the potential is there to improve the safety and cost-effectiveness of current and new vehicles by linking operation, maintenance, and logistics to the present and future health of the vehicle.

8.8 Conclusions

In this chapter we have described a subset of the techniques and approaches that can be employed in a VLRS, shown the underlying computational architecture of a VLRS, and discussed potential use cases. We have also described the technical approach to reasoning and compared and contrasted it with traditional diagnostic systems. The combination of built-in test and enhanced monitors can help the reasoner discover rare anomalous events in complex aerospace systems. We have also shown that the integration of data mining algorithms based on internal models of nominal behavior can discover unknown anomalies from historical data sets. We have presented several VLRS case studies to illustrate the potential impact on real-world safety issues. Moreover, we have described how a VLRS can reduce the need for frequent inspection, provide advanced notification of maintenance needs, reduce fault ambiguity, and reduce time for repairs. This can lead to reduced flight delays, reduced operational and business cost, and better fleet management. The scope of a VLRS is not an all-or-nothing approach, and it can vary and include different functions and subsystems. VLRS capabilities can also be implemented incrementally in modular fashion to facilitate system integration. The greatest benefits are attained when VLRS technologies are selectively applied to areas that have the most critical impact on vehicle performance and support costs.

8.9 References

Aerospace-Technology.com. July 20, 2011. "Fault-Free Flying—Aerospace Technology," Retrieved from http://www.aerospace-technology.com/features/feature124809.

ARINC. 1991. ARINC 624 Design Guidance for Onboard Maintenance System.

Bay, S. D., and M. Schwabacher. 2003. "Mining distance-based outliers in near linear time with randomization and a simple pruning rule," KDD 2003: *Proceedings of the 9th ACM SIGKDD International Conference on Knowledge Discovery and Data Mining*, 29–38.

Benedettini, O., T. S. Baines, H. W. Lightfoot, and R. M. Greenough. 2009. "State-of-the-art in integrated vehicle health management," *Proc. IMechE Vol. 223 Part G: J. Aerospace Engineering.*

Bharadwaj, R., et al. 2012. "Vehicle Level Prognostic Reasoning System," talk at SAE 2012 Aerospace Electronics and Avionics System Conference, Oct. 30–Nov. 1, 2012. Phoenix, AZ.

Chandola, V., A. Banerjee, and V. Kumar. 2009. "Anomaly Detection: A Survey," ACM Computer Survey, 41(3)(2009).

Chien, S. "Autonomous Sciencecraft Experiment," Retrieved from http://ase.jpl.nasa.gov, Accessed December 15, 2012.

Das, S., B. L. Matthews, A. N. Srivastava, and N. C. Oza. 2010. "Multiple kernel learning for heterogeneous anomaly detection: algorithm and aviation safety case study," In *Proceedings of SIGKDD 2010*, 47–56.

Felke, T., G. D. Hadden, D. Miller, and D. Mylaraswamy. 2010. "Architectures for Integrated Vehicle Health Management," AIAA-2010-3433.

Friedman, N., D. Geiger, and M. Goldszmidt. 1997. "Bayesian network classifiers," *Machine Learning*, 29:131–163.

Gorinevsky, D., A. Smotrich, R. Mah, A. Srivastava, K. Keller, and T. Felke. 2010. "Open Architecture for Integrated Health Management," AIAA Infotech@Aerospace, Atlanta, GA.

Hadden, G. D., D. Mylaraswamy, C. Schimmel, G. Biswas, X. Koutsoukos, and D. L. C. Mack. 2011. "Vehicle Integrated Prognostic Reasoner (VIPR) 2010 Annual Final Report," NASA/CR–2011-217147, http://ntrs.nasa.gov/archive/nasa/casi.ntrs.nasa.gov/20110012171_2011012606.pdf.

IEEE Std 1232-2010, IEEE Standard for Artificial Intelligence Exchange and Service Tie to All Test Environments (AI-ESTATE).

ISO 13374 (all parts), Condition monitoring and diagnostics of machines—Data processing, communication and presentation.

Iverson, D., and A. Hillin. "Inductive Monitoring System Assists in ISS Anomaly Detection Analysis," Retrieved from http://ti.arc.nasa.gov/news/ims-anomaly-detection, December 15, 2012.

Iverson, D., and M. Stop. 2004. "Inductive System Health Monitoring," in *Proceedings of the 2004 International Conference on Artificial Intelligence*.

Mack, D. L. C., G. Biswas, X. D. Koutsoukos, and D. Mylaraswamy. 2011. "Using Tree Augmented Naive Bayesian Classifiers to Improve Engine Fault Models." Presented at the 8th Bayesian Modeling Applications Workshop in Barcelona, Spain on July 14th, 2011.

Manders, E. J., S. Narasimhan, and G. Biswas. 2000. "A combined qualitative/quantitative approach for fault isolation in continuous dynamic systems," Presented at the 4th Symposium on Fault Detection Supervision and Safety of Technical Processes (SafeProcess), Budapest, Hungary, 14–16 June 2000.

Mendel, J. M. 1995. "Fuzzy logic systems for engineering: A tutorial," *IEEE Proceeding, 83(3):345–377*, Mar. 1995.

Miles, K., and C. F. Peters II. 1999. "Remote Agent does the Job for NASA," Retrieved from http://starryskies.com/articles/dln/6-99/remote.html.

Mylaraswamy, D. 2010. "Vehicle Level Reasoning and Data Mining," CIDU.

Mylaraswamy, D., et al. 2011. "Vehicle Integrated Prognostics Reasoning," NASA Aviation Safety Annual Meeting, St Louis.

Pecht, M. G. 2008. *Prognostics and Health Management of Electronics*, Wiley, ISBN 978-0-470-27802-4.

Srivastava, A. N., D. Mylaraswamy, R. W. Mah, and E. G. Cooper. 2011. "Vehicle-level Reasoning Systems" in *IVHM Perspectives on an Emerging Field*, Ed. I. K. Jennions, SAE International, Warrendale, PA.

Uluyol, O., K. Kim, and C. Hickenbottom. 2010. "A Systematic Approach to Bearing Health Monitoring," American Helicopter Society 66th Annual Forum, Phoenix Arizona, May 11–13, 2010.

CHAPTER 9 IVHM System Design

Sony Mathew, *CALCE, University of Maryland*
Michael Pecht, *CALCE, University of Maryland*

"Design is the conscious effort to impose a meaningful order."

—Victor Papanek

9.1 Introduction

What is systems design? Systems design is the process of developing a system based on customer requirements, available technology, engineering constraints, and business expectations, to satisfy market needs. Systems design involves a systematic definition of the framework of the system: the component and subsystems, the interfaces, and the required inputs and outputs of the system. The system design process usually evolves with time wherein hours of research, analysis, engineering calculations, prototyping, testing, failure analysis, and design updated are conducted. Thus, systems design is a culmination of interdisciplinary effort to define the system on the basis of requirements and engineering trade-offs. Systems design also includes strategies to meet and exceed set reliability constraints. Prognostics and Health Management (PHM) is an enabling discipline consisting of technologies and methods to assess the reliability of a system in its actual life-cycle conditions to determine the advent of failure and mitigate system risk [Pecht 2008]. PHM applied to aerospace vehicles is popularly known as Integrated Vehicle Health Management (IVHM), a term coined by NASA. The IVHM system consists of both hardware (including sensors, interconnections (wires), computers, enclosures, and fittings) and software (including programs that control the operation of the hardware, transmit data, process and analyze data, and provide warning of impending failure) needed to monitor, predict, and manage the health of an aerospace vehicle, its systems and subsystems. The systems design process presented in this chapter for the IVHM system can be applied to a PHM system developed for any given application.

Different engineering expertise is essential to predict the health of any given system and take corrective actions to improve the safety, availability, and reliability of the system under its operational life-cycle conditions. IVHM systems design begins with identifying the requirements of the IVHM system and applying a strict selection procedure for the components or subsystems, to make it robust. Having generated the requirements and specifications of the IVHM system, a framework should be developed. Next, the critical components and parameters of the IVHM system are identified, along with the prognostic approach, type of sensors, and data collection and processing techniques that will be used. This helps the design team narrow down the choices for the final IVHM system. Finally, the steps to integrate the IVHM system with the system under consideration are investigated. This chapter presents the activities that contribute toward a robustly designed system.

9.2 Requirements and Specification Generation

9.2.1 Introduction

In developing an IVHM system for monitoring the state of health of any given aerospace vehicle's system, the first step is to define the proper requirements of the IVHM system, which can then be translated into a design specification. The requirements of the IVHM system may be derived directly from the customer requirements and/or the system manufacturing constraints. The process of transforming broad goals and vague concepts into realizable, concrete requirements is a necessary step in the successful implementation of IVHM. While the company's core competencies, cultures, goals, and customers all influence the requirements, the IVHM system's definition results from a combination of business-driven product requirements, design and manufacturing constraints, and other external influences.

Poorly defined requirements could lead to a design that, if implemented, causes more issues than benefits. For example, in the case of avionics Line-Replaceable Units (LRUs), if the IVHM system includes sensors that interfere with the signals of the avionics system, then the IVHM system has failed. Understanding the system to be monitored and clearly defining the required outcomes of the IVHM system are necessary for its success. Sometimes a PHM system must be designed with a minimum set of "generic" requirements and then customized using each customer's specific requirements, which in our case refers to requirements for an aerospace system. Figure 9.1 shows the requirements that form a general PHM system's attributes.

The proper requirements for an IVHM system can be generated through a combination of customer interactions, regulatory requirements, and engineering estimates. The following subsections briefly describe the sources of IVHM system requirements.

9.2.2 Regulatory Requirements

System designers need to be aware of regulations that affect each subsystem of the aircraft. To avoid legal liabilities, the most important factors for a system designer are preparation, analysis, and documentation. For example, in the case of aircraft systems, passenger and crew safety is of primary concern. Any action that jeopardizes the safety of the aircraft or the people on board cannot be tolerated. When designing an IVHM system for aircraft systems, several alternatives should be considered prior to use. The IVHM system should comply with regulations, which are agreed to and enforced by the national and international regulatory bodies, which apply where the system is operating. Within the United States, any IVHM system to be implemented on an aircraft system must clear aviation authority

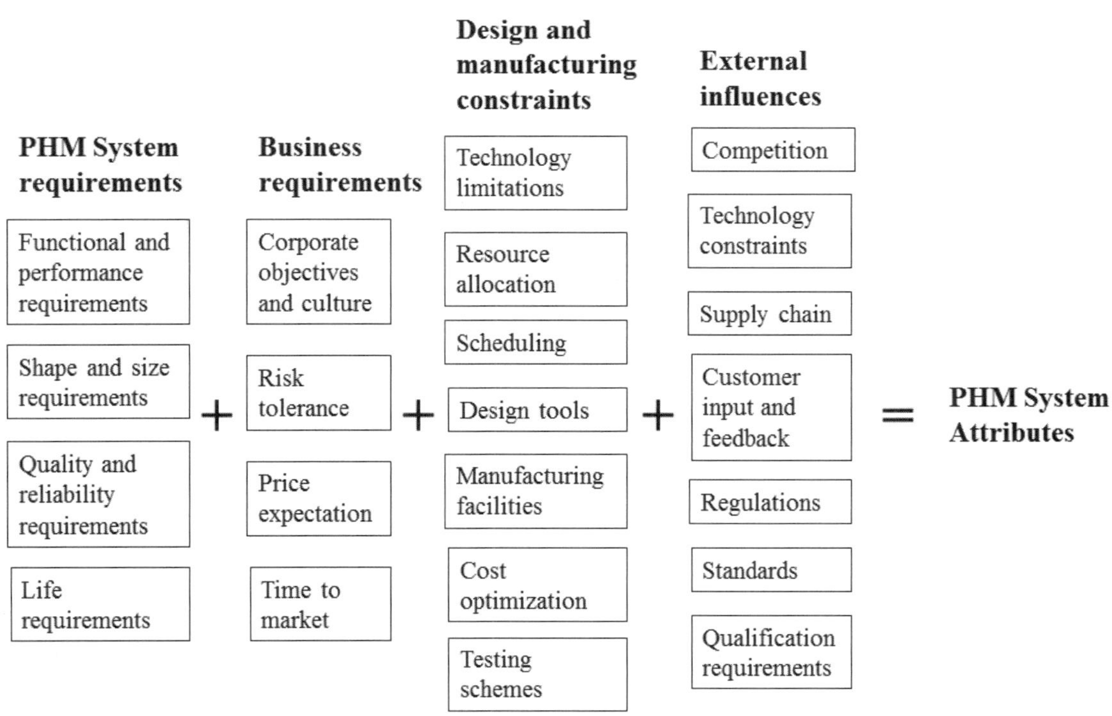

Figure 9.1 Requirements for attributes.

requirements and regulations (e.g., civil aviation—U. S. Federal Aviation Administration (FAA) and military—U.S. Military Aviation Authority (MAA). Airworthiness requirements for systems and subsystems on an aircraft necessitate evaluation of any new or modified system/subsystem for impact on structural and operational integrity [Army 2007]. If Commercial Off-The-Shelf (COTS) components are used for the IVHM system, these components have to be assessed for airworthiness impact.

Another example of a regulation that will influence the design, especially with regard to using electronics components, is that the European Union passed the Restriction of Hazardous Substances (RoHS), which bans the use of some chemicals in electronic components. Other countries including the United States, China, Japan, India, and Korea have passed similar legislation. While aviation and military systems are exempt under the provisions of the RoHS law, electronics manufacturers have migrated to lead-free electronic parts. Companies must anticipate potential product limitations, educate their workforce to make informed decisions, and follow industry and company standards.

9.2.3 Customer Interactions

It is important to consider the requirements of the customer. Customers may use a given system in many different ways, depending on the application. For example, in the case of aircraft, military and civil aviation requirements may differ, depending on the type of system. Customers may be able to help the development of the IVHM system by providing information about the most common problems they encounter with the system. In the case of systems that have multiple customer bases, the requirements of an IVHM system are defined to address the most common system problems, thereby appealing to the largest possible customer base. It is a good practice to interact with actual users of the system to get an idea of the usage-related issues that need to be tackled. For example, in the case of aircraft, having a dialog with maintenance personnel and pilots will provide helpful information regarding usage-related issues. Ultimately, business requirements of the customer play a big role in the IVHM system requirement. Goals may vary, including improving safety, increasing mission readiness, increasing availability, reducing maintenance cost, reducing inventory, and reducing warranty costs. IVHM system designers need to get a clear picture of the customer's requirements and constraints to create a system that satisfies the customer.

9.2.4 Engineering Estimate

Requirements or specifications for an IVHM system can be obtained from engineering estimates. Knowledge of the system to be monitored, and factors including the shape, size, and weight allowances of the system; available sensors; relative ease of attaching sensors to the system; and ability to transmit and process data, provides information to define the attributes of the IVHM system. An examination of the functional requirements of the system to be monitored and the outputs of the system can provide information to put together a preliminary list of parameters that the IVHM system will monitor. The ways to attach sensors in a system to gather information must be identified. For example, to monitor the solder interconnects of components on a printed circuit board within an LRU of the avionics system, designers can monitor direct current (dc) resistance, radio frequency (RF) impedance through time domain reflectometry, or output signals from the component. The current market availability of sensors that monitor the parameters, the ease of operation and integration into the system, and the cost of the sensor system need to be considered when making final decisions.

The reliability of the IVHM system should satisfy the life requirements of the system to be monitored. Inventory, part obsolescence, and supply chain and sustainment issues also have to be considered before finalizing the IVHM systems specifications. The potential production issues, resource allocation issues, and personnel issues including training for assembly, operation, and maintenance of the IVHM system should be taken into account. Manufacturing engineers may determine that purchasing certain components is better than making them in-house. The design team needs to generate a conceptual estimate of the IVHM system and then apply engineering trade-offs to update the specifications of the system.

9.3 Development of Prognostics Framework

An IVHM system is composed of several elements including sensors, computing hardware, and software algorithms for fault detection, diagnostics, prognostics, and decision support [Kunche, Chen, and Pecht 2012]. A framework for the IVHM system serves as a blueprint for designers for faster system development and deployment and provides the overall interactions between the different subsystems. A framework provides a holistic view of the system for system designers and serves as a guide for system modeling, integration, and testing. The frameworks can be categorized as either functional or physical.

A functional framework represents the functionalities of the subsystems and the data flow between them. Here, the IVHM system is decomposed into blocks of different functionalities with relations between these various blocks shown as inputs and outputs to a particular block. The physical framework represents the relative positioning of the subsystems that form the IVHM system. It shows the components in the different layers of the framework and the physical distribution of the components in the system.

The functional and physical frameworks are developed in parallel on the basis of the requirements of the PHM system [Kunche, Chen, and Pecht 2012]. The framework should cover all aspects including the communications or data flow between subsystems, inputs, and outputs of each subsystem; the operational flow; and control functions of the system to achieve IVHM. A framework provides a better understanding of the system components and enables the system designers to integrate and qualify the IVHM system before deployment. A good framework should be modular and flexible to allow seamless integration of all the IVHM subsystems, thereby facilitating the interoperability of the IVHM system with various applications. Figure 9.2 shows a functional framework for a general PHM scheme. The process starts with in-situ monitoring of the system of concern. Various operational, environmental, and system parameters may be simultaneously monitored depending on the criticality of the system. Parameter values are collected by the sensors and transmitted for further processing. The next phase of design needs to consider the method of data transmission and location of data processing. The processed data must be analyzed using specialized algorithms to identify clusters of data or classify the data as healthy or unhealthy. The data need to be further analyzed to identify if the parameters are trending into a region where the system will start to behave anomalously. If there is no anomaly detected, the cycle of monitoring for anomaly detection continues.

If an anomaly is detected, then an alarm is generated to inform the end user that the system is behaving in an anomalous manner. The data are further analyzed to isolate the faulty component or subsystem. In the diagnostics phase, the cause of the anomalous behavior is identified. At this stage another alarm is generated with the relevant information for the end user and maintenance personnel. This information can be used by system designers to improve/update the next iteration of the system. Data obtained from monitoring the system can be used in conjunction with damage or Physics-of-Failure (PoF) models to estimate the damage accumulated due to operational and environmental loads. Based on the damage to date, the Remaining Useful Life (RUL) of the system can be estimated. Alternatively, using regression analysis the RUL can be estimated. Once the RUL is estimated, the final alarm is generated with relevant information for maintenance personnel. Then the PHM system generates an action plan for managing the health of the system based on available logistical information.

Developing a framework forms the basis of the IVHM system development. The next phase in systems design involves finalizing the details regarding the overall IVHM system. The next section presents the design considerations that will help narrow the design to a more systematic and functional state.

9.4 Design Considerations

9.4.1 Introduction

Designing the IVHM system for an aerospace system is a complicated process. The design phase generally involves selection of

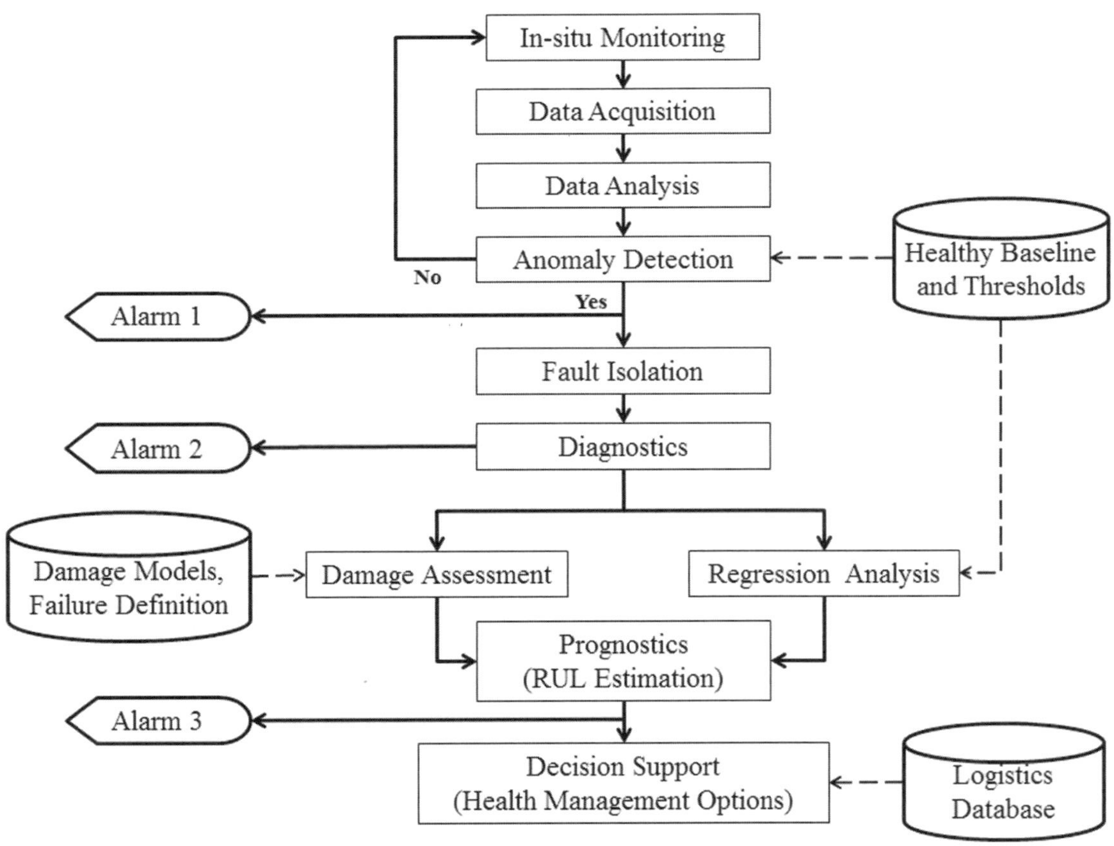

Figure 9.2 Functional framework for general PHM system.

hardware, prognostics methodology, and data processing. The design phase of the IVHM system uses information from the prognostic framework developed in the previous section. Before building the IVHM system, the critical components of the system under consideration and the parameters to monitor need to be identified. The prognostic approach to be implemented, the sensors, and the data transmission and processing techniques must be selected. The following subsections present the factors to consider during the design phase.

9.4.2 Identification of Critical Components and Parameters

To provide an accurate prediction of the remaining useful life of a product, it is essential to understand what is causing damage to the product and how the damage is manifested in the product. There are different failure signatures for different failure mechanisms, and without knowledge of the failure mechanisms the accuracy of the prognosis is questionable. Without the knowledge of failure mechanisms and modes, the sensors used for monitoring, the location of monitoring, and the models to analyze the collected data, may be selected erroneously. If the precursor parameter is erroneously selected, then monitoring that parameter will not provide accurate prediction and will therefore cause improper or delayed corrective actions. Methods that can be used to correctly identify the critical components and parameters to monitor are discussed below.

i) FMMEA: Failure Modes, Mechanisms, and Effects Analysis (FMMEA) is a method based on assessing the root cause failure mechanisms of a given product [Pecht and Dasgupta 1995]. A potential failure mode is the manner in which a failure manifests itself in a product. Failure mechanisms are the processes by which physical, electrical, chemical, and mechanical stresses induce failures individually or in combination. FMMEA is based on an understanding of the relationships between product requirements and the physical characteristics of the product (and their variation in the production process), the interactions of product materials with loads (stresses at application conditions), and their influences on the product's susceptibility to failure with respect to the use conditions. A schematic diagram showing the steps in FMMEA is presented in Figure 9.3.

Ganesan et al. [2005] provided a detailed description of the FMMEA methodology. FMMEA uses life-cycle environmental and operating conditions and the duration of the intended application with knowledge of the active stresses and potential failure mechanisms. The purpose of FMMEA is to identify the potential failure mechanisms and models for all potential failures modes and to prioritize the failure mechanisms. To ascertain the criticality of the failure mechanisms, a Risk Priority Number (RPN) is calculated for each mechanism. The higher the RPN, the higher it is ranked among the failure mechanisms. Figure 9.4 shows the prioritization of failure

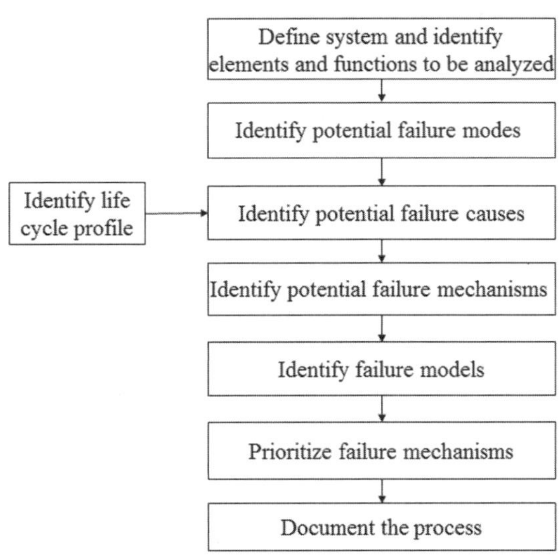

Figure 9.3 FMMEA methodology [Ganesan et al. 2005].

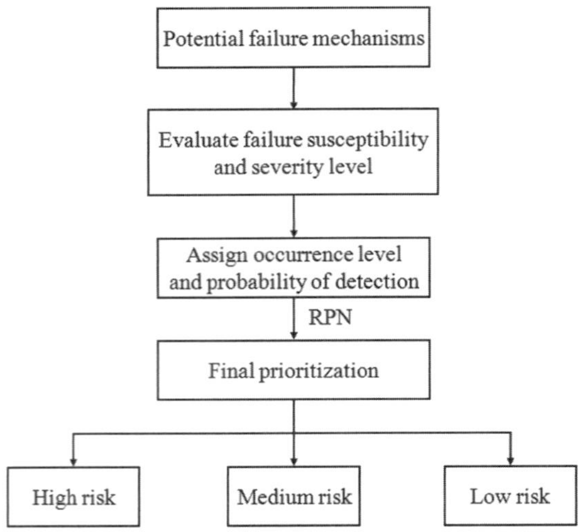

Figure 9.4 Prioritization [Mathew, Alam, and Pecht 2012].

mechanisms based on the calculated RPN for each mechanism.

The RPN is the product of the probability of detection, occurrence, and severity of each mechanism [Mathew, Alam, and Pecht 2012]. Occurrence describes how frequently a failure mechanism is expected to result in failure. Severity describes the seriousness of the effect of the failure caused by a mechanism. Detection describes the probability of detecting the failure modes associated with the failure mechanism. Figure 9.5 shows the axis of a three-dimension risk matrix.

ii) Expert knowledge and past experience: One way of narrowing down the list of potential components to be monitored in a system is to rely on expert knowledge and past experience. New systems are often designed as upgrades to existing systems. The knowledge obtained during the design and testing phase of the existing system is useful in identifying potential critical components, and failure mechanisms and modes in the new system. System designers involved with the design and manufacturing process of existing systems will try to incorporate monitoring systems in the new design to monitor the weak links. Sometimes information obtained from similar systems is also useful in narrowing down the critical components in a given system. For example, if a new version of an aircraft uses some of the technologies available in the previous version, including the engine, the knowledge gained during the

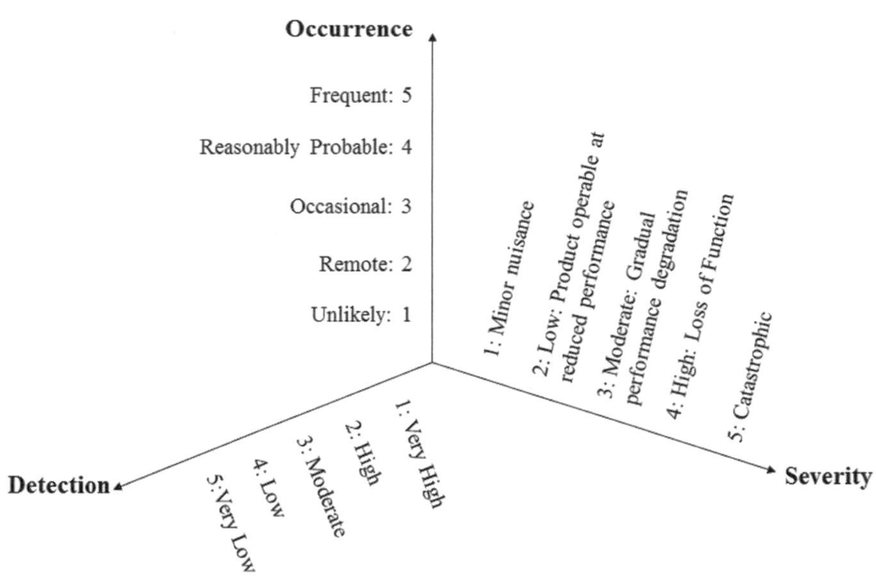

Figure 9.5 Risk matrix [Mathew, Alam, and Pecht 2012].

prototype development and testing phase of the previous version will be useful in identifying the potential faulty components in the newer version.

iii) Maintenance logs: Another source of information for system designers is maintenance logs for fielded systems. Maintenance logs are used by maintenance personnel to record and keep track of the maintenance activities conducted on a system. The information can be compiled to identify the components of the system that had to be repaired or replaced periodically. It is a record of the type of problems associated with the system during operation. The maintenance log also includes information about "intermittent" or no fault found type of incidents associated with the system. Though the data obtained are in the raw form, IVHM system designers can process this information to develop a successful monitoring system.

9.4.3 Selection of Prognostic Approach

Once the critical components and parameters to monitor have been identified, the next step is to select a prognostic approach for the system under consideration; see chapter 4. The approach selection will also depend upon the criticality of the system, and safety, regulatory, and customer requirements. This section presents in brief the different prognostics approaches.

i) Physics of failure: The PoF-based prognostics approach uses knowledge of a system's life-cycle loading and failure mechanisms to assess the systems's RUL [Pecht 2008]. PoF methodology is based on the identification of potential failure mechanisms and failure sites for a system. The first step in PoF-based PHM involves FMMEA, where design data, expected life-cycle conditions, and PoF models are the inputs for damage and RUL assessment. Based on the FMMEA, the critical failure mechanisms are selected and the life-cycle operating and environmental conditions that propagate those mechanisms are monitored. The life-cycle environment of a product consists of manufacturing, transportation, storage, handling, operating, and non-operating conditions. Life-cycle loading can be mechanical, thermal, chemical, or electrical in nature. The life-cycle loading data are used in failure models to derive the extent of degradation under the given conditions and from which the RUL of the product can be estimated.

Besides using PoF models for RUL estimation, "canary devices" designed using PoF information can also be used to estimate the RUL of a component. A canary device behaves similarly to a "canary bird" in a coal mine. Canary devices are designed to fail by the same failure mechanism that the component would fail by if the component were subjected to extended life-cycle loads [Mathew and Pecht 2011]. Under the same environmental and operational loading conditions, the canary devices are designed to fail faster than the actual component [Mathew and Pecht 2011]. Canary devices embedded in a product provide advance warning of failure due to specific wear-out failure mechanisms. A canary device fails earlier than the actual component. By knowing the acceleration factor between the canary device and the component, the time-to-failure of the component can be computed.

ii) Data driven: A data-driven approach for prognostics uses statistics and probability for analyzing current and historical data to estimate RUL. This approach is preferred when physics-based or empirical models are not available. This approach is primarily based on signal processing, pattern recognition, and state estimation. The prognosis is based on state awareness of the product, which is obtained from analyzing the monitored performance parameters and forecasting total degradation and time-to-failure. Features can be extracted directly from routinely monitored product operating data or performance data. The measured input/output data facilitate a deeper understanding of system degradation behavior [Mathew and Pecht 2011]. Machine learning techniques (covered in chapter 3) are used to process the data features to establish the healthy state of the given product. These data are used to identify performance deviation due to the presence of a fault. To be effective, it is necessary that the training data for the machine learning algorithms span the entire set of system faults and operational conditions.

Following the health-state estimation, the next step is detection of anomalous behavior in a fielded product and isolation of the fault. In the data-driven approach, the system parameters are continuously monitored, features are extracted from the data, and the prognostics algorithm compares the expected healthy state to the detected state of the system [Mathew and Pecht 2011]. Threshold limits on the data features are set to define the start of the performance degradation that will lead to operational failure. Trending of features to a defined failure criteria provides the fault or damage progression over time. This information is used to estimate the RUL of the system.

iii) Fusion: The fusion approach, detailed in chapter 7, combines the benefits of the PoF and data-driven techniques to provide an estimate of RUL [Mathew and Pecht 2011]. The fusion approach utilizes data-driven techniques for anomaly detection and classification to detect early degradation of a system. The faulty parameters are then isolated by incorporating the results from the anomaly detection and the PoF information. Knowledge of the physics-of-failure of the product helps to identifying the root cause of system failure. PoF models are used to estimate the thresholds beyond which the system will show faulty behavior. PoF models are also used to calculate the damage caused by operational and environmental loading conditions.

The time to failure is estimated based on the trend of the parameters and the failure threshold. A conservative value of the RUL estimate from the data-driven technique and the PoF model is reported. Alarms are provided after detection of anomalous behavior and after reaching the RUL estimate as specified for the system under consideration. These alarms provide adequate time for repair or replacement of the system.

9.4.4 Sensor Selection

Data collection for prognostics often requires the use of sensor systems (see chapter 2) to measure system-related parameters, including outputs, performance parameters, loading, environmental, and operational parameters. A sensor is defined as a device that provides a usable output signal in response to a specified measurand [Carstens 1993]. An IVHM sensor system typically has sensors, signal conditioning, onboard analog-to-digital converters, onboard memory, embedded computational capabilities, data transmission, and a power source or supply. To accurately assess the health and predict the remaining life of a system, monitoring may be needed during all stages of the system's life cycle. However, not all sensor systems are suitable for the implementation of IVHM. Cheng, Azarian, and Pecht [2010] presented a detailed sensor selection methodology for PHM. This section introduces some attributes of sensor systems to be considered for selection before implementation in an IVHM system.

i) Sensor requirements: While identifying sensor requirements, system designers primarily consider sensor performance features including the type of parameters the sensor can monitor, measurement range, sensitivity, accuracy, precision, resolution, sampling rate, response time, and stabilization time [Cheng, Azarian, and Pecht 2010]. However, designers also need to consider physical features such as size, weight, and shape of the sensor, as well as the number of sensors required to monitor the system under consideration. In addition, system designers need to understand the power requirements of the sensor system, including the expected power consumption and type of power source (e.g., powered by host, AC power, battery, and other power sources). Other attributes to be considered include the availability of onboard memory, capacity of onboard memory, programmability of sampling modes and rate, sampling rate, and whether the sensor conducts active or passive sampling.

ii) Sensor reliability: Sensor systems for IVHM should be reliable. Sensor systems are generally limited to some degree by noise and the surrounding environment, which vary with operating and environmental conditions. To reduce the risk of sensor system failure, the user must consider the sensor's environmental and operating range and determine if the sensor is suitable for the particular application. The packaging of the sensor system should be examined to estimate the risk due to humidity, dust, chemicals, and other mechanical and environmental conditions. Sensor validation is used to assess the integrity of the sensor system and adjust or correct it . This functionality checks the sensor performance and ensures that the sensor system is working correctly by detecting and eliminating the influence of systematic errors. Self-diagnostics, self-calibration, and sensor fusion methods can be applied to achieve this functionality.

A strategy to improve the reliability of the information gathering process is to employ sensor redundancy. It is necessary to consider the effects of the sensor system on the reliability of the system it is intended to monitor. For example, sensor systems that are heavy may reduce the reliability of circuit boards over time when attached to the surface. In addition, the method of attachment (soldering, glue, screws) can reduce the reliability of the product if the attachment material is incompatible with the product's materials of construction.

iii) Cost: Selection of a proper sensor system for IVHM application must include an evaluation of the costs. The cost evaluation should address its total cost of ownership, including the purchase, maintenance, and replacement of sensor systems. The costs of integrating the sensor system onto the system to be monitored can be categorized as recurring, nonrecurring, or infrastructural depending on the frequency and role of the corresponding activities. If the cost of the sensors exceeds the cost benefits obtained from the IVHM activity, then the effort is wasted. This may not be the case with systems that, if they fail, can cause catastrophic events (such as loss of life), serious injuries, or economic and national security issues.

9.4.5 Selection of Data Transmission and Preprocessing

After selection of the type of sensor that is suitable for monitoring the system, the next step is to decide how the data will be transmitted and processed. Data can be transmitted via a wired connection or wirelessly. Designers need to select the mode of data transmission, including active (continuous) or passive (upon request) transmission. Factors such as transmission range (for wireless), protocols, transmission rate, and security strategy need to be considered. Wireless monitoring (see chapter 11) has emerged as a promising technology that can enhance real-time IVHM capabilities. For wireless transmission, the distances involved may be short (a few meters, as in a television remote control) or very long (thousands or even millions of kilometers for radio communications). Wireless sensor nodes can be used to monitor inhospitable and toxic environments remotely. Wired data transmission can offer high-speed transmission, but it is limited by the need for transmission wires. Wireless transmission can offer very convenient data communication, eliminating the need for a wire, but the transmission rate is lower than that for wired transmission. Therefore, there are trade-offs for a given application.

The data collected must be processed in a manner that provides IVHM algorithms with the right inputs. The raw data collected from the sensors may not be suited for health assessment. After the raw data are captured, the data must be processed to be transformed into the proper form for use in diagnostic and prognostic algorithms. Problems with raw data may arise including collection of "not-applicable" data value, IVHM hardware problems (faulty sensors, wires, data logging equipment), IVHM software problems (bugs in data collection and data logging software, errors in data transmission), human-induced problems (faulty or wrong connections), wrong data entry, multiple data sources for same parameters, duplicate records, and noise in the system and environment.

The data preprocessing includes such steps as data cleaning, data integration, data transformation, data reduction, data discretization, data denoising, and feature extraction. Depending on the need, one or more of the techniques just mentioned may be applied to the raw data. Data cleaning involves such actions as filling in missing values and resolving the inconsistencies in the data. Data integration involves putting together data with different representations and from different sources. Data transformation includes actions such as normalization of data, aggregation of data, and generalization of data. Data reduction seeks to minimize the collected data without losing the valuable information regarding the system. Data discretization involves partitioning the continuous data into discrete intervals, where each discrete set provides a snapshot of the system health at the given instance. Denoising is the act of smoothing or removing noise from data. Feature extraction involves techniques that take specific attributes of the raw data that are significant in some particular context. There is a vast variety of algorithms for data-preprocessing available to systems designers for implementation into the IVHM system.

The data may be preprocessed on board the IVHM sensor system or offsite depending on the type of sensor features selected. A sensor equipped with onboard memory and a programmable interface can preprocess the data on board before transmitting the processed data for further analysis. Onboard preprocessing can significantly reduce the number of data points and thus free up memory for more data storage. This, in turn, reduces the volume of data that must be transmitted out to a base station or computer. A large number of sensor systems working in a network would allow decentralization of computational power and facilitate efficient parallel processing of data. Embedded computations can be set to provide real-time updates for taking immediate action, such as powering off the equipment to avoid accidents or catastrophic failures, and provide a sufficient time frame for conducting future repair and maintenance activities . The abilities of the onboard processor are limited by some physical constraints. One constraint is the available power. Extended calculation and high calculation speeds involve an increased consumption of power by the sensor system. The other constraint is onboard memory capacity. Running complex software requires a lot of memory. These two constraints make it challenging to embed complex algorithms in onboard processors. However, even using simple algorithms and routines to process the raw sensor data can provide valuable information for in-situ analysis.

9.4.6 Algorithm Selection

Predicting future behavior of a given system is tied to the ability to learn from the system's past and present data. A number of techniques exist for analyzing the sensed system data and making inferences for fault detection and failure prediction. If the physics behind the system functioning and failure is known, then the PoF method is adopted. PoF methods employ domain knowledge about a system and its components, and how they fail through the use of damage assessment models that capture the underlying phenomena [Pecht 2008]. Appropriate models need to be selected for specific failure mechanisms. In some cases, PoF models are already available for individual components of the system. Examples of PoF models include the Engelmaier model for solder joint fatigue under temperature cycling loads, Rudra model for conductive filament formation in printed circuit boards, and Eyring models for corrosion and electro-migration. There are situations in which individual models for the components/subsystems are not available. However, physics/knowledge-based models are derivable based on system functions; for example, Saha et al. [2009] shows a physics-based model constructed from an equivalent circuit to predict the remaining useful life of a battery system. Daigel and Goebel [2011] constructed a model for predicting the failure of pneumatic valves based on the valve's working principle.

Data-driven methods are employed for modeling complex systems for which the physics are not well understood, or for which PoF models are very difficult to develop [Bonissone, Xue, and Subbu 2011]. Data-driven approaches exploit statistical techniques that depend on the available past and present observed data, and statistical models. These techniques do not rely on the knowledge of material properties, structures, or failure mechanisms. Statistical techniques could either be parametric or nonparametric 0. Parametric statistical techniques assume that the system's data fit a certain type of distribution (for example, the Gaussian distribution), and critical parameters (such as the mean and the standard deviation) of the distribution are easily calculated from the data. In many applications, the information regarding the distribution of the data may not be known. In these cases, statistical approaches that are nonparametric in nature are used. Nonparametric methods do not make any assumptions regarding the underlying distribution of data.

In the PHM community, machine learning is a widely used technique that is based on statistical learning theory, uses algorithms that enable computer systems to learn from system's performance data [Poggio and Smale 2003]. Classification or clustering techniques are used for fault diagnostics (i.e., fault detection and isolation, chapter 3). The selection of classification or clustering algorithm is based on multiple factors such as data type, amount of test and validation data, computational resources, experience, and allowable risk level. The difference between classification and clustering is the existence of class labels. Classification is the task of arranging data into specific classes according to a sorting criterion, where each class has its own label. Patterns of the collected data are analyzed for identification of the health condition of a system by means of a classification algorithm. With classification, different kinds of labeled classes represent different health conditions of the system. Matching a specific abnormal condition with a specific state of the system with the help of an expert, or historical data on the system, is possible. A specific health condition of a system can be defined with the concept of classification. Clustering is different from classification in that the label of a class cannot be identified.

In clustering, various health classes are identified based on the segregation of extracted features (obtained from system performance data) in the multidimensional feature space. Classification algorithms are either supervised or unsupervised. In "supervised" methods, the data are labeled and the algorithm learns from the distinctions among the labeled data. The data in unsupervised learning have no predefined classes and do not include any labeled data. The system using unsupervised learning finds clusters from its own data automatically. Both supervised and unsupervised learning can be divided into two categories based on the estimation method of the posterior probability. The discriminative approach directly models the conditional probability of y given x, $p(y|x)$, using probability theory, where x is the input data (mostly features extracted from system's performance data) and y is the associated class label. The generative approach models the prior probability density $p(x|y)$ for each class, y, and then chooses the class that best fits the observed data, x, based on optimization algorithms. The generative approach produces a different probability density model for each class and describes the overall probability over all the variables.

Data-driven prediction techniques are essentially regression techniques that estimate the future health based on either the previous health states or based on extracted features. Regression is further classified as linear or nonlinear regression. Within these classes, there could be cases of simple regression and multiple-regression. Regression is used for prediction and is based on estimating the dependence of the RUL upon monitored system parameters. The type of regression analysis to be used depends on the correlation between the monitored parameters of the system under consideration. Overall selection of a data-driven algorithm is a complicated process but is a must for designing the PHM system.

9.5 Testing

9.5.1 Introduction

After selection of the sensors, and data acquisition and data processing hardware and software, the next step is to test the IVHM system hardware and software to assess the reliability under operating conditions. The purpose of testing or qualifying the IVHM system is to eliminate mistakes in the design selections, identify the significant input factors (environment, loads, physical characteristics), assess the reliability under operating conditions, improve the robustness of the system, and balance the trade-offs in the overall system's design. The technical feasibility of the design can be gauged based on the testing results. This section presents the steps involved in the qualification of hardware and testing and validation of software.

9.5.2 Qualification of IVHM Hardware

Qualification includes all activities that ensure that the nominal design and manufacturing specifications will meet or exceed the desired reliability targets. Qualification validates the ability of the nominal design and manufacturing specifications of the product to meet the customer's expectations, and assesses the probability of survival of the system over its complete life cycle [Wang, Azarian, and Pecht 2008]. The purpose of qualification is to define the acceptable range of variability for all critical product parameters affected by design and manufacturing, such as geometric dimensions, material properties, and operating environmental limits.

i) Virtual Qualification (VQ): Virtual qualification is a simulation-based methodology used to identify the dominant failure mechanisms associated with a given part or component under the lifecycle loads and to determine the corresponding time-to-failure distributions. VQ provides preliminary results for qualification of the hardware and can be used to update the design, if necessary, for cases in which certain components or parts of the IVHM system fail to survive the required life estimates. VQ requires significantly less time and money than physical testing. The VQ process consists of the following steps: 1) design capture, 2) load history characterization, 3) load transformation, 4) damage assessment, and 5) calculation of the remaining life of the system [Mathew et al. 2007]. The software used to calculate the damage and RUL contains models that can incorporate "curve fitting" to describe failure data or can be based on stress-damage models that relate the fundamental mechanical, thermal, electrical, radiation, and chemical failure processes arising from the environmental and operational lifecycle loads to the time-to-failure of the component. Figure 9.6 shows an example process flow for the simulation of a time-to-failure calculation for a system subjected to thermal and vibratory lifecycle loads.

The accuracy of the results using VQ depends on the accuracy of the inputs to the process, such as the accuracy of the life-cycle loads, the choice of the failure models used, the choice of the analysis method (e.g., 2-D, pseudo-3-D, or full 3-D), the constants in the failure model, and the material properties. Hence, to obtain a reliable prediction, the variability in the inputs should be specified using distribution functions, and the validity of the failure models should be tested by conducting accelerated tests. Although the data obtained from virtual qualification cannot fully replace those obtained from physical tests, they can increase the efficiency of physical tests by indicating the potential failure modes and mechanisms that the operator can expect to encounter.

ii) Design Of Experiments (DOE): A DOE is a systematic planning method of research wherein controlled experiments are conducted for purposes of comparison of factors that affect the systems performance. The idea behind the DOE is to have a detailed plan before conducting the actual test. A DOE helps the engineer identify which input factors will affect the reliability or RUL of

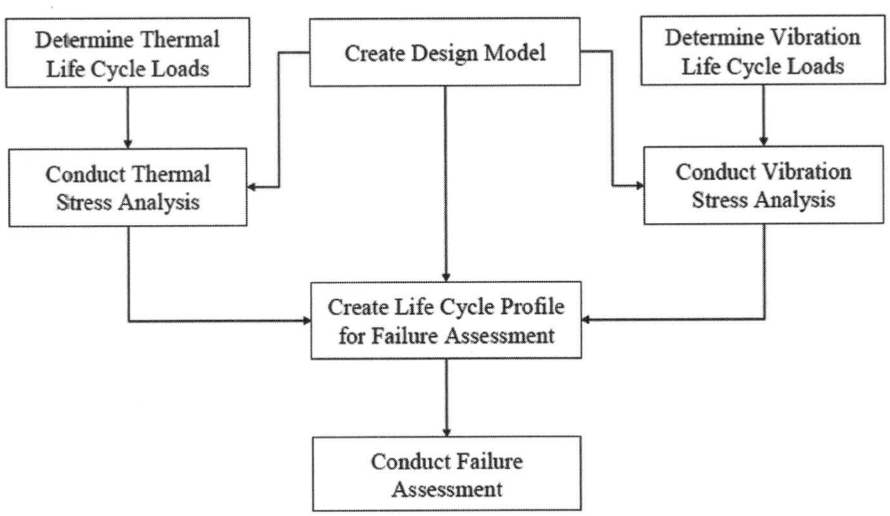

Figure 9.6 Example of process flow for simulation [Mathew et al. 2007].

the IVHM system. A DOE is planning of experiments so that the test results can be analyzed to yield valid and objective conclusions. The DOE is a process to develop test program(s) that focuses on the potential failure mechanism(s) and mode(s) that were identified in the FMMEA phase as the weak links under life-cycle loads. The first step in creating a DOE is to determine the objective of each test. The next is to list the inputs of the test and the outputs to be monitored. The relationship between the input factors and the monitored outputs should provide information about the reliability of the component or system in actual operational conditions.

It is important to decide what kind of test will be conducted, including the loading conditions during the test. In the case of an accelerated test, the acceleration factor between the accelerated test and the actual operating conditions needs to be estimated to make sense of the testing results. To perform testing of the IVHM system efficiently, engineers need to develop a DOE matrix that contains information about the component (s) to be tested, the environmental and operational loading conditions, and the parameters that will be monitored for each test case.

Figure 9.7 shows examples of input factors, test conditions, and output parameters. The test engineer needs to clearly identify the potential relations between the inputs, test conditions, and outputs so that the testing results can be used for assessing the system's reliability under tested conditions. A well-designed experiment provides the required information in a simple, reproducible, and cost-effective manner.

iii) Accelerated testing: Accelerated testing measures the performance of the test product at loads or stresses that are more severe than would normally be encountered. The goal of accelerated testing is to accelerate the time-dependent failure mechanisms and the damage accumulation rate to reduce the time to failure. The failure mechanisms and modes in the accelerated environment must be the same as those observed under actual usage conditions. Accelerated testing begins by identifying all the possible overstress and wear-out failure mechanisms [Hu et al. 1992]. The load parameter that directly causes the time-dependent failure is selected as the acceleration parameter. Common accelerated loads include thermal loads such as temperature, temperature cycling, and rates of

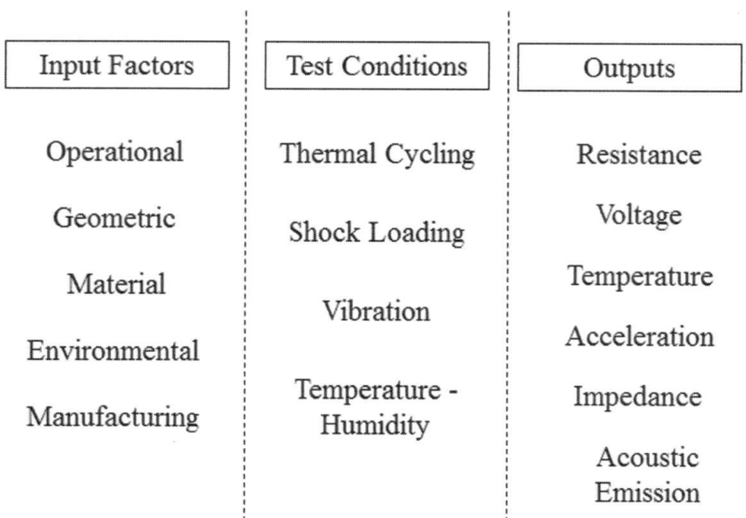

Figure 9.7 Examples of considerations for DOE.

temperature change; chemical loads such as humidity, corrosives, acid, and salt; electrical loads such as voltage or power; and mechanical loads such as vibration, mechanical load cycles, strain cycles, and shock. Accelerated tests require careful planning to represent the actual usage environments and operating conditions without introducing extraneous failure mechanisms or nonrepresentative physical or material behavior. The degree of stress acceleration is usually controlled by an acceleration factor, defined as the ratio of the life under normal use conditions to that under the accelerated condition.

For accelerated testing, it is necessary to select the appropriate acceleration load; determine the test procedures and the stress levels; determine the test method, such as constant stress acceleration or step-stress acceleration; perform the tests; and interpret the test data, which includes extrapolating the accelerated test results to normal operating conditions. The test results provide qualitative failure information for improving the hardware through design and/or process changes. Accelerated life testing can provide comparisons between alternate designs if the same failure mechanism is precipitated. Failure due to a particular mechanism can be induced by several acceleration parameters. For example, corrosion can be accelerated by both temperature and humidity, and creep can be accelerated by both mechanical stress and temperature. The accelerated environment may include a combination of these loads. Interpretation of results for combined loads requires a quantitative understanding of their relative interactions and the contribution of each load to the overall damage.

iv) Failure Analysis (FA): Detailed FA of failed samples is an essential step in the qualification and validation program. Without FA and feedback to the design team for corrective action, the goal of the qualification program is defeated. It is not adequate simply to collect failure data from testing; instead, the test results must be used to provide insights into the relevant failure mechanisms and to cost-effectively prevent failures. FA techniques include nondestructive and destructive techniques. Nondestructive techniques include visual observation, observations under optical microscope, x-ray, and acoustic microscopy. Destructive techniques include cross-sectioning of samples, and de-capsulation. FA is used to identify the locations at which failures occur and the fundamental mechanisms by which failures occur. FA will be successful if it is approached systematically, staring with nondestructive examinations of the failed test samples and then moving on to more advanced destructive examinations.

9.5.3 Testing and Validation of PHM Software

During development of an embedded system, it is necessary to determine if specifications are met and the outputs are correct [Tran 1999]. Verification and Validation (V & V) of IVHM algorithms is important. The V&V process provides tools and analysis procedures appropriate to each phase of the software development life cycle. V&V identifies deficiencies in the software's test planning, allows early error identification and correction, increases awareness of the requirements of the IVHM algorithm, increases compliance between specification and actual output, and increases the probability of successful deployment.

"Verification is a process that is used to evaluate whether or not a product, service, or system complies with regulations, specifications, or conditions imposed at the start of a development phase" [Brat 2010]. The question to answer here is: are we building the product right? Verification techniques include dynamic testing, functional testing, structural testing, random testing, static testing, consistency testing, and testing of properties of the algorithm. "Validation is the process of establishing evidence that provides a high degree of assurance that a product, service, or system accomplishes its intended requirements" [Brat 2010]. The question here is are we building the right product? Validation techniques include formal methods, fault injection, and dependability analysis. Validation usually takes place at the end of the development cycle, and looks at the complete system as opposed to verification, which focuses on smaller subsystems.

Usually the algorithms selected for anomaly detection, fault isolation, diagnostics, and prognostics have to be written using a computer programming language. There are many different programming languages and platforms upon which algorithms are coded. It is also possible that different modules of the overall software are obtained from different vendors. Software debugging aims to locate the mistakes in the code and fix them. For in-house software development, the difficult task of examining the code as well as executing it under different scenarios has to be conducted.

Unlike software debugging, in which each line of code is examined for mistakes and then corrected, testing seeks to verify the logical interactions of different modules/functions within the entire software. Testing of the software in an IVHM system is as crucial as qualifying the hardware components. Whether it is in-house or purchased from outside vendors, the software should be tested to avoid false alarms in operation. Software testing is a detailed investigation of a program's ability to work as expected under different data input scenarios. The goal of testing is to identify if the coding is correct, and complete, and find errors so that they can be corrected before operation. In software testing, the output of the software is matched to known values of parameter obtained from physical testing. As the complexity of the software increases, testing the software becomes challenging, and it may not be possible to test all scenarios.

9.6 Integration

After testing the subsystems/components of the IVHM system, the next step is to update the system design to improve its reliability and robustness. After design updates are applied, the design of the system is finalized. At this stage, although the design is complete, the IVHM system prototype has to be assembled and integrated into the system under consideration. A working prototype brings the design to life. It helps to identify flaws in the design, material selection, and overall concept of the IVHM system. The prototype has to be tested to evaluate the overall performance of the IVHM system in conjunction with the system to be monitored under the actual operational environment. The performance of the prototype under field conditions should be evaluated on the basis of performance metrics that have been developed for the IVHM system. The features of the prototype may be updated to match the user specifications and expected outputs for IVHM.

Appropriate life sustainment strategies should be developed for the IVHM system, so that continuous monitoring is enabled. The inventory needs of the IVHM system should be estimated and proper maintenance and logistical strategies need to be devised. Procedures for operating and conducting maintenance should be documented in manuals. Training procedures should also be documented. The end users and maintenance personnel should be trained to use the IVHM system. A clear guide should be presented to the end users and maintenance personnel to interpret the alarms and information obtained from the IVHM system.

9.7 Conclusions

Producing a reliable IVHM system begins at the system design phase and is contingent on a systematic definition of the framework of the IVHM system, and its components, subsystems, interfaces, and required inputs and outputs. The systems design helps in the process of moving from a concept to a physical entity. Designing an IVHM system requires the system designer to engage in a detailed process of concept design, framework development, design selections, testing and qualification, prototyping, design updates, and developing detailed operational and support procedures. This requires inputs and coordination between personnel from inter-disciplinary fields.

Defining the proper requirements of the IVHM system helps in generating the design specifications. A framework for the IVHM system serves as a blueprint for system designers for faster system development and deployment. Conducting FMMEA of the system whose health is to be monitored provides designers with knowledge of root causes of failure and the failure modes. This knowledge helps in identifying the parameters to monitor and the sensors available for monitoring. Selection of a prognostics approach and the algorithm(s) to process the data to obtain meaningful information regarding the system's state of health is beneficial in furthering the IVHM design. Testing the hardware of the IVHM system helps eliminate mistakes in the design selections. V&V of IVHM algorithms need to be conducted to ensure reliable operation of the IVHM system.

9.8 References

Army Regulation 70–62. 2007. Airworthiness Qualification of Aircraft Systems, U.S. Army, Washington, DC.

Bonissone, P., F. Xue, and R. Subbu. 2011. "Fast Meta-Models for Local Fusion of Multiple Predictive Models," *Applied Soft Computing*, Vol. 11, pp. 1529–1539.

Brat, G. 2010. "Verification & Validation for PHM," NASA ARC, Tutorial, Annual Conference of PHM Society, Portland, OR.

Carstens, J. R. 1993. *"Electrical Sensors and Transducers,"* Regents/Prentice-Hall, Englewood Cliffs, New Jersey.

Cheng, S., M. Azarian, and M. Pecht. 2010. "Sensor Systems for Prognostics and Health Management," *Sensors*, No. 10, pp. 5774–5797.

Daigel, M., and K. Goebel. 2011. "A Model-based Prognostics Approach Applied to Pneumatic Valves," *International Journal of Prognostics and Health Management*, pp. 1–16.

Ganesan, S., V. Eveloy, D. Das, and M. Pecht. 2005. "Identification and Utilization of Failure Mechanisms to Enhance FMEA and FMECA," *Proceedings of the IEEE Workshop on Accelerated Stress Testing & Reliability (ASTR)*, Austin, Texas, October 2–5, 2005.

Hu, J., D. Barker, A. Dasgupta, and A. Arora. 1992. "Role of Failure-mechanism Identification in Accelerated Testing," *Proceedings of Annual Reliability and Maintainability Symposium*, pp. 181–188, January 1992.

Kunche, S., C. Chen, and M. Pecht. 2012. "A Review of PHM System's Architectural Frameworks," *MFPT 2012 Proceedings: The Prognostics and Health Management Solutions Conference*, Dayton, OH, April 24–26, 2012.

Mathew, S., and M. Pecht. 2011. "Prognostics of Systems: Approaches and Applications," *Proceedings of the 24th International Congress on Condition Monitoring and Diagnostic Engineering Management (COMADEM 2011)*, Norway, May 31–June 2, 2011.

Mathew, S., D. Das, M. Osterman, M. Pecht, J. Clayton, and R. Ferebee. 2007. "Virtual Remaining Life Assessment of Electronic

Hardware Subjected to Shock and Random Vibration Life Cycle Loads," *Journal of the IEST*, Vol. 50, No. 1, pp. 86–97, April 2007.

Mathew, S., M. Alam, and M. Pecht. 2012. "Identification of Failure Mechanisms to Enhance Prognostic Outcomes," ASM *Journal of Failure Analysis and Prevention*: Vol. 12, Issue 1, pp. 66–73.

Pecht, M. 2008. *Prognostics and Health Management of Electronics*, Wiley-Interscience, New York, New York.

Pecht, M., and A. Dasgupta. 1995. "Physics-of-Failure: An Approach to Reliable Product Development," *Journal of the Institute of Environmental Sciences*, Vol. 38, pp. 30–34, September/October 1995.

Poggio, T., and S. Smale. 2003. "The mathematics of learning: dealing with data," *Notices of the American Mathematical Society*, Vol. 50, No. 5, pp. 537–544.

Saha, B., K. Goebel, S. Poll, and J. Christophersen. 2009. "Prognostics Methods for Battery Health Monitoring using a Bayesian Framework," *IEEE Transactions on Instrumentation and Measurement*, Vol. 58, No. 2.

Si, X. S., W. Wang, C. H. Hu, and D. H. Zhou. 2011. "Remaining Useful Life Estimation—A Review on the Statistical Data-Driven Approaches," *European Journal of Operational Research*, Vol. 213, pp. 1–14.

Tran, E. 1999. "Verification/Validation/Certification," *Dependable Embedded Systems*, Carnegie Mellon University.

Wang, W., M. Azarian, and M. Pecht. 2008. "Qualification for Product Development," International Conference on Electronic Packaging Technology and the International Symposium on High Density Packaging (ICEPT-HDP), Shanghai, China, July 28–31, 2008.

CHAPTER 10 Applications

Peter Foote, *Cranfield University*
Charlie Dibsdale, *OSys*
Keith Jackson, *Meggitt*
Ranjith Kumar, Sreenilayam Raveendran, Michael H. Azarian, and Michael Pecht, *Center for Advanced Life Cycle Engineering, Univ. of Maryland*

> "The obvious is that which is never seen until someone expresses it simply."
>
> —Kahlil Gibran

10.1 Introduction

In this chapter, four very different examples of IVHM application are given, each revealing a different side to the subject and the ease, or not, of applying it in different sectors. The first two sections deal with aerospace-based issues of Structural Health Monitoring (SHM) and rotating equipment, while the next two deal with the very different sectors of motorsport and wind turbines.

The chapter opens with a view on SHM in which applications to current aircraft are described and emerging technologies discussed. The military has, for a long time, routinely monitored operational loads to deduce life but, as the technology becomes more widely used, or is taken over to civil aircraft, the business case becomes more difficult. There emerges a Catch-22 in which customers want robust, accurate, SHM technologies, but these will not be achieved until the technology has been proven and matured: hence the dilemma. The next application deals with a wide variety of rotating equipment, from electrical motors to steam turbines. This equipment can be monitored by a number of, relatively few, techniques quite successfully. Monitoring oil, looking at static and dynamic data, and using control system signals as health indicators all feed into the overall picture. The added dimension that IVHM brings is the fusion of all the data to make the best possible diagnosis up to, and including, the platform level.

The third application area of motorsport is one in which the acronym IVHM is not recognized, a fact true of most sectors outside of aerospace, but the sensing, data processing, and learning happen nonetheless. This area is rich in lessons for IVHM as, due to rules and regulations that are changing at an alarming rate, the sport (and its application of IVHM) is very agile. The section ends with a list of factors to take into account when considering an IVHM system. This list is as applicable to trains and planes as it is to motorsport and indicates a field from which much could be learned. The last, and very comprehensive application, is that of IVHM to wind turbines. Types of wind turbine are discussed, along with their failure rates; 95% of downtime occurs due to gearbox and generator problems. Condition monitoring of these, and other, components is then discussed and brought together in health management implementation strategies.

10.2 SHM for Aerospace Applications

Peter Foote, Cranfield University

“Damage control is easy. Reading Klingon—that's hard.”

—Montogomery "Scotty" Scott, chief engineer, Starship Enterprise

10.2.1 SHM and Its Uses

Structural Heath Monitoring (SHM) is the "the process of acquiring and analyzing data from onboard sensors to evaluate the health of a structure" [SAE International] and is a means of verifying that a structure is fit for its intended purpose and safe to operate.

Although monitoring of structures is not a new activity (it has been an integral, mandated component in aircraft maintenance programs during most of the period of powered flight), the means to achieve this using permanently installed sensors rather than manual inspections is a new and emerging technology (see [Boller, Chang, and Yozo Fujino 2009] for a detailed overview).

The industry and stakeholder community linked to SHM is still relatively unstructured, and no standards or guidelines on the nature and use of SHM exist yet. The SAE International Aerospace Recommended Practices [SAE International] and recent additions to the procedures for designing aircraft maintenance programs owned by the Airlines For America (formerly the Air Transport Association: ATA) are a step forward in this respect [MSG-3 2009].

In any discussion of the technology, an understanding of the intended use of the sensors and data is essential. SHM technologies fall into two main types, depending on their intended functions:

1. Operational monitoring. Use of sensors to monitor:
   - Fatigue
   - Exceedance
   - Environment
2. Damage detection and monitoring. Use of sensors to detect and monitor:
   - Fatigue damage
   - Environmental damage
   - Accidental damage

Operational monitoring is an indirect means of assessing and predicting structural integrity by monitoring how the structure is used. This includes measurement of the static and dynamic stresses it has experienced and also the environment to which it has been exposed. Dynamic and static stress monitoring (fatigue monitoring) of an airframe usually takes place at key load-bearing features or control points. The data is acquired either by strain measurement (e.g., strain gauges) or is calculated parametrically by using flight control data to infer loads from a dynamic model of the airframe or load cases [Hunt and Hebden 2001]. The parametric approach constitutes a virtual sensor system.

Monitoring of this type is currently found mostly on military aircraft, where the usage patterns are less predictable than commercial aircraft. It also underpins design assumptions made by some military design authorities, such as the UK Ministry of Defence (MoD), who employ "safe-life" structural design philosophies.

Environmental monitoring is achieved using sensors or other records and data to track the exposure of an airframe to factors in the environment that can cause structural degradation, such as salt water that will cause aluminium alloy structures to corrode if their protective layers are deficient or damaged. Humid atmospheres will also accelerate aging in composite materials in which any moisture penetration of the fiber/polymer matrix will cause loss of strength.

In both cases, sensors can provide measurements of the key parameters such as duration of exposure, wetness/humidity, acidity/alkalinity (Ph), and temperature.

For fatigue and environmental monitoring, the sensor and other parametric data are used in physical models to estimate likely damage. In the case of fatigue monitoring, an understanding of fracture mechanics is used to link measurement data to the fatigue state of a structural item. Continuous monitoring allows the consumption of fatigue life to be measured as well as an estimate of remaining useful life to be made. Similarly, the environment data can be used in corrosion prediction models developed from knowledge of underlying physics and chemistry or empirically, and from the observed behavior of test specimens exposed to similar conditions.

Damage detection and monitoring is a means of measuring damage directly by sensors. These perform a function identical to current, NonDestructive Inspection (NDI) methods in which instruments are used by a human inspector to probe regions of the structure to find cracks, regions of corrosion, delamination of composite plies, missing material, etc. The essential distinction between SHM and NDI is the role of human factors. In the latter, the detection capability inherent in the process relies not only on equipment but also on the ability of inspectors to perform their tasks [DOT 1993].

SHM for fatigue monitoring is already established as part of structural health management programs for military aircraft, with several national air forces implementing flight fatigue loads tracking. Environment monitoring is a more developmental capability but is of increasing importance as the legacy of aging aluminium alloy-based fleets of aircraft grows.

Damage detection SHM techniques are still early in development. Although NDI methods are well established, the transition to automated forms of inspection is technically challenging, not least because of the high degree of reliability required and the consequent high barrier to acceptance.

Both operational monitoring and damage detection monitoring technologies will be explored in more detail in the following sections.

10.2.2 Rationale of SHM in the Context of Aircraft Maintenance

Aircraft structural requirements, whether military or commercial, are governed by rigorous regulatory and airworthiness requirements such as those embodied in Airworthiness Circular AC25 571. A broad description of current approaches to ensuring integrity of structures can be found in [Goranson 1997], but to recap: the prevailing methods adopted for commercial aircraft follow the damage-tolerant design approach in which structures are assumed to enter service with damage (e.g., cracks) that are just below the threshold of detection. The safe operation is then guaranteed by careful analysis at the design stage of how such cracks would grow under operational conditions and lead to reduced structural strength. Inspection intervals are assigned to ensure that if such cracks exist, they will be easily detectable by inspection well before structural strength falls to the regulatory fail-safe load requirement (see Figure 10.2.1). Hence, the design and inspection interval is linked by the factors of minimum detectable damage limits (such as crack length), anticipated usage, reliability, and confidence that an inspection will detect damage during inspections. Guidance on this analysis is held in the A4A MSG-3 process [MSG-3 2009], which is the industry accepted practice and methodology from which all commercial airplane maintenance programs are built.

The damage-tolerant approach is more than a design philosophy: it is a design *system* because it embodies the through-life support and maintenance programs for the structures. It is illustrated in Figure 10.2.1.

Underpinning this system is the means to detect damage that could reduce the strength of the structure. This is currently achieved by manual inspection processes categorized by MSG-3 as

1. General Visual Inspection (GVI)
2. DETailed Inspection (DET)
3. Special Detailed Inspection (SDI)

Recently, a fourth category was added to the MSG-3 document (revision 2009-1): it is referred to as Scheduled-SHM (S-SHM) and defined as *"the act to use/run/read out a SHM device at an interval set at a fixed schedule."* The use and interpretation of this newly added form of inspection is discussed further in the SAE International Aerospace Recommended Practices [SAE International].

The inclusion of SHM is in recognition of the advantages it could offer over manual inspection, in which sensors installed deep in a structure could relay data on structural integrity without the need for an intrusive manual inspection.

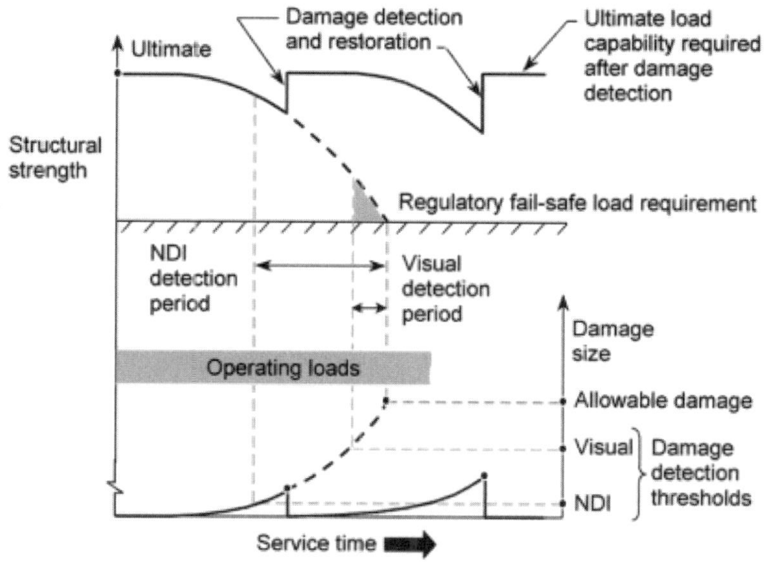

Figure 10.2.1 Damage-tolerant design system that links design parameters to a regime of structural inspections at scheduled intervals [Goranson 1997].

The quality of the "inspection" performed by the sensors must at least equal the reliability of the current, manual inspection methods. These are quantified using a probability of detection of 90% with confidence level of 95% (see [DOT 1993]).

A further revision to MSG-3, which has been proposed (but not yet adopted), is called Automated-SHM (A-SHM) and is described as *technology which does not have a pre-determined interval at which maintenance action must take place, but instead relies on the system to inform maintenance personnel that action must take place.*

There is a subtle distinction between the techniques of S-SHM and A-SHM. They may well share exactly the same technology, but the manner in which the SHM data is used is profoundly different. Scheduled SHM takes place as part of a conventional scheduled maintenance and inspection process. As such, its use requires no special certification consideration. It is simply an alternative means of fulfilling a requirement for inspection. On the other hand, A-SHM replaces a scheduled task and instead relies on the SHM technology to alert maintainers to the presence of structural damage that warrants a maintenance action. This latter application of SHM would have certification implications because it has the potential to adjust the intervals built into the damage tolerance design assumptions.

10.2.3 Examples of SHM Applications and Benefits

The previous descriptions of SHM apply to both operational and damage detection types. SHM could be used as a direct replacement for NDI, when a specific inspection is called for. An example may be a Service Instruction or Service Bulletin, issued by a design authority, which requires inspection of an internal bay within an aircraft structure following an issue or arising within the operating fleet. The instruction may call for repeated inspections at scheduled intervals that would require repeated intrusion into the structure to perform a manual inspection. S-SHM would allow a suitable sensor system to be fitted at the initial inspection, which can be read out from a remote terminal without the need for the structure to be opened in future inspections. The sensor would perform the equivalent inspection task more rapidly and at lower cost during subsequent maintenance. This is an Alternative Means of Compliance (AMOC) that employs a damage detection form of SHM technology.

Another example could be an inspection that is triggered by a pilot report of a hard landing. European Aviation Safety Agency (EASA) guidelines for events started by such a report include phased inspection, from superficial to more in-depth inspection depending on outcomes. SHM could play a dual role in this kind of event. First, operational monitoring could replace the pilot report by automatically monitoring loads and detecting if a prescribed exceedance threshold has been passed. Second, damage detection sensors placed at structural fuses (attached strips of material designed to break at predetermined strain levels to indicate a local exceedance of stress) and areas known to be early or typical indicators in these events could be used to diagnose further. With an integrated system, this information could be available immediately on download of the data. The potential benefit is the avoidance of a grounded aircraft following a false report in which a pilot suspects a hard landing, but subsequent inspections reveal that no damage has occurred. A further benefit in the event that damage has occurred is the capability afforded by SHM to better identify the zones requiring maintenance action.

10.2.4 Military Applications

Military airworthiness regulations are tailored to national priorities and as such do not have the same global consensus as those for commercial aviation. The U.S. Air Force, for example, operates its aircraft within an Aircraft Structural Integrity Program (ASIP) [Wells and Troy 1968]. ASIP is a damage-tolerant approach, as with commercial aircraft, and is the primary basis for aircraft structure maintenance programs for the past 40+ years. Inspection requirements are based on rogue flaw assumptions (slow crack growth) and NDI capability. By contrast, however, the UK MoD designs and operates its aircraft on a "safe-life" principal, in which designs incorporate worst-case assumptions about operational loads and so, by over-engineering, ensures that a minimum design goal objective measured in flying hours can be met without significant fatigue damage. This design approach is underpinned by a Major Airframe Fatigue Test (MAFT), in which a complete airframe undergoes dynamic fatigue testing in simulation of operating conditions across the aircraft's performance spectrum. The test will run into multiples of the vehicle's design life and begin before the aircraft enters service. The fatigue life consumed by the MAFT will always be in advance of any of the aircraft type in service, with the objective of revealing any structural issues before they appear in the active fleet. The variations in approach between military and civil requirements reflect the less-predictable demands placed on military aircraft. For example, combat aircraft routinely encounter high-*g* maneuvers and buffeting from-low level flight, whereas commercial aircraft experience predominantly high-altitude, steady flight with few unpredictable events (with the exception of hard and heavy landings).

Military operators now routinely employ operational monitoring as a means of tracking aircraft fatigue life consumption. The UK MoD [MoD 2003] states, "Every aircraft in the fleet shall be provided with instrumentation for the purpose of estimating the fatigue life consumption of fatigue-critical structure . . . . Provision shall be made for this instrumentation during production."

A design factor penalty of 1.5 can result from an absence of monitoring. "Thus unmetered flying consumes the safe life of the aeroplane 50% faster than metered flying." In other words, the requirement not only calls for monitoring to measure fatigue consumption but also offers a design factor credit as a result. This means that airframe monitoring is used to produce more efficient structures *by design*.

For SHM, the safe-life design would appear to obviate the need for any inspection and hence render damage-detecting forms of SHM irrelevant. However, in practice, even safe-life aircraft undergo some degree of scheduled inspection, especially when unforeseen, in-service issues emerge (known as arisings), either from the MAFT or other origin. Military aircraft are also frequently modified or reconfigured to adapt to changing roles, which may take them outside of their original design envelope. This would include re-lifing, when a safe-life aircraft reaches the end of its nominal design life but for economic or tactical reasons is required to remain flying.

Key issues for all SHM applications are the veracity of the SHM technologies, coupled with the value they deliver in terms of maintenance credit (the ability to eliminate a maintenance requirement) and maintenance benefit (the ability to adjust a maintenance requirement). Potentially, A-SHM could pave the way to Condition-Based Maintenance (CBM), with all the associated economic and operational benefits.

10.2.5 Snapshot of Some Emerging Technologies

The major types of SHM technologies are shown in Tables 10.2.1 and 10.2.2. This is not comprehensive, but it illustrates the range

Operational Monitoring Parameter	Measurement Technique Application	Sensor/ Measurement Data Source
Fatigue	Operational loads monitoring	Electrical resistive strain gauges Fiber-optic strain sensors • Discrete fiber-optic Bragg gratings, intrinsic interferometers • Continues distributed back scatter methods Parametric (e.g., flight control parameters, *g*-meters)
Exceedance	Overload detection	As above plus: Mechanical fuses Impact detection systems (high-bandwidth strain sensors) Accelerometers
Environment	Corrosion Moisture ingress Overheat	Corrosion parameter sensors: • Time of wetness • Multi-parameter: Ph, Temperature, H_2O • Sacrificial or witness (canary) sensors. (Pseudo alloy conductors corrode and give early warning) Parametric (e.g., location, date); time linked to environment databases

Table 10.2.1 SHM Sensors for Operational and Environmental Monitoring

Damage Detection/ Monitoring Parameter	Measurement Technique Application	Sensor/ Measurement Data Source
Fatigue damage	Metal crack detection Disbond initiation and growth	Surface breaking cracks: • Electrical conductive track crack gauges • Vacuum gallery crack gauges Fixed eddy current patches Phased array acousto-ultrasound (fixed transducers producing imaging ultrasound) Lamb wave acoustic transducer/sensors Pitch-catch acousto-ultrasound Acoustic emission detection
Accidental damage	Impact detection (composites) Void/delamination detection (composites)	Phased array acousto-ultrasound (fixed transducers producing imaging ultrasound) Lamb wave acoustic transducer /sensors Pitch-catch acousto-ultrasound Acoustic emission detection Electrical intrinsic self-sensing (carbon fiber composites)
Environment induced damage	Corrosion Moisture ingress Overheat	Corrosion parameter Sensors: • Time of wetness • Multi-parameter: Ph, Temperature, H_2O • Sacrificial or witness (canary) sensors. (Pseudo alloy conductors corrode and give early warning) • Other sacrificial material-based sensors Phased array acousto-ultrasound (fixed transducers producing imaging ultrasound) Lamb wave acoustic transducer/sensors Pitch-catch acousto-ultrasound Acoustic emission detection Electrical intrinsic self-sensing (carbon fiber composites)

Table 10.2.2 SHM Sensors for Damage Detection

and nature of the techniques available for both operational monitoring and damage detection.

A detailed account of each technology type is beyond the scope of this book. Instead, a selection of techniques will be described qualitatively as representative of emerging SHM technologies.

10.2.5.1 Pitch-Catch Transducer Sensor Arrays

These systems consist of arrays of acoustic transducers capable of transmitting and receiving sound [Qing et al. 2007]. The transducers are bonded to the structure in a geometry that defines the monitored region. Typically, the devices are piezoelectric

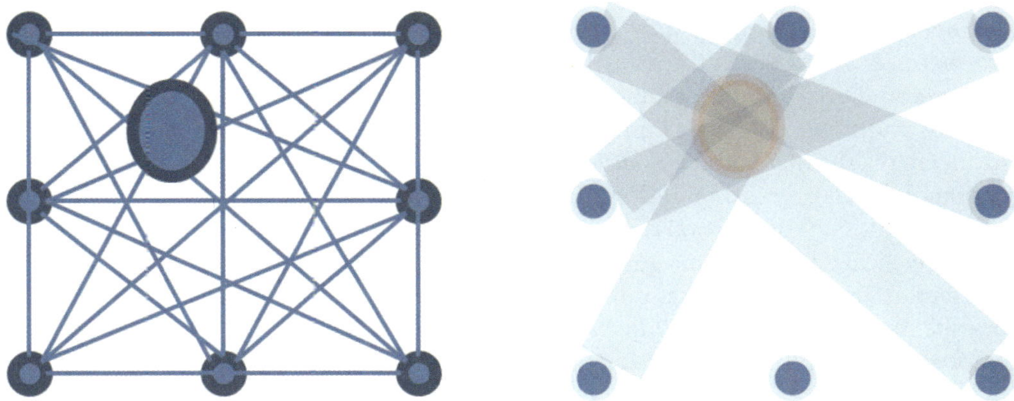

Figure 10.2.2 Notional array of transmitter/receiver acoustic transducers bonded to a structure. The system sequentially maps pathways in the structure by transmitting acoustic bursts from each transducer, in turn, while the others receive. When damage occurs (represented by the off-center oval patch), a difference is detected in the signals from transducer pairs that intercept the damaged region

discs of less than 10 mm diameter and 0.5 mm thickness, held in a polymer film (e.g., Kapton—the same substrate material from which resistive foil strain gauges are made).

A high-frequency voltage in a pulse envelope is applied to each transducer, in turn. The piezoelectric material in the transducer converts the voltage to a similarly modulated acoustic signal, which is transmitted into the structure. The acoustic pulse propagates away from the transducer toward the other elements in the array. While a transducer is energized, the others are in receiving mode, each detecting the acoustic burst as it arrives (hence the "pitch-catch" connotation). After each transducer has been energized, in turn, several times, an acoustic map of the structure is developed. The map is simply a record of each permutation of received signals from all transducers. The essentials of the process are that any receiver sensor will detect the acoustic signal from the active transducer on a direct path (see Figure 10.2.2). The acoustic map is, therefore, a mesh of such direct signal paths

To detect damage, the undamaged structure is first mapped. This is called a baseline image or data set. The structure is interrogated periodically in the same manner, and the collected signals are compared with the corresponding baselines signals. The comparison could be simple signal subtracting with a residual value characterizing any change of a particular transmit-receive signal. If damage occurs, those signal paths that intercept the damaged region will experience the largest change. This change would include attenuation, reflections, and other features that generally distort the signal in comparison with the baseline case.

If only those signal paths affected are plotted, a coarse indication of the presence and location of damage emerges (see Figure 10.2.2).

The ability of the technique to locate and measure the extent of the damage depends on a number of factors, dominated by the total number of transducers in the array for a given monitored area and the stability of the baseline conditions. A greater number of transducers results in a better resolution. Application of signal processing methods based on the physics of interaction of sound waves, with typical damage forms such as cracks in metals and voids and delaminations in composite materials, can be used to extract more information from the collected data.

10.2.5.2 Acoustic Emission Detection

Materials as they fracture can emit small bursts of acoustic waves across a broad range of frequencies, from audible to ultrasound (up to 1 MHz). The energies of these bursts are tiny in comparison to audible sound but are nevertheless detectable using piezoelectric sensors. Hence, a means of detecting the presence and growth of cracks in both metals and other materials that are subject to cracking, such as the epoxy matrix material in fiber-reinforced composite materials, is possible using these sensors.

The emissions occur in bursts of broadband acoustic waves, which propagate through a structure away from the source at predetermined velocities. If three or more sensors are close enough to the source to detect these bursts, then not only the presence of a growing crack but also its location can be determined. The relative time delay between the discrete bursts arriving at each sensor allows the point of origin to be determined via a triangulation (see Figure 10.2.3).

Relatively few sensors are required to monitor an area of structure (for example, typically less than 10 sensors for an area of 1 m^2). Metallic structures transmit ultrasound with very little attenuation. Composite

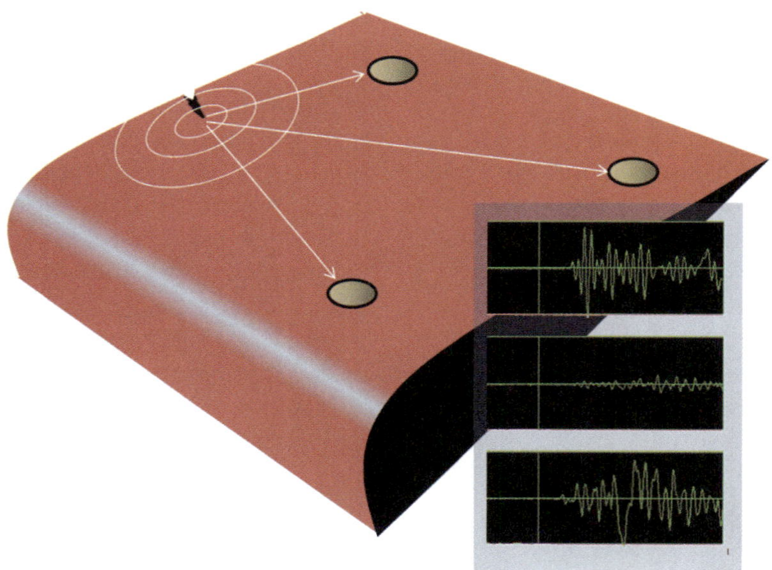

Figure 10.2.3 Principles of damage detection by monitoring of acoustic emission. A group of sensors detects acoustic bursts resulting from a growing crack. By measuring the relative times of arrival of the signals, the location of the source and hence the crack can be deduced using a triangulation method.

material such as carbon-fiber-reinforced plastic will also transmit sound, but with higher attenuation than, for example, aluminum alloy.

State-of-the-art acoustic emission detection sensors are extremely sensitive, with sensitivities of 50–100 eV or an energy input of approximately 10^{-17} J. By comparison, the motion of a single, atomic scale vacancy and a single dislocation emits a signal of about 0.1–0.05 eV. So these sensors can detect the disruptions in metals at near-atomic scale, revealing crack growth invisible to other (e.g., assisted visual) forms of detection.

The major sources of acoustic emission in metals are crack jumps, fracturing, and loosening of hard inclusions and plastic deformation development. There are other microscopic sources, such as dislocation movement. Sources in composite material are matrix cracking and fretting (for example, associated with delamination in plies following impact damage) and fiber breakage.

Importantly, emissions originate not just from crack tips but also from the fretting surfaces along the length of a crack as it opens and closes under cyclic loading. This is important for damage detection, because acoustic emission can be used to monitor the extent of a crack as well as its active tip as long as the coordinates of the acoustic bursts can be determined accurately enough.

Figure 10.2.4 shows a sequence of two-dimensional plots of acoustic emissions generated by a growing crack in a flat aluminum plate undergoing constant-amplitude cyclic loading. The crack grew from a notch in the center of the right-hand edge and grew from the edge (plot (a)) to approximately 120 mm (plot (e)) before the specimen failed. Four sensors were used to cover the area of the specimen. This illustrates a capability of acoustic emission to plot the progress of a developing crack from initiation to critical length.

Acoustic emission has been in use for many years in mechanical test environments, especially for proof testing of pressure vessels, but its adoption as an in-situ damage detection method for aerospace applications is still at the experimental stage. Flight trials [Read et al. 2008] have demonstrated that the technique can be used to detect crack growth in metal structure, in real time on board a fast military jet. Such trials have demonstrated:

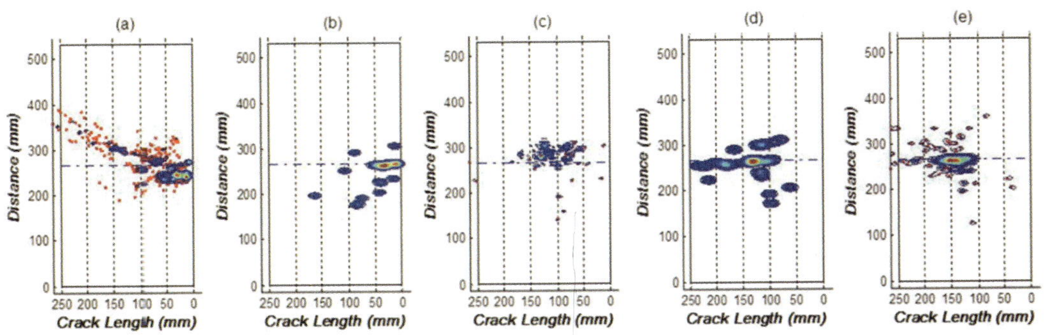

Figure 10.2.4 Sequence of maps of acoustic emissions generated by a crack growing from the center of the right-hand edge of an aluminum plate. The contours indicate the intensity of acoustic emissions (burst rate), with brightest being the most intense.

- The technique is sufficiently sensitive and discriminating to detect crack growth against high-amplitude acoustic noise backgrounds (jet engine noise levels of 160 dB, and near-supersonic aerodynamic airflows)
- The equipment is robust enough to function in the fast jet environment (high-*g* maneuvers, rapid temperature excursions between ground ambient and −60°C, high vibration, and electromagnetic interference)
- False signals from electromagnetic interference can be detected, isolated, and rejected.

10.2.5.3 Comparison of Pitch-Catch and Acoustic Emission Detection Techniques

The two preceding examples both use ultrasound as a means of detecting and locating damage, and they represent two distinct categories of detection methods, giving rise to associated advantages and disadvantages.

The pitch-catch techniques are representative of methods that require some form of baseline comparison in which damage is detected as a change in conditions. The quality of detection, therefore, depends on the stability of the baseline condition, which is the signature of the undamaged structure. It is possible, however, for baseline drift to occur during operation and use of the structure as a result of configuration changes (e.g., added components, repainting) and also changes in environment such as temperature, humidity, and contamination that are not necessarily signs of damage. These factors could lead to false positive indications of damage.

Another disadvantage of this technique is the number of sensors required to locate the damage. As can be appreciated from Figure 10.2.2, a region of damage can only be located to within one "cell" of overlapping signal paths from three transducers. Depending on the flaw size detection criteria, a suitably high density of sensors will be required. So, for example, to detect a flaw size of 10 cm anywhere within a region of 0.5 m^2 may require 36 sensors, each with their own connecting wires.

A major advantage of the technique for aerospace application is that the instrumentation to power and interrogate the sensor arrays does not need to be on board the aircraft. An inspection using this method could be performed using ground-based instrumentation, which is connected to the sensors as part of a maintenance process. It is not necessary to produce airworthy equipment, thus avoiding the associated system design complexity and cost.

Acoustic emission detection does not rely on a baseline measurement. A crack or other form of damage generates its own signature that the sensors, acting in a purely passive or listening mode, are able to detect. Moreover, the triangulation method of detection requires only three sensors to monitor a region. The limitation on damage location accuracy and number of required sensors are determined ultimately by trade-offs in bandwidth of the sensors, and the velocity and attenuation of sound waves in the structural material. In a structure made from metal alloy, areas of several square meters can be monitored using just three or four sensors. Even for carbon fiber composites, in which the sound is more readily attenuated, four sensors can typically monitor 0.5 to 1.0 m^2.

The disadvantages of acoustic emission are the requirement of some dynamic load input to the structure to trigger emissions. In practice, for an aerospace application this means that the sensor instruments must be on board in order to catch the emissions as the structure undergoes cyclic loading. A second disadvantage is that the phenomenon is still not well understood in terms of the relationship between crack size and growth characteristics, and the nature of the structural loads. At best, it can detect the extent and position of a crack. As a minimum capability, the technique can indicate that a crack is present and growing somewhere within a monitored zone.

10.2.5.4 Fiber-Optic Sensors

The emergence of fiber-optic technology for telecommunications has spawned a range of sensor techniques that exploit the interaction of physical parameters such as stress, acoustic energy, and temperature on the light-transmitting properties of optical fibers.

Distributed temperature and strain measurement systems are used for monitoring pipelines and security perimeters where the low-loss transmission properties of the fibers (as optimized for telecommunications) make it ideally suited for monitoring over large distances.

Optical fiber sensors are especially attractive for aerospace applications because of their extreme low mass and small cross-section (125 μm diameter), their immunity to electrical interference, their multiplexing capability (thousands of sensing points can be located in a single fiber, requiring just one cable and connection), and their lack of conducting wires and electrical power (no short-circuiting, ignition, and electromagnetic compatibility issues). Additionally, their small cross-section means that they can be embedded in structural material such as glass and carbon reinforced plastic with little or no impact on the material strength and durability. For this reason, the use of embedded optical sensors to create "smart structures" has been a topic of much research during the last two decades.

Of the many types of fiber-optic sensors, the most successful, in terms of deployment in structural monitoring applications and commercial success, is the fiber Bragg grating. These devices are intrinsic to the optical fiber and are discrete zones created by an ultraviolet light exposure process. These devices reflect light only in a narrow range of wavelengths (typically less than 1 nm bandwidth). The devices comprise a ripple of refractive index modulation in the fiber, which stretches and compresses along with the fiber when subject to strain. The physical change in dimensions of the gratings causes their reflection wavelengths to shift in proportion to the local strain. If broadband light is shined along the fiber, the reflections from these gratings can be captured and analyzed back at the fiber's end. The "color" reflected by each grating will indicate the strain at that grating's location. A string of these gratings, all with slightly differing and known nominal reflection wavelengths, can be identified in a light spectrum analyzer and their state of strain measured. In this way, they act as a distribution of strain sensors (see Figure 10.2.5).

The technology has so far found application in civil engineering and the renewable energy sector (wind turbines) but has not progressed beyond research trials in aerospace. The factors affecting its introduction to aircraft structures monitoring are:

1. The case for introducing new technology into established monitoring programs needs to be established (e.g., military operational loads monitoring programs that already use electrical strain gauge and parametric systems).

   Optical sensors of any kind require specialized optoelectronic interfaces to power-up the fiber and perform the optical parameter measurements. The supply chain for optoelectronic technology, although mature for the telecommunications industry and for other aircraft avionics such as high-bandwidth data buses, is not sufficiently developed for sensors applications to

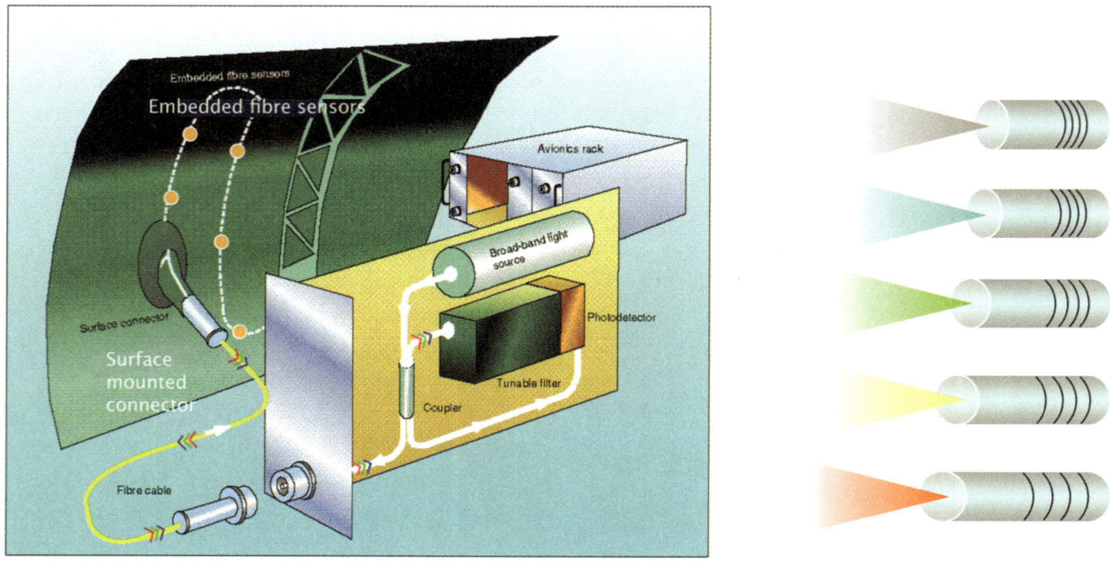

Figure 10.2.5 Schematic fiber Bragg grating smart structure concept. A single fiber containing multiple Bragg grating strain sensors is embedded within a structure and interrogated by an external optoelectronic module. The gratings work as strain sensors by shifting their reflection wavelengths in response to tension and compression. (Courtesy of BAE Systems)

meet most manufacturers' and systems integrators' requirements.

2. If optical sensors are to be embedded in composite aircraft structures, a new certification of the material would probably be required. The optical fibers, although small, are sufficient to represent an inclusion (technically a defect) in the material. To verify that their inclusion does not compromise strength, durability, and damage tolerance, extensive testing and evaluation will be required. This includes not just the sensor fiber, but also the means for connection, which must be partially amalgamated with the structural material. Once again, a robust case is needed to justify the costs of such a process.

Nevertheless, the background engineering and evidence to support this case is accumulating [Read and Foote 2001]. One such activity was the development of manufacturing compatible optical connector designs, as shown in Figure 10.2.6. These connectors were purpose engineered in a design for manufacture approach for embedding in high-quality, prepreg carbon-fiber layups used in the aerospace industry. The prototype connectors were designed to endure the temperatures and pressures of autoclave curing, followed by machining operations used for net shape recovery of structural items. Designed for embedding in panels and sheets (e.g., wing skins and fuselage panels), these connectors were for both surface egress and edge egress geometries.

10.2.5.5 SHM as an Element in Improved Structural Design

As mentioned earlier, the current UK MoD defense standards for safe-life aircraft require monitoring for fatigue life consumption as a design consideration. This effectively "buys" flying hours for safe-life airframes, while a lack of the monitoring capability incurs

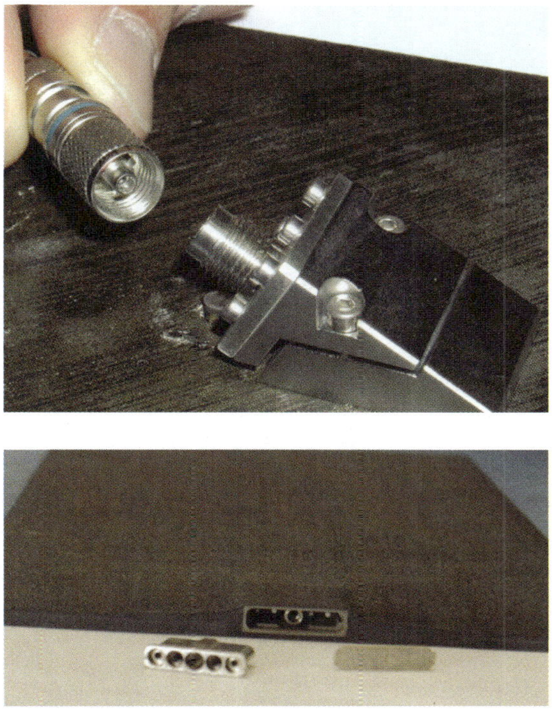

Figure 10.2.6 Prototype fiber-optic connector designs for optical sensors embedded in aerospace grade carbon composite panels: panel edge emerging geometry (left) and surface emerging geometry (right). (Courtesy of BAE Systems)

an operating life penalty. From a designer's perspective, this already means that a structure can be designed to be slightly lighter, with lower design factors due to the presence of a monitoring system.

This metered operation principle can be extended further into the damage-tolerant structural design system, which is intimately bound with inspectability and opportunities for inspection. Damage-tolerant design assumes that intrinsic damage is present, just below the threshold of detection, even in a new structure. The structure must be designed with an inspection threshold in mind so that the assumed damage cannot grow to a critical size before the first inspection threshold, at which there must be high confidence that the damage will be detected and rectified. Potentially, SHM can test that assumption of intrinsic flaws. A monitored aircraft could have inspection intervals adjusted based on the actual state of damage (CBM), but also could be designed without features currently included to satisfy inspectability requirements.

An example is considered by Telgkamp and Schmidt [2003], in which an analysis of the damage-tolerant crack growth criteria as applied to airframe stringer stiffeners is performed. If such a stringer is currently not inspectable, then it must be assumed to be broken. The inspection interval is determined such that the crack will not reach critical crack length (by some margin) before it is visible to inspection at the scheduled interval. The structure is, therefore, designed with sufficient thickness of skin to ensure that stress levels are low enough to reach that criteria for crack growth. If instead, the stringer is monitored continuously using SHM, then the assumption of a broken stringer can be excluded. Still adhering to the fixed inspection interval requirement, the designer can afford higher stress levels and hence reduced thickness in skin material while still preserving the crack growth criteria.

Other examples stem from current design with composite structures. Currently, the full advantages of bonded joints are not realized due to uncertainties in joint qualities coupled with the ability to inspect them. To overcome this limitation, composite airframes often include "chicken fasteners" that supplement joint strength using conventional bolted arrangements. This adds weight, not just from the metallic fasteners themselves, but also from the additional composite ply build-up and reinforcing around the fastener holes. If the bonded joints are monitored continuously with SHM it may still be possible to achieve the damage tolerance requirements without these extra measures.

Additionally, design factors to allow for reduction in compression strength following barely visible impact damage in composites are used. SHM could again be applied as a means to meet damage requirements as an alternative to conservatively strengthened structures.

10.2.6 Summary

This section has described some state-of-the-art uses of SHM in current aircraft applications. Monitoring of operational loads is now routine in military programs and has laid the path for emerging technologies with greater functionality, such as structurally embedded optical sensors and damage detection and monitoring sensors.

The evolution of the technology is slow due the (rightly) conservative nature of the aerospace industry. The prospects for the emerging technology, examples of which were described, are dependent on value propositions for their adoption and the veracity of the technology itself. The investment needed to mature the technology will only be unlocked when a strong business pull (market) from end users is established. On the other hand, end users are reluctant to adopt the technology as it is perceived as being immature. This "Catch-22" will be overcome only as the technology and its supply chain matures. Unlike previous introductions of monitoring systems such as Heath and Usage Monitoring Systems (HUMS) in commercial helicopters, which were mandated for safety reasons, no such imperative exists for SHM as described here. The drivers for SHM are operational and economic rather than the needs of safety. Nevertheless, the economics of aircraft fleet operation and ownership are now so finely balanced that even marginal increments in technology that benefit through-life processes can make the difference between profitable and loss-making enterprises. For this reason, SHM will undoubtedly be a feature of future aircraft.

10.2.7 Acknowledgments

The author gratefully acknowledges the help of BAE Systems.

10.2.8 References

Boller, C., F.-K. Chang, and Y. Yozo Fujino, eds. 2009. *Encyclopedia of Structural Health Monitoring*, Wiley.

DOT. 1993. U.S. Department of Transportation document: DOT/FAA/CT-92/12, 1, "Reliability at Airline Inspection Facilities: Vol. 1. A generic protocol for inspection reliability experiments," March 1993.

Goranson, U. 1997. "Fatigue issues in aircraft maintenance and repairs," *Int. J. Fatigue*, 20, No. 6, pp. 413–431.

Hunt, S. R., and I. G. Hebden. 2001. "Validation of the Eurofighter Typhoon structural health and usage monitoring system," *Smart Materials and Structures*, 10(3), pp. 497–503.

MoD. 2003. UK MoD Defence Standard: DEF STAN 00970, Part 1, Issue 5, section 3.

MSG-3: Operator/Manufacturer Scheduled Maintenance Development, Airlines For America publication. 2009

Qing, X. P., S. J. Beard, A. Kumar, K. Teng, T. K. Ooi, and F.-K. Chang. 2007. "Built-in Sensor Network for Structural Health Monitoring of Composite Structure," *Journal of Intelligent Material Systems and Structures*, Vol. 18, 1: pp. 39–49.

Read, I. J., P. Foote, J. McFeat, and Y. MacPheat. 2008. "Flight trials of structural damage detection systems on a fast military jet," *Proc. 5th European Conf. on Structural Health Monitoring*, Krakow, Poland.

Read, I., and P. Foote. 2001. "Sea and flight trials of optical fibre Bragg grating strain sensing systems," *Smart Mater. Struct.*, 10, pp. 1085–1094.

SAE International Aerospace Recommended Practices Document No: ARP 6461, "Guidelines for Implementation of Structural Health Monitoring on Fixed Wing Aircraft."

Telgkamp, J., and H.-J. Schmidt. 2003. "Benefits by the application of structural health monitoring systems on civil transport aircraft," *Proc. International Workshop on Structural Health Monitoring*, Stanford, CA.

Wells, H. M., and T. Troy. 1968. AS-TR-66-57, January 1968, "Air Force Aircraft structural integrity program: airplane requirements," U.S. Air Force.

10.3 Rotating Machinery

Charlie Dibsdale, OSys

"Man will never be enslaved by machinery if the man tending the machine be paid enough."

—Karel Capek (1890–1938)

10.3.1 Introduction

This section provides an overview of IVHM technology commonly applied to rotating machinery. It provides a simple view of the arrangement and some of the key types and functions of the machinery, and describes the application of existing health-monitoring techniques. The added dimension of IVHM fuses these techniques to provide a richer picture of the rotating machinery health, as well as covering the wider system the rotating machinery may be part of.

An overview of some of the emerging technology is provided, covering the introduction of ceramic bearings, how the IVHM techniques should be applied, and what the future may hold in automating some aspects of NonDestructive Examination (NDE) for rotating machinery.

10.3.2 Rotating Machinery: Scope and Diversity of Application

Rotating machinery is one of the major application areas of IVHM; it is comparable in scope to structures. Many applications of rotating machinery are configured as a "train" of equipment, consisting of a prime mover, driven equipment with a connecting shaft, and with a possible gearbox and auxiliary machinery such as lube oil. The supporting equipment is often called "balance of plant" or "skid equipment." Other rotating machinery configurations consist of one machine, such as a gas turbine engine on an aircraft, or a motor in a hydraulic or electrical system. Eighty percent of global rotating machinery applications are electrical induction motors.

10.3.2.1 Prime Movers, Balance of Plant, and Driven Equipment

A typical arrangement of a train of equipment may look like Figure 10.3.1. The industrial gas turbine is slightly different from one utilized in aerospace, in that it features a gas turbine (gas generator), using a physically separated power turbine that turns the driven equipment (typically an electrical generator). The hot exhaust gases also may be utilized to produce steam. Steam may be used to drive a steam turbine and generator, or the steam may be used in other chemical processes.

10.3.2.2 Electrical Rotating Machinery

Electrical rotating machinery includes both generators and motors, which may be Alternating Current (AC) and Direct Current (DC) types. These range in size from tiny machines to those fitted in the largest power stations, pumping or compression applications. The larger machines will have their

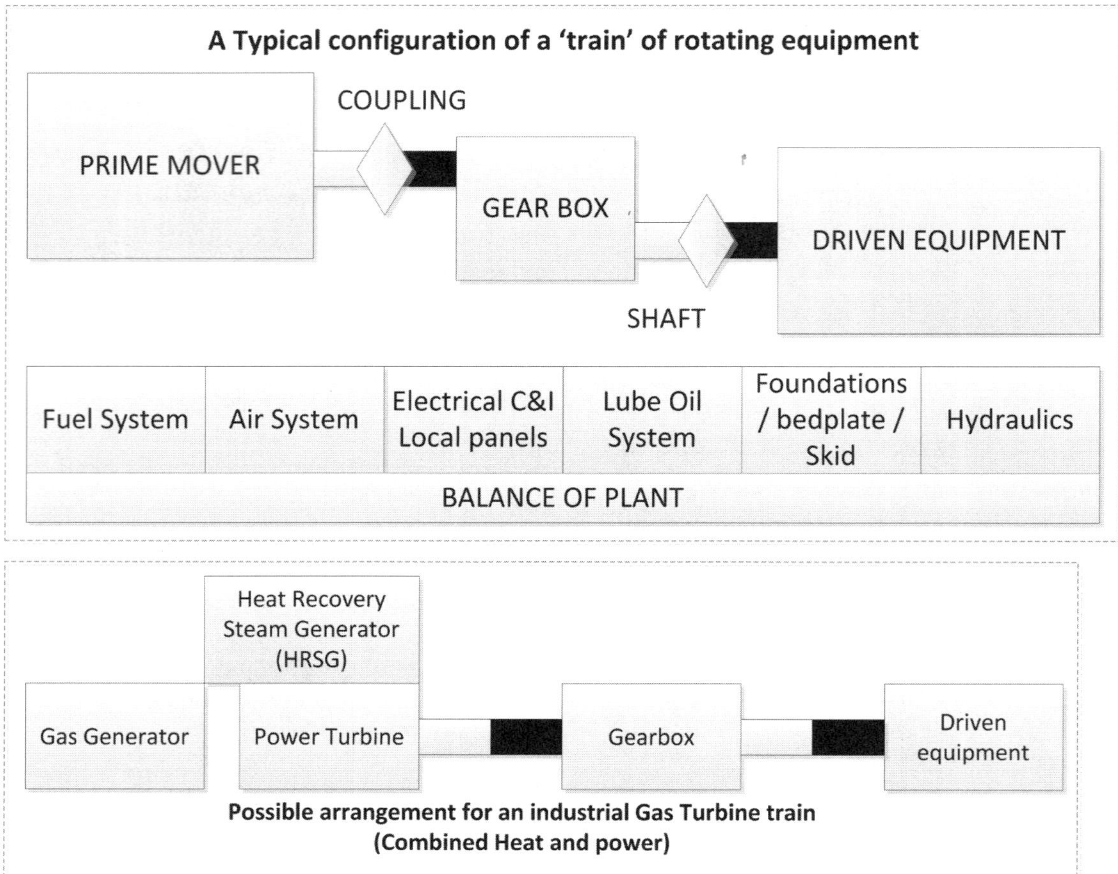

Figure 10.3.1 Typical configurations of rotating machinery trains.

own cooling and lubrication systems. AC machines may be synchronous (in which the rotational speed of the rotor matches the speed of the induced rotating magnetic field), otherwise the rotor speed exhibits "slip," dependent on load. A larger machine will have bearings (smaller machines, rolling element; larger ones, journal bearings), which are likely to have localized lube oil temperature sensors, as well as vibration sensors (usually accelerometers for rolling element bearings and proximity probes for journal bearings). Large machines often have current sensors (clip-on transformers) and voltage sensors, which provide the parameters necessary for effective IVHM. Power factor and correction are present in AC electrical systems.

10.3.2.3 Reciprocating, Centrifugal, Axial Engines, and Compressors

Reciprocating machines will include internal combustion engines (prime movers; see Figure 10.3.1), reciprocating compressors (driven equipment), and hydraulic motors or pumps. The critical functions will focus on the timing of fuel injection and combustion for engines, and pressure containment within the cylinders, along with timing of the inlet and discharge cylinder valves. Dynamic pressure sensing, head temperatures, and crank deflections; comparison of inlet temperatures, pressures, and compressibility to outlet conditions; and normalizing for any cooling are all good indicators of performance.

10.3.2.4 Gearboxes, Couplings, and Shafts

Most equipment trains are fitted with a gearbox and couplings. Gearbox monitoring is most often delivered by oil and debris analysis, vibration analysis, and inspection. If it is possible, continuous transient or dynamic monitoring of shaft torque to spot the incidence of overload and shock load condition, which may initiate damage to the gear teeth, should be carried out. Couplings may absorb instantaneous shock, but they provide a means by which machinery trains may be separated for change or major overhaul. Couplings may be clutch-type or liquid. Alignment of machinery in a shaft is critical, and misalignment is often caused by bad maintenance.

10.3.2.5 Bearings, Rolling Element, Journal, and Pad

All rotating machinery will have bearings to reduce friction in rotation. The bearings fall into three types: rolling element, journal, and pad. Although oil, debris, and vibration are effective parameters for monitoring all of these types, the features and symptoms will be different. For rolling element bearings, oil debris and spectral analysis of tracked orders are particularly effective. For journal and pad bearings, analysis of orbital and phase vibration, along with determining suspended metals in oil, are more relevant.

Pad bearings are often used for thrust bearings; individual pads are often fitted with temperature sensors, and the pads may also have wear indicators with which wear rates-of-change should be trended.

10.3.2.6 Seals and Glands

Seals and glands are essential to deliver the "containment" functional requirement for many rotating machines, as well as to ensure that the efficiency of some other machines are maintained. Containment functional failure is often safety critical, because it implies the release of stored energy, or hazardous liquids or gases. An example is a dry gas seal for a centrifugal compressor, in which the contained gas includes hydrogen sulfide in a high-pressure and -temperature environment. Most seals or glands have design leakage, which may have associated instrumentation (differential pressure, or flow cells) that enables the leakage rate to be monitored. Some seals are supplemented by the supply of pressurized/vacuum sections to ensure that migration of contained fluids are prevented. A health-monitoring system can monitor the flows and pressures normalized by the pressure of the fluid being sealed to determine that the seals are functioning correctly. An example would be a series of labyrinth seals on a high-pressure steam turbine, alternatively supplied with steam and vacuum (connected to the condenser). See Figure 10.3.2.

10.3.3 Rotating Machinery: Current Technology for IVHM

Some common IVHM techniques used for rotation machinery include

- **Performance analysis:** monitoring physical parameters such as temperatures, pressures, and flows; and for electrical machinery voltage, current, phase angles (for AC), and magnetic fields. In aeronautics, static engine performance analysis involves trending various performance parameters (temperatures, pressures, and flows) throughout the gas-path and the engine's environment, in distinct machine states such as "take off," "climb," "cruise," and "descent." Discrimination of machine state reduces data scatter and ensures

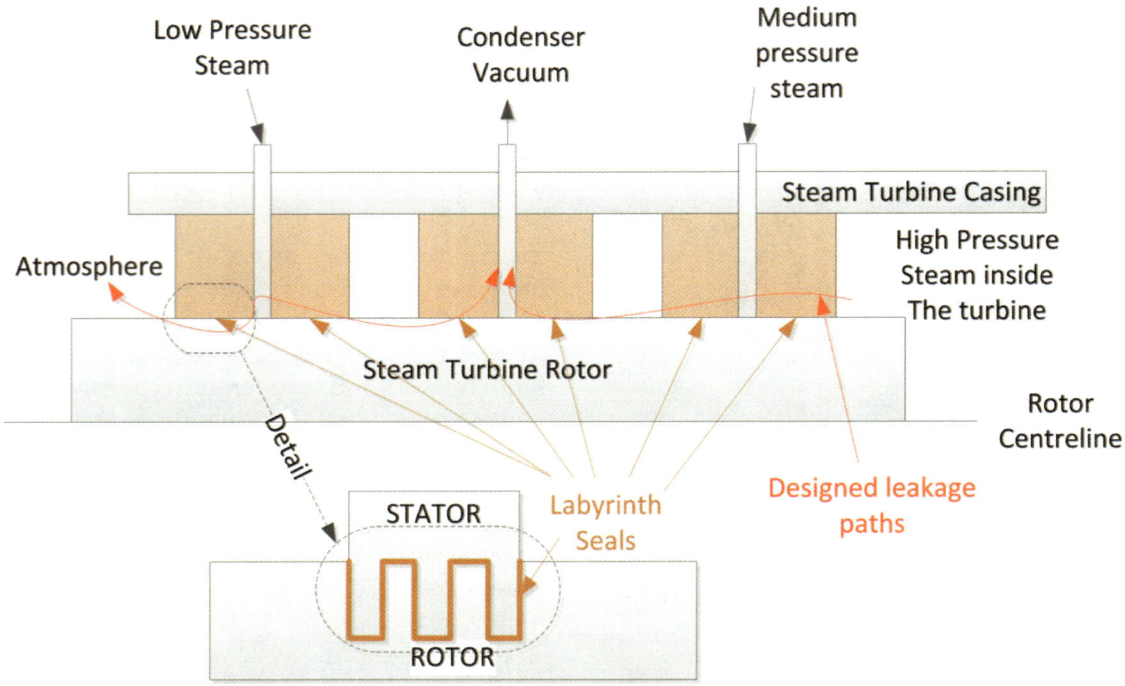

Figure 10.3.2 Typical labyrinth seal arrangement for a steam turbine.

greater certainty in detecting degradation, which may be more evident in these states compared with others.

- **Vibration analysis:** monitoring vibrations emanating from running machinery; conducting phase, balance, and orbital analysis [Bently 2002].
- **Oil analysis:** monitoring the lubricity, viscosity, contaminants, and additives (such as anti-oxidants) in the oil.
- **Oil debris analysis:** monitoring suspended particulates suspended in oil or grease, for size and chemical composition.
- **Thermography:** monitoring the heat of a rotating machine to identify areas where overheating may be occurring. Thermography is often used in monitoring static electrical machinery, and is very useful in spotting high-resistance electrical connectors in rotating machinery terminal boxes.

Each of the techniques also have differing data sampling regimes that are determined by the type of failure signatures required to be observed.

A typical system schematic [Vachtsevanos et al. 2006] for IVHM that is associated with rotating machinery, and which enables the techniques just discussed, is shown in Figure 10.3.3.

The Failure Recording, Analysis, and Corrective Action System (FRACAS) in Figure 10.3.3 is a well-known specification developed in the military domain [MIL-STD-2155, 1985] for recording and analyzing failure or deficiency reports necessary to derive reliability metrics. FRACAS may also be associated with and support Root Cause Analysis (RCA). A description of typical sensors fitted to rotating machinery is included in the preceding paragraphs. Sensors are connected to a data bus, of which there are many industrial

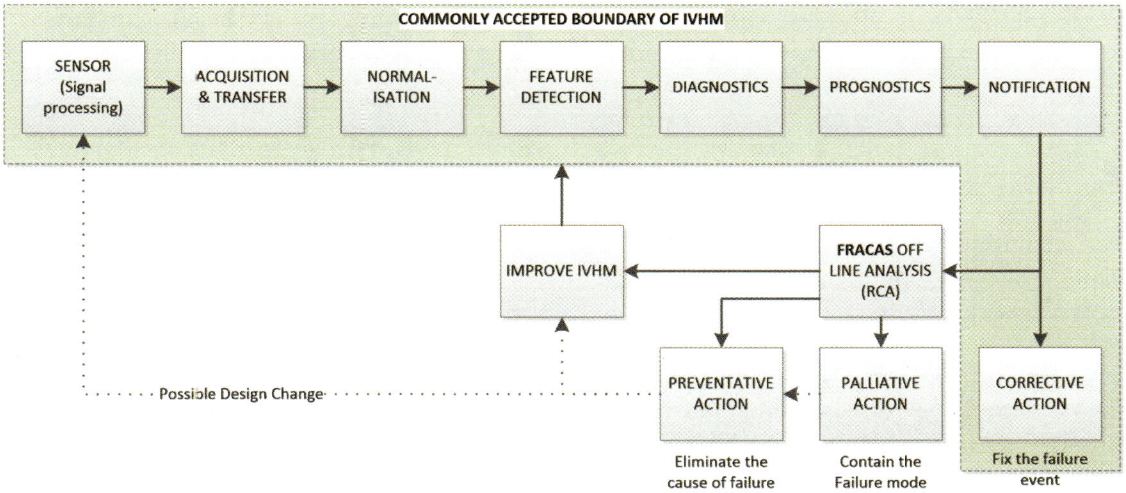

Figure 10.3.3 Typical flow through an IVHM system and the use of output data.

examples. Sensors are fitted to supply protection, control, and an operator. It is up to the engine control unit to provide most of the real-time parameters, which may be used by a centralized monitoring system for analysis, and timely detection of abnormal degradation or incipient failure.

10.3.3.1 Dynamic Analysis

Dynamic analysis typically samples data in the tens of kilohertz ranges. This is typically used for vibration analysis but may also be used for performance and electrical analysis. Examples include dynamically sampling pressure in a gas turbine to detect fan flutter, or with electrical current, which may also be used to measure vibration.

The frequency of sampling is driven by Shannon-Nyquist, in which observable "features" in the data need to be sampled at least twice over the features' underlying period.

10.3.3.2 Transient Analysis

Transient analysis typically samples data in the one to ten hertz range. This analysis is conducted when the underlying machine is in a state of change, and has not yet steadied out, or heat flux soaked to a new equilibrium. It may take many minutes for a machine to reach a steady state after a major power or speed change. The transient data sampling rate is governed by Shannon-Nyquist. A number of failure modes may only occur, or are observed with greater certainty, when a machine is in the transient state. An example may be monitoring an induction motor start-up torque to determine the heath of the squirrel cage rotor bars. Machine start-up analysis is a common use of transient state data.

10.3.3.3 Static Analysis

Static analysis is performed by sampling data at a periodic rate that is determined by the typical PF interval (Potential to Functional failure time or interval) from when a machine may be diagnosed to be in the potentially failed state. The sampling rate must be a fraction of the PF, allowing for any delays from sampling to analysis (for example, it may take some time from collecting a magnetic chip detector sample of debris, and getting that sample to a remote laboratory that will conduct the analysis).

Static analysis is the simplest approach, but care must be taken to ensure that data fed into static analysis is data sampled during the static state. Transient state data will introduce error, which is likely to increase false positive diagnosis [Nabney 2002].

Permanently fitted data acquisition systems, capable of continual monitoring, may sample data at a set rate (often once an hour). This is often the case for performance monitoring systems. Other systems take snapshots of data at repeatable points within a machine mission profile (such as a gas turbine aircraft engine, at take off, climb, cruise, descent, etc.) to minimize scatter in the data.

Static analysis is the most common approach taken and is easier to conduct compared with transient analysis.

10.3.3.4 Tribology Applied to Rotating Machinery

According to Wikipedia [Wikipedia 2013], "Tribology is the science and engineering of interacting surfaces in relative motion." Oil debris and oil analysis are two techniques that yield information about the failure involved with interacting surfaces in relative motion. Debris analysis can be as simple as periodically observing ferrous particles on a magnetic chip detector placed in line with a lube oil line; through having online particle counters; up to performing laboratory analysis of shape, size, and morphology of the debris materials, which may be used to isolate the source, failure mechanisms, and remaining useful life associated with the degrading condition. The three approaches each deliver an increasing amount of useful information within the IVHM system. Laboratory debris analysis is a very effective means of monitoring rolling element bearings, but it has a disadvantage of requiring maintenance effort to retrieve and send the debris samples to a laboratory for analysis. The periodicity of sampling and the inherent delay times for sending the samples must be a fraction of the natural PF interval of the failure mode being monitored.

Similarly, oil analysis can range from simple local checks of color, crackle tests for water content, and simple viscosity checks, to far more sophisticated laboratory analyses in which Total Acidity Number (TAN) indicates the consumption of anti-oxidization additives in the oil, and identification of suspended metal in the oil provides effective wear detection of pad or journal bearings.

10.3.3.5 State-Based Prognostics

The fusion of tribology analysis with other techniques in the IVHM system may also show changes of state in a degrading system. Consider a rolling element bearing failure in Figure 10.3.4.

Oil debris and oil laboratory analysis has the sensitivity to detect the earliest signs of bearing failure (when race cracking surfaces and spalling starts) at particle sizes of 5 to 10 μm. As the failure progresses, spectral vibration analysis picks up increases in vibration energy at discrete frequencies associated with the rolling element passing rate (which may be determined from the shaft speeds). Finally, as the failure progresses toward the functional failure point, performance monitoring may indicate that gross vibration is increasing, lube oil temperature rises, and the machine may lose marginal efficiency. Thermography may pick up hot spots in the bearing housing. As these signs are picked up in sequence, prognostics may be determined. Fusion of several different techniques in an IVHM system increases the certainty of diagnostics and delivers usable prognostics.

10.3.3.6 Reusing Digital Control System Data

Often, digital control systems produce data that is of value in an IVHM system. These data include

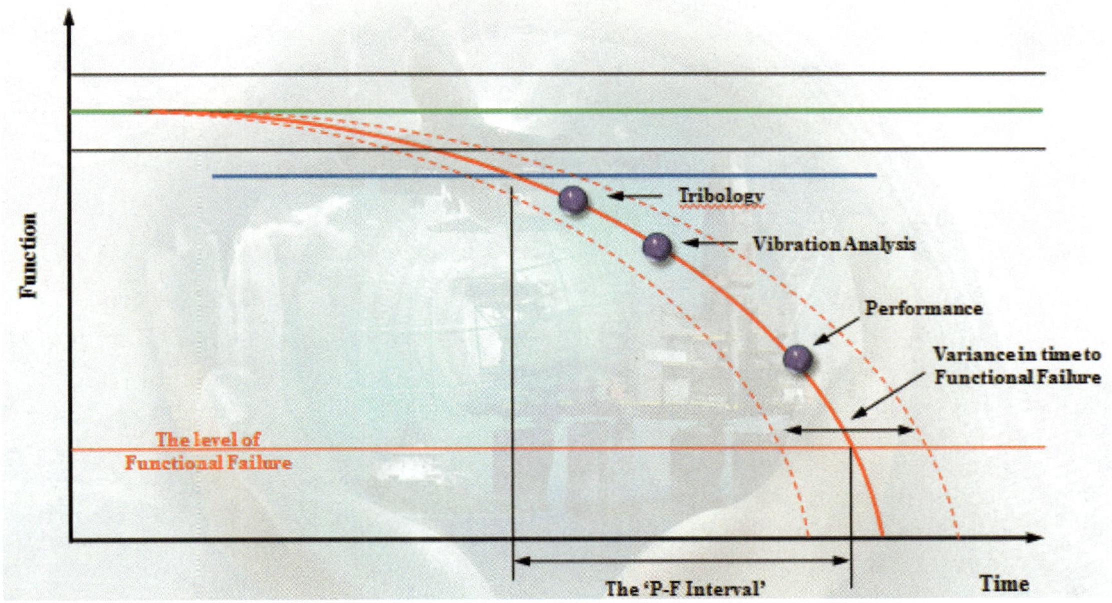

Figure 10.3.4 Fusing techniques in an IVHM system to deliver state-based prognostics.

- State data. Often, control system algorithms are built on a "state engine" and produce a readable "state flag." This flag data is potentially useful in helping distinguish when a machine is in a transient or steady state, which may be used to change the rate of data sampling.
- State data that is used to indicate that the control system may vary its control laws, which may affect "normality." These changes may be required to change how "normalization" works (see Figure 10.3.3). It may be that a different model of normality is used for differing machine states, and the state flags may trigger the use of the various normalization models.
- Error codes (or BITE codes). BITE (Built-In Test Equipment) codes are indicators of when machines breach internal thresholds, which indicate a fault or degraded condition. IVHM systems are usually designed to sense the onset of conditions before BITE or error codes, but these are still a useful source of "features" that may be used to augment IVHM diagnostics.

10.3.4 Emerging Technologies for Rotating Machines and Their Impact on IVHM

10.3.4.1 Hybrid and Full Ceramic Bearings

Hybrid ceramic bearings is an emerging technology that uses silicon nitride ceramics instead of hardened steel for the balls/rollers in rolling element bearings. The use of ceramics presents advantages:

- They may be used in high-temperature environments, with a low coefficient of thermal expansion.
- They are "harder," can withstand higher loads, are less prone to wear, and have a high "fracture toughness."
- They are not subject to corrosion.
- The material is lighter than steel.

- The lower compressiveness means friction is reduced by approximately 80% compared with steel bearings.
- They are electrical insulators, which largely avoids a bearing failure mode due to micro-arcing discharges occurring through thin oil films causing fluting on the steel ball and races surfaces.

The overall effect of using ceramics is to reduce size and increase the reliability of the bearings. The onset of failure for hybrid bearings would be to monitor for conventional signs of failure in the ball races:

- Constant monitoring of load at a transient rate of data sampling may detect shock or overload events that increase the probability of the emergence of a potential failure.
- Use acoustics to isolate signs of internal cracking (this technology is not yet mature).
- Once cracks are apparent at the surface of the races, spalling will occur, and oil debris analysis may be used to detect particles.
- As spalls grow in size, their effects may be observed in the vibration signals, where peaks of energy may be observed in vibration spectral analysis coincident at the ball passing rate.

3. Finally, in the failure life cycle, performance parameters will start to indicate bearing problems. Temperature and machine efficiency will start to change; also, if gross peak vibration across the whole spectrum is available (this parameter is considered a performance parameter) and shows a ramp increase over time, then these three symptoms may be used to diagnose bearing failure. At this point in the failure life cycle, it is very close to functional failure at which the bearing may seize or disintegrate, possibly causing extensive secondary damage to shafts or shock-loaded gearboxes.

The process of failures of full ceramic bearings is problematic, as it is highly likely that the propagation of cracking may be far faster than for steel. This may yield very little useful PF interval, and IVHM techniques may not be applicable.

10.3.5 Development

10.3.5.1 Automating Nondestructive Examination: the Application of Robotics

An aspect of on-condition monitoring is inspection, and in particular, NDE, which is predominantly applied to static machinery. NDE is also applied to rotating machinery; a well-known example is borescope examination. At present, borescoping is conducted on a scheduled basis to confirm design degradation; by exception to troubleshoot and isolate a fault condition; or to determine damage after a shock event, such as Foreign Object Damage (FOD) in a gas turbine. The image data may be analyzed immediately by the borescope operator, or transmitted to a number of experts who interpret the image. Manual intervention to conduct borescoping is relatively expensive. It would be valuable to permanently fit borescopes and have them gather data automatically over a wide range of frequencies (even using radar for measuring geometric changes caused by erosion). The borescope sensor may also be autonomously articulated robotically to gain a wider range of sample data. The image or dimensional data may then be automatically processed; and features that may be trended, extracted. The trend-able features may be fused with the IVHM system (there is nothing preventing the IVHM system from extracting the features) to provide a richer set of data when fused from other sensors, to improve the granularity and certainty of diagnosis and prognosis.

IVHM is about applying health and condition monitoring techniques: using sensor data, non-intrusively to the machinery (thereby avoiding being a possible cause of failure); using a high degree of automation (to reduce costs and improve scalability); and delivering highly reliable outcomes to avoid the consequences of machinery functional failure. The added dimension that IVHM adds to traditional rotating machinery health-monitoring techniques is in fusion. Fusing the commonly applied techniques described in this chapter, and fusing the monitoring and management of rotating machinery up to the platform level.

10.3.6 References

Bently, D. 2002. *Fundamentals of Rotating Machinery Diagnostics*, Bently Pressurized Bearing Press.

MIL-STD-2155. 1985. http://src.alionscience.com/pdf/MIL-STD-2155.pdf, Accessed Feb. 5, 2013.

Nabney, I. 2002. *Advances in Pattern Recognition*, Springer-Verlag.

Vachtsevanos, G., F. Lewis, M. Roemer, A. Hess, and B. Wu. 2006. *Intelligent Fault Diagnosis and Prognosis for Engineering Systems*, Wiley.

Wikipedia. 2013. http://en.wikipedia.org/wiki/Tribology, Accessed Feb. 5, 2013.

10.4 The Application of IVHM in Motorsport

Keith Jackson, Meggitt

"Any car which holds together for more than a race is too heavy"

—Colin Chapman, founder of Lotus Cars

10.4.1 Introduction

Most people working in the motorsport industry would not recognize the acronym of IVHM as part of their skill set. However, if you mentioned processing real-time data and engineering data, modeling and simulation systems, and diagnostic and predictive capabilities, all applied to improve the team's performance—lap to lap, race to race, and season to season to achieve their business aims for their sponsors—they would instantly identify with the discipline. Motorsport today is a highly professional sport. Generating, collecting, understanding, fusing, and using engineering data on a march to improve short-, medium-, and long-term performance; and manage and predict events as they arise, are essential skills to achieving the consistent and best results for a team.

Motorsports teams use data from multiple sources including design, simulation, testing, wind tunnels, and track data, which can be fused and processed to reduce weight, improve designs, improve models, plan race strategies, and determine tactics. Engineering data are not uniquely considered for health management. Instead, health management and IVHM is an embedded part of the total data management strategy.

For practitioners of IVHM in other industries, it is worth taking a look at the motorsport industry for initiatives and ideas; however, it must be kept in mind that motorsport operates to different business rules than any other industry and is governed by rules to make the sport exciting and to manage financial commitments. For instance, in Formula 1, telemetry communications can only operate in one direction from the car to the pit (this rule is mandated to inhibit creative uses of the technology), while it is certainly not explicitly prohibited in other industries [FIA 2012]. Because motorsport is contained and controlled, yet teams are always seeking areas of improvement, it is an industry in which new ideas are regularly tried, without strong barriers to technology introduction. Equally, if a technology is proven not to bring value, it is quickly discarded. Consequently,

Figure 10.4.1 The author at Donnington Park GP Circuit, Leicestershire, UK. "Even club and track day racers make use of advanced monitoring systems—you can see the HD video system in the windscreen, gathering context information, which is linked to a data logging system."

motorsport enjoys short development cycles that make other industries shudder; IVHM is a critical element of the data gathering, analysis, and action-taking, which enables these fast development cycles to be achieved.

Since the early 1990s, the top level of motorsport has become increasingly well structured and disciplined, and the days of inspired engineering, tested on the track with a stop watch, have moved to informed engineering. Now, designs are simulated and modeled; options are traded against each other; wind tunnels are used to validate the aerodynamics; and ultimately data, acquired from the track during testing, practice, qualifying, and the actual race, are used to validate and improve models and engineering knowledge.

The constant increase in the power of electronics and software means that the data systems used in a clubman's race car today would have been acceptable to a Formula 1 team 20 years ago. Indeed, much of the challenge for a team is to use the data that has been logged rather than the acquisition of the data itself. This is not dissimilar to other industries: it is easy to collect operating data, not too hard even to find anomalies in the data that indicate "something," but the real skill is working out what the "something" (or diagnostic) is and being able to act upon the information.

Today a clubman's race car can be equipped with a data logging system that connects into the engine and gearbox data buses and fuses their output with inertial, GPS, and sensor data (see Figure 10.4.1). When this data is off-loaded, an analysis package not only shows raw data (lap times, speeds, temperature, pressures, and track positions) but also includes a degree of intelligence to direct the users through the wealth of data to show where performance can be improved. Example information includes

- A track map, created from the speed, GPS, and inertial systems. This shows the "line" the car has driven (to varying degrees of accuracy) but not the track "ribbon" itself. However, it is useful to set context and to look at the performance of segments of the track, etc.
- Pen traces, which may be with respect to lap position and time. The most obvious traces are speed, rpm, gear, *g*-force, pedal actuation, and steering angle.
- Statistical information. Studying a race full of pen traces is not always effective, and trends or relationships may not jump out from the screen. Analysis packages have the ability to join channels using mathematical functions, which can then be plotted and displayed using a variety of statistical and display techniques.
- Alarms, faults, exceedances, and watching simple channels or derived channels for faults. Obvious examples include temperatures, pressures, fluid levels, engine overspeed and tire pressures. Note that faults may not be permanent, and so

"backing off" for a few laps may allow recovery, particularly of thermal issues.

- Comparisons. Comparing lap to lap, driver to driver, change to change, and day to day all enable the team to find the best set-up for a given situation.

Just as with other IVHM applications, the raw unprocessed data from the equipment is not very useful; it needs to be processed, converted to engineering data, cleaned, and combined with other data sources to produce useful and actionable information. As stated before, IVHM or health management data and actions are not distinct from other management information or activities, they are entwined in the total system.

A top league race car's systems (say, Formula 1) extend the basic capabilities of a lower formula car. Data capacity rises, the number of data channels (individual measurements) increases, the sophistication of the sensors becomes greater, and the ability to process the data increases: this is enabled by the technology but is primarily down to the team of engineers studying the data supplied to them. They study screens of raw and processed data, on the pit wall, in the garage, and even back at the factory, each with a specific task: race strategy, setup, lifing, design, etc.

Though it is not universal to top league motorsport, one of the major technology capabilities is the use of real-time telemetry systems that allow data from the car to be sent in real time to the pit wall.

While the width of the data pipe for telemetry may vary (with location, by regulation, or by affordability), it is always true that not all of the data being captured on the car can be sent to the pit wall in real time. However, not all of the information is useful in real time. The data acquisition systems can therefore be configured to sample and transmit data as required to make best use of the limited data pipe (but great care must be taken not to filter out critical data spikes, which may be essential diagnostics). Triggers can be set, sampling frequencies changed, data compressed, and derived or processed data produced. For instance, sampling four car corners of ride height at 100 Hz for the entire lap may not be the best use of the data pipe. Sampling the four corners of ride height for 30 m before each corner, around a corner, and through the exit is likely to be more interesting to send to the pit in real time. Once on a straight, a lower sampling rate will mostly be adequate. However, there will always be exceptions, and maybe 100-Hz sampling is useful on the straight to understand design changes in testing. To enable this, smart data logging systems are easy to configure, but not when the car is moving on the circuit (because of rule restrictions).

This is similar to other IVHM examples in which it is important to (a) sample what is useful and avoid a flood of data, and (b) ensure that the correct data reaches the correct user, in a timely manner. Algorithms are available for reducing the size of the data if there is too much data to evaluate without losing the behavior of interest. Algorithms are also available to create virtual data when insufficient engineering data exists; it will fill in the places where data is missing to create a baseline behavior from which an anomaly can be identified.

Sensing technology also becomes more sophisticated in higher leagues of racing. Strain gauges on suspension and structures to look at loads have been used for many years. These become particularly important where conventional suspension components (dampers and wishbones moving around pivots) are replaced with very rigid suspension structures and push rods, which operate through cams to remote springs and dampers in which the resulting suspension movements are small, even when resulting from very high forces. The suspension systems are also greatly adjustable, not just for the spring and damper rates but also for the geometry (wheel toe angle, run out, and camber). Therefore, to fully optimize the suspension, the strain gauge information is combined with input from other sensors (particularly laser ride height that measures the height of each corner of the car above the track) to form a detailed view of the car's expected and actual performance.

Of equal interest to ride height is slip angle. In greatly simplified terms, whenever (any) car turns, the steering wheels are turned in the requested direction of travel, but the actual turning of the car will normally be slightly less than the requested turning angle. This is known as the slip angle, the delta between the requested and the actual turning angle. The effect is reported as "understeer" by the driver—the car not turning by as much as expected. Generally, in normal road usage, the understeer is not detected by the driver. However, the higher the speed, the greater the steering angle, and the more the slip increases. In fact, with a rear-wheel-drive sports car, it is easy for the slip angle to be so great that additional lock on the steering has no effect (it just adds to the slip angle). If the chassis is unsettled (by driver action, surface change, or other factors), the opposite effect of oversteer (where the back of the car comes around) can be experienced. Understanding slip angle is a critical performance parameter, as a car that is sliding is slowed more than one that is not; a car that is too twitchy unsettles the driver; and a car that is too forgiving (as most road cars have to be) is not agile enough for racing [Milliken and Milliken 1994].

As seen throughout IVHM applications, a great deal of use can come from basic sensors and processing, but there are often some problems that need highly specialized sensors. This is better illustrated in the next few paragraphs.

A phrase used in racing is "keep it on the black stuff," meaning keep it on the track and not in the gravel, grass, or barriers. Most of the forces created by the car are transmitted through the tires, so an understanding of tires is critical. Performance data for tires is guarded by the tire manufacturers and its distribution restricted. The actual performance is influenced by the operating context (forces—x,y, and z axis and geometry) and the environment (track temperature, surface type, tire temperature, and pressure). Modeling of tires (e.g., Pacejka models) can provide a predicative indicator, which then has to be validated against the actual performance experienced on the track, and the final model is complex. Consider a complex corner with a change in elevation, where the entry is a descent, the apex is level, and the exit is uphill. The x, y, and z forces are all changing, and if the driver takes a different line, this will affect the force combination and sequence. Therefore, it is not a surprise that race cars can be instrumented with pressure and temperature sensing systems. Ultimately, the engineer wants to understand the "contact patch" (the part of the tire touching the track) and the limits of the forces that can be put through it—which keep the car on the black stuff [Pacejka 2012].

Tire performance is clearly a critical element, but is complex in its makeup and thus difficult to measure. It is also guarded by the tire manufacturers and understandably affected by environmental and operational factors, which make it difficult to perfect the models. This is not unlike the situation with complex items of equipment (say a gas turbine) in which the basic principles are understood by many, but the ability to extract useful information with the fidelity required for an accurate monitoring system is often largely restricted to the OEMs.

It is well understood that in IVHM systems, environmental factors have to be normalized and compensations made to obtain accurate and sensitive measurements from the data. This is equally true for motorsport, in which the weather, driver, and operating context have major effects on the data being sampled from the car. A good driver, even in the lower leagues of motorsport, is a highly skilled and repetitive machine. Given the same conditions, drivers will lap within 1% of the same time; shift gear, brake, and accelerate at almost identical points; and turn in and apex on corners, on their chosen race line, within a fraction of a second of repeatability and a few centimeters of positional accuracy.

However, under real-world conditions, not everything remains constant. First, race cars literally guzzle fuel, and so the associated weight reduction means that a car with a low fuel load will accelerate faster and, just as importantly, slow down more quickly and corner faster than a fully fueled car. As discussed before, tire condition and behavior also changes significantly during the race, depending on the conditions. Heat absorption by the tires (and brakes) improves the grip, and eventually they reach peak performance. But, tires also wear, overheat, and can become damaged through their lives.

The weather is also a major influence: a wet track offers less grip than a dry track, and a cold track offers less grip than a hot track; getting off the racing line can feel like moving from super-grip to ice. Touching a painted white line, which is wet, can be enough to lose a car's grip, and it will exit the track sharply.

Hence, environmental effects are very significant and must either be normalized from the data or data samples deleted to avoid building an unrepresentative model of the car's performance.

Another variable (of this inexhaustive list) is the race itself. Lapping fast on an empty track is one thing, whereas responding to the challenges of passing other cars; avoiding being passed; holding position against a car with less fuel, better tires, better driver, or different aero package and getting the best for the whole race is a planning challenge. It may be better to be slower at the start of a race, with a heavy fuel load, maybe losing places, but then recovering them later in the race when the remaining extra fuel can be used to advantage, and attrition may have thinned the field of competitors. Here, scenario and game playing come into effect to determine the best race strategy of when to come into the pits, when to change tires, and when to attempt to recover position. This is all with the understanding that the other teams are trying to do the same thing and predict your actions in a competitive manner. Today, some teams have unique "codes" that help the game strategy decision-making process. Maybe, one day, the motorsport industry will contribute an example to game theory that is as famous as *Prisoner's Dilemma: John Von Neumann, Game Theory and the Puzzle of the Bomb* [Poundstone 1993].

Given this background into the use of IVHM in motorsport, it can be seen that it represents a good, high-speed example of an IVHM system and, in particular, the system is used for much more than just condition and fault monitoring. Explicitly, IVHM in motorsport is probably one of the broadest and most complete applications of the discipline. The remaining paragraphs of this chapter look at the major IVHM disciplines exercised in motorsport.

10.4.2 Condition Monitoring

This is the most traditional application of IVHM. For many people, IVHM is condition monitoring applied to a whole vehicle. The control and monitoring systems in a race car will warn the driver and the team (via telemetry) of a simple-to-detect fault (e.g., overheat, low levels, punctured tire). The systems are also extended to provide predictive indicators or trends (for instance, the ramping thermal trend). Even in these circumstances, setting fixed limits could be inappropriate, as temperatures will rise with ambient temperature changes or the driving conditions, varying between open track and in traffic where the aerodynamics will change. Consequently, operating context information is used to qualify warnings.

10.4.3 Lifing

Race cars may be rebuilt after major races or have components changed at will by maintenance crews but, increasingly, choices are regulated and locked down (for instance, even in a modest race series an engine may be "sealed" and marked to avoid changes). Lifing to work out when engines, suspension, and other components need to be replaced avoids wasting residual life in components and manages the risk of in-use failures. Many teams are budget constrained and may choose to take the risk of performance loss and manage the rebuild and replacement program during the season. This is a common technique enabled by IVHM (i.e., manage life against risk of failure, which may have primary and secondary consequences).

10.4.4 Set-up

A race car will be optimized at a race for the conditions, the condition of the vehicle, and the expected conditions on race day. During practice and qualification, the suspension, tire pressures, and aerodynamics of the car will be adjusted to maximize the desired performance. In simple cases, it will just be a case of exploring that option A is faster than option B, but it may also be used to validate the simulation models with live data of a track for its conditions. Normally, a car's set-up is a trade-off between options. Straight line speed vs. corner speed (aerodynamics), maximum speed vs. acceleration (gear ratios), tire wear vs. grip (suspension geometry), and suitability of the car for the driver. The optimum set-up is the best set of compromises for the driver and the car on the day.

10.4.5 Driver Training

In common with other industries where IVHM is applied, the effectiveness of using IVHM data to train and improve a driver varies with the driver's willingness to accept the information being provided. The now widespread use of video, audio, and GPS context data along with the sensed data from the car itself and the ability to compare lap to lap, set-up to set-up, and driver to driver, makes it harder to deny the data when it is presented appropriately.

10.4.6 Design

The results on the race track are the ultimate feedback for the design engineers for insight on the effectiveness of their designs. In the design process, they will have used Computational Fluid Dynamics (CFD) codes, suspension models, stiffness models, performance models, tire models, and an experience base to design the car for which they will have an expected performance. The models are never perfect, and the race data showing actual performance against predicted performance provides critical information to improve both the car's design, and the design information and correlations between simulated, wind tunnel, and actual measures.

10.4.7 Strategy

Motorsport is, by definition, competitive and having the fastest car with the best driver will not give consistent results if a weak strategy is applied to the race. Today, the team manager and their ensemble of race engineers use the live data feeds, weather forecasts, and even knowledge of the activities of other pit crews to determine how to adapt their race strategy (e.g., should you try to overtake someone who is coming in to the pit's next lap; can your teammate lose time holding up a competitor so you rejoin the track in front of them; how much life do you want to take out of components at this stage of a race or the season of racing, etc.?) Today, race strategy is a mental process, but the increase in use of gaming theory in business and in commerce could see motorsport adopt and develop the techniques for its application.

10.4.8 Summary and Conclusions

This brief description of the application of IVHM in motorsport (even if motorsport professionals would not call it by that name) shows some very clear examples of what it takes to gain maximum advantage from the capability of a complex system. Practitioners of IVHM may want to consider these factors when considering their IVHM systems and applications:

- Rarely can a business case be made for an IVHM system if the data is just used for condition monitoring, and even if it can, there may be greater value in collecting IVHM data for other purposes (lifing, design improvement, game strategy).
- Environmental and context factors will influence what is sensed from a system being monitored. It may be possible to normalize the data mathematically, or data may have to be discarded where the operating context is unusual. Temperatures rising when a race car is following another car may not be an indication of a problem.
- Even with the ability to collect, transmit, and store vast quantities of data, finding areas of interest in the data can be an exhaustive search. Thinking about the sampling, compression, and analysis in advance is likely to increase the effectiveness of the system greatly.
- Decision-making as to what actions to take on an asset is often impossible if only the single asset is considered. For a race car, the strategy for tire changes and pit stops will depend on the other car in the team, environmental factors, and what the other teams are doing. Decision-making requires an operating and business context that is unknown by the car itself or even the driver.
- Splitting the functionality between the onboard (on-car) and offboard (on pit, in garage, in factory, in design) and making sure each recipient gets the right data, in the right format, in a timely manner, enables effectiveness. The team manager on the pit wall needs real-time telemetry, whereas the design engineers need more data which may be off-loaded at the end of the race.
- Even though numerical methods and models continually improve, there is still reliance on a person to make the final executive decisions. Maybe as gaming theory, knowledge bases, and expert systems become more advanced, this will become less of the case. However, today people and experts are still in the loop.
- Using the sampled IVHM data to validate, correlate, and improve expectation models and the engineering process creates value.
- Highly sophisticated sensors can produce data that are very valuable, but considerable value and information can also be created from simple data sets. In other words, measuring times for a car over the same segment of a circuit, with accuracy, is easy to accomplish, understand, and use. Gathering slip angle information is more complex and requires sophisticated sensors. A rookie team, even if armed with slip data, may struggle to use it effectively.
- The information is only useful if the recipients are prepared not only to believe it but also to use it. This means presenting it in a format that is understandable and acceptable, that it is correct, and that judgement is exercised in understanding the data—which is why so many people and racks of equipment are still required to process the data.

Motorsport is a somewhat unusual industry, and the insight into how it operates and exploits IVHM, as well as the pace at which new and valuable technologies can be developed to suit the needs of the industry, make it a useful reference for other industries.

10.4.9 References

FIA. 2012. Formula One Technical Regulations: Section 8.5. http://www.fia.com/sport/regulations.

Milliken, William F., and Douglas L. Milliken. 1994. *Race Car Vehicle Dynamics*, http://books.sae.org/book-r-146/, SAE International, Warrendale, PA.

Pacejka, Hans B. 2012. *Tire and Vehicle Dynamics*, http://books.sae.org/book-r-418/, SAE International, Warrendale, PA.

Poundstone, William. 1993. *Prisoner's Dilemma: John Von Neumann, Game Theory and the Puzzle of the Bomb*, Anchor Books.

10.5 Prognostics and Health Management for Wind Turbines

Ranjith Kumar, Sreenilayam Raveendran, Michael H. Azarian, and Michael Pecht, Center for Advanced Life Cycle Engineering, Univ. of Maryland

“But to truly transform our economy, protect our security, and save our planet from the ravages of climate change, we need to ultimately make clean, renewable energy the profitable kind of energy.”

—President Barack Obama

10.5.1 Introduction

Wind energy is a rapidly growing renewable energy source and is an alternative to fossil fuels for power generation. Wind energy is harnessed by means of wind turbines, which capture the energy in the wind that is generated by pressure differences arising from the uneven heating of the earth's surface by solar radiation. This energy capture is achieved by the rotation of an aerodynamic rotor, which converts wind energy into rotational energy. The rotational energy is converted into electrical energy by means of an electrical generator. Since the generation of electrical energy by a wind turbine depends on the wind flow occurring in its vicinity, the electrical power output of a wind turbine inherently fluctuates. Hence, wind turbines are located in areas with near-constant flow of non-turbulent air throughout the year.

It is estimated that the global wind energy potential is between 20,000 Tera Watt hours (TWh) and 50,000 TWh of electricity each year, and that 8 to 25% of total global electricity demand by 2030 will be delivered by wind energy [India 2012]. The driving force for adoption of wind energy includes reduced dependence on imported fuel, reduction in CO_2 emissions resulting in improved air quality, and economic development as a result of increased investments and job growth. In 2011, wind turbines generated 500 TWh of electricity, which constituted 3% of the total global electricity demand. The turnover of the wind sector worldwide reached U.S. $65 billion in the year 2011, compared with U.S. $50 billion in 2010. Asia and the United States accounted for 74% of new installed capacity, and Germany doubled its installed capacity of offshore wind turbines from 108 MW to 215 MW [World 2012].

10.5.2 Wind Turbine Components

10.5.2.1 Introduction

Wind turbines are classified into two categories, based on the direction of the rotational axis. Wind turbines whose rotational axis is parallel to the direction of the wind are known as Horizontal-Axis Wind Turbines (HAWTs). The rotational axis of Vertical-Axis

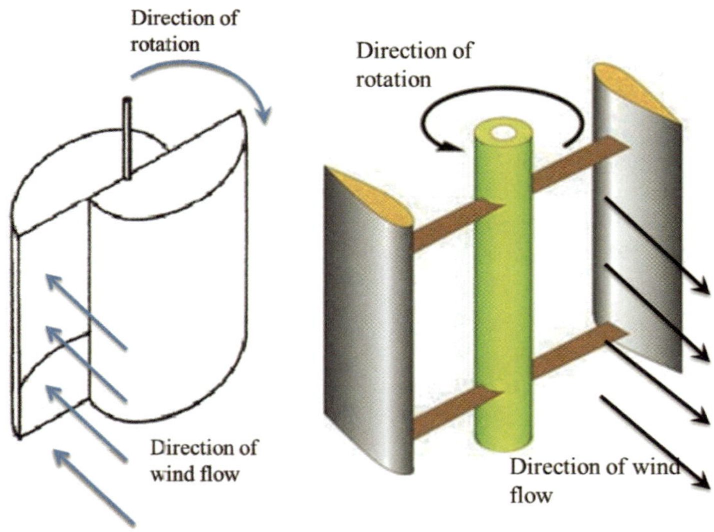

Figure 10.5.1 Schematic images of the Savonius (left) and the Darrieus (right) type VAWTs [Wikipedia 2013].

Wind Turbines (VAWTs) is perpendicular to the direction of wind flow. VAWTs can be further classified as the Savonius rotor-type VAWT and the Darrieus type VAWT. The Savonius rotor-type wind turbines are also called the H-type VAWT. These two types of VAWTs are shown in Figure 10.5.1.

HAWTs can be classified based on the number of blades: single-bladed, double-bladed, or triple-bladed. Even though both categories of wind turbines differ in their basic construction, they are made up of the same general components: a rotor to convert the energy in the wind into rotational energy consisting of blades and a supporting hub; a drivetrain that includes the rotating parts of the wind turbine (other than the rotor): the shafts, gearbox, couplings, mechanical brake system, and the generator; the nacelle and the main frame, including the pitch and yaw system, which acts as a housing for the drivetrain; a tower structure and foundation that supports the rotor and the drivetrain; and the electrical and electronic controls, including the cables, switchgear, transformers, and power converters.

10.5.2.2 Drivetrain

The drivetrain of a wind turbine typically consists of a low-speed shaft connected to the slowly rotating rotor side of the wind turbine; a gearbox, which increases the rotor speed to that required by the generator; and a high-speed shaft, which is connected to the generator of the wind turbine. The gearbox consists of planetary gears and parallel gear stages that are supported by intermediate shafts. Other drivetrain components include the supporting bearings, couplings, brake system, and the rotating parts of the generator of the wind turbine. Direct-drive wind turbines are a special category of wind turbines that use specially designed low-speed generators that can generate electricity at low operating speeds. These designs eliminate the need for a gearbox to increase the low operating speed of the wind turbine. Hence, they are called direct-drive wind turbines. These wind turbines are generally equipped with synchronous-type or induction-type generators to convert the mechanical energy to electrical energy, along with power converters between the generator and connection to the grid to enable the operation of the turbine at a variable operating speed.

10.5.2.3 Generators

Generators convert rotational energy to electrical power; motors convert electrical power to rotational energy. A generator can function as a motor and vice versa. The generator and motor are referred to as electrical machines. The electrical machines used in wind turbines are generators. The two most common types of generators are induction generators and synchronous generators. Both types of generators are used in wind turbines. The application of these two types of induction motors in a wind turbine depends on the availability of a constant or nearly constant rotational speed when the generator is directly connected to the power grid. For wind turbines directly connected to the grid, Squirrel Cage Induction Generators (SCIG) are used. The range of variation in the speed of an SCIG is a narrow range, slightly higher than its synchronous speed (a four-pole generator operating in a 60-Hz grid has a synchronous speed of 1800 rpm). SCIGs are preferred when a rugged and inexpensive design as well as ease of connectivity with the grid is desired in a wind turbine. For greater flexibility in the variation of speeds, generators are connected to the grid by means of power electronic converters. The requirement of power electronics in a variable-speed operation is to ensure the generation of AC electrical current at a constant frequency such that it can be fed into the grid. A commonly used type of induction generator for variable-speed operation is the Doubly Fed Induction Generator (DFIG). For an induction machine to generate electricity, speed during generation should be greater than the synchronous speed of the generator (super-synchronous). This is because the AC frequency connected to the stator of the generator is a fixed frequency. To generate power during sub-synchronous operation, the rotor is also fed with electrical power at a certain frequency such that it compensates for the lower rotating speed of the rotor. Hence, in a DFIG, the stator is directly connected to the grid, while the rotor of the generator is connected to the grid through electronic converters and slip rings. Therefore, a part of the power produced by the generator is fed through the power electronic converter to the rotor. The nominal power of the converter can consume about 30% of the wind turbine power, enabling a rotor speed variation in the range of about ±30% of the nominal speed [Chen, Guerrero, and Blaabjerg 2009]. Thus, by controlling the active power of the converter, the rotational speed of the generator can be varied, and thus power generation can be achieved during variable-speed operation of the wind turbine rotor.

10.5.2.4 Nacelle

The nacelle of a wind turbine consists of the housing, the machine bedplate or main frame, and the yaw orientation system. The purpose of the main frame is to provide mounting and proper alignment of the drivetrain components of the wind turbine. The housing protects the contents of the nacelle of the wind turbine from the weather. The purpose of the yaw orientation system is to keep the wind turbine blades properly aligned with the direction of wind flow. The main component of the yaw orientation system is the yaw bearing that connects the main frame to the tower. The rotation of the main frame is achieved by means of one or more yaw motors, which drives a pinion gear against a bull gear attached to the yaw bearing. The rotation of the bull gear positions the drivetrain in the direction of the wind flow. The exact position of the yaw system is controlled by an automatic yaw control system with a wind direction sensor mounted on the nacelle of the wind turbine. Brakes are also used to hold the position of the main frame in the direction that is required by the yaw orientation system.

10.5.2.5 Tower Structure and Foundation

The tower structure and foundation supports the static and dynamic loads of all the components in a wind turbine. The types of wind turbine supporting structures are the free-standing type that uses steel tubes, lattice (or truss) towers, and concrete towers for on-shore wind turbines. For smaller land-based wind turbines, guyed towers are also used. The design of the foundation for offshore wind turbines depends on the depth of the water at the place where the wind turbine is installed. For wind turbines installed in shallow water (5 to 18 m), land-based wind turbine foundation designs can be used. The most typical on-shore wind turbine foundation is the monopile design. The gravity-based design and suction bucket design have been developed based on the monopile design to adapt it to shallow water conditions [Musial, Butterfield, and Ram 2006]. For water depths from 60 m and greater, different designs have been studied and are in the research stages. Designs include the tripod tower design, the guyed monopole design, the full-height jacket (truss) design, the submerged jacket design with transition to tube tower, and the enhanced suction bucket or gravity-based design. Floating foundation-based designs also have been explored in recent studies. Some of the floating foundation designs are the Dutch Tri-floater design, barge design, spar-buoy with two tiers of guy-wires, three-arm mono-hull tension leg platform design, concrete tension leg platform with gravity anchor, and a deep water spar design [Musial, Butterfield, and Ram 2006].

Selection of the tower structure and foundation design is greatly influenced by the characteristics of the site at which the wind turbine is installed. Different factors play a role in the design of the tower; they include wind turbine system dynamics such as coupled vibrations between the rotor and tower, the effect of tower shadow (the wake created by air flow around a tower) on turbine dynamics, power fluctuations, and noise generation. For example, the tower shadow effect on downwind turbines leads to their noisier operation compared to their upwind counterparts [Manwell, McGowan, and Rogers 2009].

10.5.2.6 Control System

The control system of a wind turbine monitors the operation of the wind turbine to maximize power production. The components of a control system typically include sensors, controller, power amplifiers, actuators, and the computing systems. Sensors monitor parameters such as wind flow, temperature, current, voltage, speed, and direction the blades are pointed. Controllers consist of the mechanical systems and the electrical circuits which carry out the commands of the computing systems. The computing systems consist of the computers and microprocessors in the wind turbine. The controllers operate the actuators including motors, pistons, magnets, and solenoids. Power amplifiers such as hydraulic pumps, electrical amplifiers, switches, and valves help in the operation of the actuating system.

The main function of control systems on variable-speed wind turbines is to adjust blade pitch settings and generator torque dynamically to control the power produced under high winds. There are two types of control systems in wind turbines: supervisory control and dynamic control. Monitoring turbine operation and sequencing of the control actions such as brake release and contactor closing are carried out by the supervisory control system. During wind turbine operation, actions carried out such as changing blade pitch in response to turbulent winds, which can affect the operation of the turbine, is performed by the dynamic control system. Wind turbine control systems operate at three functional levels: a controller for all wind turbines in a wind farm, a supervisory controller for each individual wind turbine, and separate dynamic controllers for the various turbine subsystems in each wind turbine. These three levels of control are interrelated and are carried out in a hierarchical order.

Supervisory Control and Data Acquisition (SCADA) systems are often called wind farm controllers. These systems can initiate and shut down wind turbine operation. They also coordinate the operation of all the wind turbines in the wind farm. This is carried out by the linkage between the SCADA systems and the supervisory controllers on each wind turbine. Control actions related to the medium- and long-term changes in environmental and operating conditions are communicated to the supervisory controllers on each wind turbine by the SCADA system; hence the interval between successive control actions by the SCADA system to the supervisory control system could take a relatively long time. The functions of the supervisory control system include switching wind turbine operating states such as from power production to low-wind shutdown; monitoring wind flow and severe operating conditions such as high load on the wind turbine and the operational limit conditions of the wind turbine; starting and stopping the wind turbine in a sequential order; and providing control action commands to the wind turbine dynamic controllers, such as adjusting the desired blade tip speed ratio. Control action commands sent by the dynamic controllers on each wind turbine are carried out frequently, compared to the other two levels of the control system. Some of the control commands include continuous dynamic adjustments to wind turbine actuators and components based on the dynamic changes in external environmental conditions. Hence, each wind turbine might contain several dynamic controllers, because each dynamic controller manages only one specific subsystem of the wind turbine at a time. The coordination between each dynamic controller is carried out by the supervisory control system. Some of the functions of the dynamic control systems include adjusting blade pitch to limit the torque on the drivetrain to be within the design limit of the wind turbine, controlling the power supplied to the power electronic converters on the wind turbine, and controlling the operation of the actuators that carry out the actual physical operation.

A dynamic controller, therefore, operates actuators or switches influencing each wind turbine subsystem and hence controls the overall operation of the wind turbine. The actions carried out by the actuators or switches are also monitored by the dynamic control system and are used as inputs for further control actions by the system. This monitoring is carried out by means of sensors that communicate the wind turbine operation parameters to the control system. Some of these measured parameters include speeds such as generator speed, rotor speed, wind speed, yaw rate, and direction of rotation; temperatures such as gearbox oil, hydraulic oil, gearbox bearing, generator bearing, generator winding, ambient air, and electronic temperatures; directional position of wind turbine components such as blade pitch, teeter angle, aileron position, blade azimuth, yaw position, yaw error, tilt angle, and wind direction; electrical characteristics such as grid power, current, power factor, voltage, grid frequency, ground faults, and converter operation; fluid flow parameters such as hydraulic or pneumatic pressures, hydraulic oil level, hydraulic oil flow; motion, stresses, and strain parameters such as tower top acceleration, tower strain, shaft torque, gearbox vibration, and blade root bending moment; and environmental conditions such as turbine or sensor icing, humidity, and lightning [Manwell, McGowan, and Rogers 2009]. These measured parameters are translated into control actions to the actuators or switches by the dynamic control system.

Some of these actuators or switches controlled by the dynamic control system include mechanical systems such as tail rotors, linkages, springs, and fly ball governors, which control blade pitch, yaw position, and rotor speed; electrical systems conduct the desired control action based on the output of the sensor signals such as energizing coils in relays or switches. Computers in the dynamic control system handle the digital and analog inputs and outputs and carry out complicated logic computations to provide dynamic responses to the control system. The advantage of computer control systems is the ability to change the control code by reprogramming the computer. Amplifiers present in the control system amplify the power of the control actions if the control signals from the controller are not powerful enough to power the actuator. Some of the switches in the wind turbine include relays, contactors, power electronic switches such as transistors and Silicon-Controlled Rectifiers (SCRs), and hydraulic valves. Mechanical actuators in a wind turbine include electromechanical devices such as DC motors, stepper motors, AC motors with solid-state controllers, linear actuators, magnets, and solid-state switching components; hydraulic pistons for positioning systems; hydraulic pumps for providing high-pressure fluid to control valves; resistance heaters and fans to control the temperature in the subsystems of the wind turbine; and mechanical actuators such as gears, linkages, and other machine elements that modify the actuating force or direction.

10.5.2.7 Power Electronics and Electrical Systems

The rest of the wind turbine includes miscellaneous components such as cables for interconnection with electrical networks, switchgear, transformers, power electronic converters, and power factor correction capacitors. These components play an important part as well in providing safety aspects and enabling the functioning of the principal components of the wind turbine discussed above. For example, while land-based wind turbines are connected to the medium-voltage system, large offshore wind turbines are required to be connected to the high-voltage and extra high-voltage systems. Hence, the transformer is normally located close to the wind turbines to avoid high current flowing in long low-voltage cables connecting the offshore wind turbine with the land-based power grid [Chen, Guerrero, and Blaabjerg 2009].

Power electronic converters in wind turbines include semiconductor devices for driving, protecting, and enabling control circuits functions such as voltage magnitude and frequency conversion. Some of the semiconductor devices include self-commutated devices such as Insulated Gate Bipolar Transistors (IGBTs), Metal Oxide Semiconductor Field-Effect Transistors (MOSFETs), Integrated Gate Commutated Thyristors (IGCTs), MOS-gate thyristors, and silicon carbide FETs [Chen, Guerrero, and Blaabjerg 2009]. Power electronic converters are used in grid-commutated and self-commutated converter systems. Grid-commutated converters consist of thyristor-based converters, which require high-power capacities. This is possible in thyristor-based converters because they consume inductive reactive power and hence are unable to control the reactive power. This property is preferred in high-voltage and -power applications, such as conventional high-voltage DC systems. Pulse Width-Modulated (PWM) control systems are used in self-commutated converter systems for switching function in semiconductors such as turning them on or off. IGBTs are typically used in self-commutated converter systems. The advantage of these converters is that they can control both active and reactive power, even under operating conditions such as ac-dc or dc-ac conversion [Chen and Spooner 2001]. Hence, these control systems can provide the reactive power demand by means of a PWM converter. This high-frequency switching of a PWM converter could lead to the generation of harmonics components in the kilohertz range of frequencies. Thus, small-size filters are required to be used in converters to remove the interference problems due to these harmonics.

10.5.3 Wind Turbine Failures

The amount of energy in the wind that is captured by the wind turbines is proportional to the area swept by the rotor during its rotation and the cubic average wind speed at the center of the rotor. To increase the power-generation capacity of a wind turbine, wind turbines have become bigger and taller, as shown in Figure 10.5.2.

Thus, there is a need for continuous research and development in wind turbine design, manufacturing, and maintenance to improve the power-generation capacity. Some of the difficulties with respect to the enormity of size of the wind turbines include the difficulty in performing inspection and maintenance work at the elevated heights of the wind turbine; mishaps are prone to happen, some which have led to fatalities; availability of better wind resources in remote locations such as mountainous or rough sea regions is a challenge for carrying out maintenance and repair actions; the requirement of specialized mechanical equipment for lifting and handling heavy wind turbine components such as cranes and forklifts; increase in capacity of the wind turbine increases the price of the wind turbine as well as maintenance costs; and finally, the requirement of an efficient monitoring system to improve safety, reduce down time, and reduce frequency of breakdowns [Ciang, Lee, and Bang 2008].

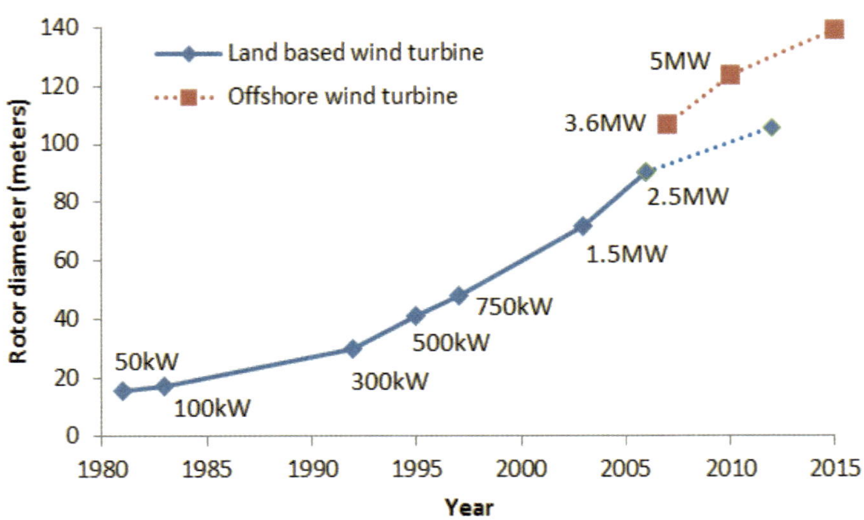

Figure 10.5.2 Increase in the capacity of horizontal-axis wind turbines and rotor diameter over time [DOE 2008].

Wind turbine maintenance costs can be significant, as wind turbines are typically difficult to access. For example, replacement of a $5000 bearing can turn into a $250,000 project involving cranes and a service crew, in addition to the loss of power generation [Hatch 2004]. For a turbine with a 20-year operating life, the Operation and Maintenance (O&M) and part replacement costs have been estimated to be at least 15% of the total income from power generation [Walford 2006]. For off-shore wind turbines, the O&M costs have been estimated to be as high as 30% of overall costs. Failures occurring in wind turbines are categorized as 1) failures occurring at high frequencies but that can be rectified with short downtime and 2) failures that rarely occur but involve longer downtimes. Studies have shown that 75% of all failures occurring in onshore wind turbines are due to failures with a high frequency of occurrence, and they cause up to 5% of the annual downtime. Figure 10.5.3 shows failure rates and downtimes per failure type in onshore wind turbines in Germany. Electrical system failures account for frequently occurring failures. Ninety-five percent of downtime occurring in wind turbines is due to rarely occurring failures such as gearbox and generator failures [Faulstich et al. 2009]. An increase in failure rates has also been identified with increasing wind turbine capacities [Spinato et al. 2009].

Condition-Monitoring Systems (CMS) are currently used in wind turbines for fault diagnosis and scheduling maintenance actions. Some of the wind turbine failures reported in the literature is discussed in this section. Wind turbine structural failure have been reported extensively and have been attributed to various causes, from a failure in the concrete base to a failure of the blades due to a bolt shear or the buckling of a load-bearing brace [Ashley et al. 2007]. Structural failures in wind turbines have been reported

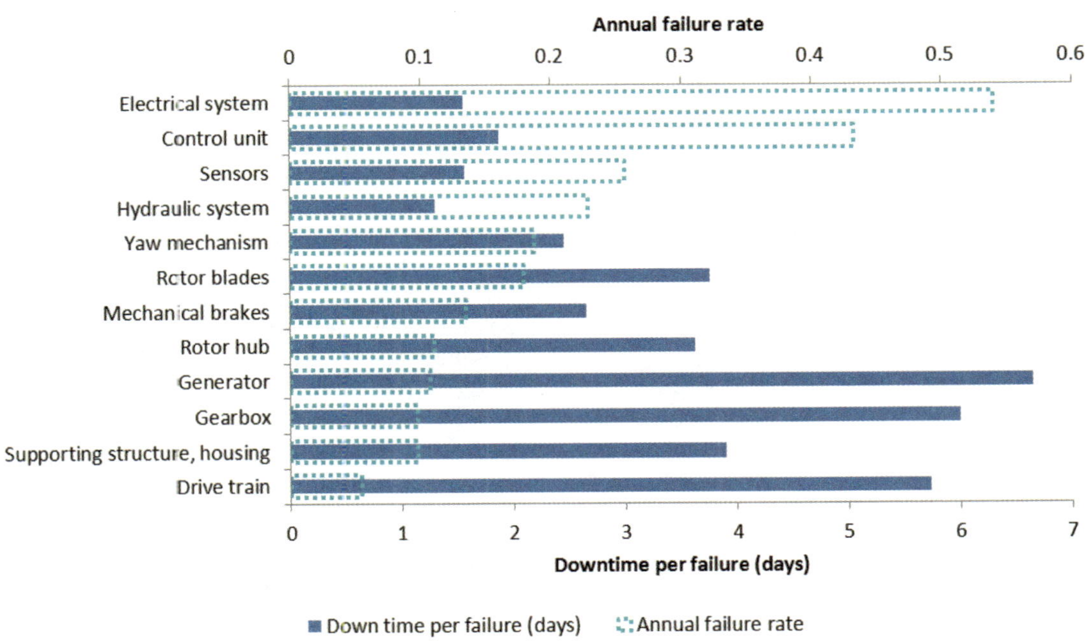

Figure 10.5.3 Failure rates and downtimes per failure type in onshore wind turbines in Germany [Windmonitor 2012].

in several wind farms in Wales, Scotland, Spain, Germany, France, Denmark, Japan, and New Zealand [Rosenbloom 2013].
For example, a wind turbine incident was reported from Germany in the year 2002 due the failure of a blade in the wind turbine, leading to debris being strewn on surrounding fields [Ashley et al. 2007]. In a similar incident, blade failure led to airborne debris, which flew up to 8 km and landed through the window of a house [Caithness 2013]. Among the structural failures, blade tip breakage and yaw bearing failures have the highest failure rate for a typical wind turbine system [Khan, Iqbal, and Khan 2004]. The failure rate of the different wind turbine components is shown in Table 10.5.1 from [Khan, Iqbal, and Khan 2004].

The most common structural component affected by damage to the wind turbine structure is the rotor or blade and main tower structure [Caithness 2013]. Hence, extensive research has been carried out in the health monitoring of the blades of wind turbines. Failure of any one of the blades of a wind turbine can lead to the replacement of either one or all of the blades, which can cost up to 15–20% of the total wind turbine cost [Larsen and Sorensen 2003]. Therefore, blade failures can increase maintenance costs as well as downtime for repair [Larsen and Sorensen 2003]. Blade failures can cause additional damage to the wind turbine as well; for example, a minor blade damage can lead to a unbalance in the rotating masses of the wind turbine, which can lead to damage to the wind turbine structure and eventually to collapse of the whole tower [Rosenbloom 2013]. Some of the most commonly occurring wind turbine blade failures are listed in Table 10.5.2 [Ciang, Lee, and Bang 2008].

The second component that is susceptible to damage is the main tower structure of the

Component	Failure rate (h^{-1})
Blade tip breakage	$1 * 10^{-4}$
Yaw bearing failure	$1.15 * 10^{-5}$
Blade failure	$1.116 * 10^{-5}$
Bolt failure	$1.116 * 10^{-5}$
Hub failure	$1.116 * 10^{-5}$
Generator failure	$0.769 * 10^{-6}$
Gearbox failure	$0.63 * 10^{-6}$
Brake failure	$2.16 * 10^{-6}$
Tower and anchor bolt failures	$1 * 10^{-7}$

Table 10.5.1 Wind Turbine Component Failure Rate

Description
Damage formation and growth in the adhesive layer joining skin and main spar flanges (skin/adhesive debonding and/or main spar/adhesive layer debonding)
Damage formation and growth in the adhesive layer joining the up- and downwind skins along leading and/or trailing edges (adhesive joint failure between skins)
Damage formation and growth at the interface between face and core in sandwich panels in skins and main spar web (sandwich panel face/core debonding)
Internal damage formation and growth in laminates in skin and/or main spar flanges, under a tensile or compression load (delamination driven by a tensional or a buckling load)
Splitting and fracture of separate fibers in laminates of the skin and main spar (fiber failure in tension; laminate failure in compression)
Buckling of the skin due to damage formation and growth in the bond between skin and main spar under compressive load (skin/adhesive debonding induced by buckling, a specific type 1 case)
Formation and growth of cracks in the gel-coat; debonding of the gel-coat from the skin (gel-coat cracking and gel-coat/skin debonding)

Table 10.5.2 Wind Turbine Blade Failures [Sundaresan, Schulz, and Ghoshal 2002; Sørensen et al. 2002, 2004]

wind turbine. Wind turbine structure designs are of two types: lattice design and cylindrical tube design. Lattice designs are made of steel structural components, which have the risk of corrosion at the junctions or connections in the structure. Corrosion can weaken the main diagonal structural elements that maintain the shape of the structure and are sensitive to wind excitation. This can pose a risk in higher-capacity wind turbines that are supported by a structure of this type of design. Cylindrical tube-based wind turbine towers are made of steel or concrete. Concrete cylindrical tube towers are used in wind turbines exceeding 85 m in height because the steel-based cylindrical structures are not able to balance the vibration excitation of these tall wind turbines during operation [Harte and Van Zijl 2007]. However, concrete towers are not perfect and are prone to problems caused by thermal expansion, leading to cracking and thereby increasing the natural frequency and reducing the stiffness properties. Environmental problems such as storm damage can also cause damage to the blades or a collapse of the entire structure [Caithness 2013]. A turbine tower failure in France during a storm in 2000 led to the toppling of an entire wind turbine, which was the first in a series of such incidents that led to a formal investigation [Ashley et al. 2007]. A similar incident in Germany resulted in a sudden and total collapse of an entire wind turbine in 2000 [Ashley et al. 2007]. The failure of this wind turbine was attributed to concrete damage at the base of the wind turbine. Wind turbine tower structure failures have been reported in Germany (as recently as 2002) and the United States (e.g., in Oklahoma, on May 6, 2005) [Rosenbloom 2013; Caithness 2013]. Some failures observed in newly installed wind turbines have been attributed to poor quality control practices, improper installation or assembly of components, and individual component failures [Ashley et al. 2007; Caithness 2013]. Lightning damage results in the peeling of blade coatings, leading to a reduction in strength and causing serious structural damage.

Fire and strong winds are other environmental factors that can damage a wind turbine. The highest load that the blade is subjected to is the flap-wise (chord-wise) bending load when the turbine has been brought to a standstill due to high wind, and a blade is hit by an extreme gust of wind [Thomsen 2006]. This load can be increased further when the braking system fails during high wind failure. The braking system in a turbine is designed to stop the rotors if the wind is too strong. This brake failure can lead to out-of-control spinning of the blades as well as hitting of the blades on the wind turbine structure at very high speeds. Such instances have been reported in Germany, on numerous occasions during 1999, 2000, and 2003. Such a failure can result in considerable damage, ranging from damage to parts of the blade up to uprooting of the entire nacelle (rotors attached) from the tower structure. Blades and other substantial parts have landed as far away as 500 m in some cases. These incidents indicate that there is an urgent need for an effective CMS to prevent such catastrophic failures in the wind turbine. For example, if the failure of the brake system can be detected at an earlier stage, the problem of an increase in the chord-wise bending load on the blades due to high wind can be avoided. The following section discusses the condition-monitoring systems in current practice.

10.5.4 Condition-Monitoring Systems in Wind Turbines

10.5.4.1 Introduction

There are two main approaches to condition monitoring of wind turbines: condition-based maintenance and scheduled maintenance. Scheduled maintenance is carried out at regular intervals by skilled technicians using portable tools to assess the health condition of the different components in the wind turbine.

Condition-based maintenance approaches involve component repairs or part replacements being performed based on the actual health condition that is estimated based on remote monitoring data carried out on each wind turbine. CMS are more desirable than scheduled maintenance because they avoid the replacement of healthy components. Condition monitoring in wind turbines is carried out at both the system level and the subsystem level. Subsystem-level CMS are beneficial for gearboxes, generators, rotors, and the blades, because these components dominate the total machine cost in a wind turbine, and the failure of one component can influence the failure of others as well. For example, the failure of a $1500 bearing could lead to $250,000 in maintenance costs if subsystem-level condition monitoring is not implemented. This increased maintenance cost due to the failure of the bearing causes a gearbox replacement worth $100,000; a rewinding of the generator worth $50,000; miscellaneous replacement of failed components worth $70,000; and additional lost production costs [Hyers et al. 2006].

As discussed in section 10.5.2.6, SCADAs are used in wind farms for controlling their operation. SCADAs communicate with individual turbines, substations, and meteorological stations through a communication network. Wind turbine parameters are collected and stored every 5–10 minutes by the SCADA system. Some of these parameters include wind-based parameters (wind speed, wind direction, wind intensity, and wind turbulence), energy conversion-based parameters (active power output, blade pitch angle, generator torque, rotor speed, phase currents, reactive power, and power factor), vibration-based parameters (drivetrain acceleration and tower acceleration), and temperature parameters (gearbox bearing temperature, gearbox lubricant temperature, and nacelle temperature).

10.5.4.2 Condition Monitoring of Rotors and Blades

Wind turbines are subjected to fatigue stresses during their lifetime. Rotors are required to survive more than 10∧8 stress cycles during a 20-year lifetime [Hyers et al. 2006]. In addition to fatigue stresses, these components are also subjected to creep fatigue and corrosion fatigue. The most common failure modes observed in rotors and blades are cracks and delaminations in the composite material used in the blades. The effects observed due to blade or rotor failures are rotor imbalance, aerodynamic asymmetry, moisture uptake, icing, and change in surface roughness [Caselitz and Giebhardt 2005]. Rotor imbalance and aerodynamic asymmetry cause additional vibrations of the nacelle and tower structure as well as variations in rotor speed. This also increases the stress amplitudes in the fatigue cycles, thereby shortening the fatigue life of the blades. Moisture uptake creates an additional imbalance, which also degrades the composite matrix properties. Icing can further aggravate the fatigue crack growth process, while changes in the surface roughness can reduce the aerodynamic performance of the blades.

Condition monitoring of the operating conditions of a wind turbine is carried out by detecting the occurrence of periodic variations in rotor speed by means of frequency domain analysis techniques [Caselitz et al. 1996]. This effect is also observed in tower vibration [Knudsen, Bak, and Tabatabaeipour 2012] as well as the electrical current signal. The occurrence of a peak at the rotor rotational frequency in the frequency spectrum of the electrical current signal is attributed to the presence of rotor imbalance or aerodynamic asymmetry. Bicoherence of the electrical power output of a 45-kW wind turbine was carried out when the wind turbine had a blade with less stiffness compared to the other blades [Jeffries, Chambers, and Infield 1998]. The frequencies that are excited in the spectrum of the electrical current due to the presence of rotor asymmetries were also identified in [Jeffries, Chambers, and Infield 1998].

Changes in surface roughness on the blades of a wind turbine have also been identified due to erosion, icing, and insects in the vicinity of a wind turbine. A study by Corten and Veldkamp [2001] has shown that the power characteristic of the wind turbine is affected by these causes. The power characteristic of a wind turbine is a functional relationship between its wind speed and output power and is established by the wind turbine manufacturer for each particular design of the wind turbine.

Commercial blade monitoring systems have also been developed by blade manufacturers such as LM and Aerpac [Verbruggen 2003]. These technologies use accelerometers and proportional proximity sensors for condition monitoring of the blades. Fiber-optic Bragg Gratings (FBG) strain gauges have also been reported in the literature as being used for monitoring the load history of blades [Rademakers et al. 2004]. CMS based on optical fiber networks have been developed by NREL/Sandia Laboratories, Smart Fibres Limited, and LM Blade Monitoring [Elsevier 2005].

10.5.4.3 Condition Monitoring of Drivetrain

The drivetrain of a wind turbine consists of the shafts, couplings, bearings, gearbox, generator, and braking systems. The primary reliability concern pertaining to the drivetrain in a wind turbine is wear of the gear tooth and bearings in the gearbox. The reasons for this operating condition are the presence of particulate material in the lubricating oil due to contamination during assembly, corrosion, or wear products occurring during operation of the wind turbine. The wear particles are generated due to variations in the wind speed that lead to imbalance, causing chatter in the gear teeth and fretting. Wear particles can also be generated from stress concentrations in the gear tooth and bearings, leading to wear, mechanical interference problems, or the loss of lubrication in the gearbox.

Condition monitoring of gearboxes and bearings in a wind turbine has been carried out using electrical current signals, wear debris in the oil, and by vibration monitoring. The usage of electrical power for condition monitoring of the rotor misalignment and bearing failure was demonstrated by Watson et al. [2010].

A survey of commercial wind turbine CMS indicates that most systems are based on vibration monitoring using accelerometers. Commercial systems also use oil debris particle counters, torque and strain measuring systems, and displacement and acoustic sensors [Verbruggen 2003; GE 2012].

10.5.4.4 Condition Monitoring of Electrical and Electronic Systems

The cost of electronic controls accounts for only about 1% of the total cost of a wind turbine, but electronic controls cause 13% of all failures. The reasons for failures in electronic control systems have been attributed to vibration-induced fatigue of printed circuits, corrosion, water ingress, condensation, and dust. Power electronics failures in wind turbines have been attributed to thermal management issues due to overload or loss of cooling as well as thermo-mechanical fatigue due to variations in load [Hyers et al. 2006].

Thermography is a useful tool that can be used for condition monitoring of electrical systems such as transformers, contacting points, and switching gears in wind turbines [Hyers et al. 2006; Hameed et al. 2009]. This method enables the detection of hot spots. However, it requires the interpretation of results by visual inspection, which is the inherent drawback of this method and limits its implementation in actual wind turbines.

Motor Current Signature Analysis (MCSA) has been implemented in wind turbines for the detection of faults in a wind turbine generator. This method is implemented for the detection of stator turn-to-turn faults as a result of shorted turns or electrical influence on the generator due to grid faults. This is carried out by analyzing the frequency spectrum of the stator electrical current. The frequency that is monitored for identifying stator turn-to-turn faults is shown in Eq. 10.5.1.

$$f_{ault} = f1\left[\frac{n}{p}(1-s) \pm k\right] \qquad (10.5.1)$$

where f_1 is the supply frequency; $n = 1, 2, 3, \ldots$; $k = 1, 3, 5, \ldots$; p is the number of pole pairs; and s is the slip [Hameed et al. 2009; Kliman and Stein 1992].

10.5.4.5 Condition Monitoring of Pitch and Yaw Systems

Blade pitch and yaw control systems are critical components of a wind turbine because they help to enhance the energy captured by the blades, reduce the operating load on the wind turbine, and enable stalling and aerodynamic braking. These systems consist of either hydraulic actuators or electric motors. Electric motor systems have a faster response compared to hydraulic systems, whereas hydraulic systems have better stiffness, little backlash, and higher reliability [Lu et al. 2009]. In a Swedish wind farm between 2000 and 2004, 13.3% of all the failures were attributed to the hydraulic system failure of the pitch and yaw system [Ribrant and Bertling 2007].

Simulation results of pitch errors indicate that the acceleration signals in the X- and Z-directions as well as the electrical power signal are useful for the detection of faults in the pitch system [Baines and Way 2007]. Although the detection in the electrical power signal is less significant, a 3–4 dB change in the acceleration signal has been observed for pitch error faults of 0.5°.

Yaw system faults in a wind turbine can be caused by significant loads on the blades (20°) or due to unusual wind changes (10°). A change of up to 5° is observed in the yaw system of a wind turbine during normal operating conditions [Holst-Jensen and Aarhus 2001]. These errors have also been detected in the electrical power spectrum by means of simulation as well as by measurements carried out on a test wind turbine under normal and faulty operating conditions.

Commercial implementations of CMS for pitching systems in a wind turbine have been reported by MLS Electrosystem [MLS 2007] and GE [Maughan 2007]. These control systems used features such as excessive backlash through control signals as a diagnostic feature to detect blade angle asymmetry.

10.5.5 Survey of Commercial Condition-Monitoring Systems for a Wind Turbine

In addition to the above parameters, some additional parameters that are measured using commercial CMS are the dynamic load on structural components such as the roots of the rotor blades using strain gauges, shaft torque using strain sensors, and oil conductivity and pH measurement to evaluate the wear of bearings and gears.

Wind turbine faults	Reason for occurrence	CMS parameter used for detection
Blade surface roughness	Dirt, insects, manufacturing defects (cracks and blowholes) and icing weather conditions	Reduction in power output due to reduced aerodynamic performance of blade profile
Rotor mass imbalance	Icing weather condition and penetration of material in blades	Periodic transverse vibrations
Aerodynamic asymmetries	Blade pitch drive faults	Axial and torsional oscillations creates bending moments
Bearing faults	Wear-out process	Time domain vibration features (rms, kurtosis, and crest factor) and frequency domain features (Fourier transforms and envelope analysis)
Gear faults	Wear-out process	Side band analysis in frequency domain
Generator faults (winding faults, rotor cage faults, and faulty slip rings or bushings)	Wear-out process	Electrical current monitoring, temperature monitoring
Transformer and switching gear faults	Wear-out process	Resistance and temperature monitoring

Table 10.5.3 Wind Turbine Failures, Causes, and Detection Strategies

Table 10.5.3 summarizes the commonly occurring faults in wind turbines and the CMS parameters that are used to identify these faults.

Supporting components for a CMS include data acquisition using analog-to-digital converters with varying data acquisition rates from 0.1 Hz to 20 kHz, data preprocessing systems such as Fourier analysis and digital filtering systems, data analysis systems for fault detection and providing alarms to wind farm operators, databases for trend analysis and investigation of fault causes, and communication channels for transferring data from wind turbines to the SCADA system.

There are several commercially available wind turbine CMS. The Brüel & Kjær Vibro CMS monitors vibration signals while extracting features from the signal such as peak value, harmonics, and sidebands, as well as speed and phase information. This system monitors the bearings, shaft, gearbox, and generator of a wind turbine as well as communicates with monitoring centers through Local Area Network (LAN), wireless, General Packet Radio Service (GPRS), or Universal Mobile Telecommunications System (UMTS) cellular modems [Brüel & Kjær 2012]. The faults in a wind turbine that are monitored include gear wear, gear tooth faults, bearing defects, shaft imbalance and misalignment, rotor unbalance, bearing misalignment, bent shafts, and gearbox suspension faults.

The CMS developed by Nordex monitors the vibration of the main bearing, the gearbox, and the generator by means of acceleration sensors mounted on these components. Frequency, envelope, and order analysis of the vibration signals is carried out to send warning signals to the operators once they detect a fault in the system [Nordex 2013]. GE's Bently Nevada ADAPT.wind system monitors the main bearing, gearbox, generator, and the tower sway of a wind turbine. This is carried out by means of accelerometers, proximity sensors, and oil particle sensors collecting signals at the rate of 102,400 samples per second [GE Bently 2012; GE ADAPT 2012]. Features such as peak, rms, rotor frequency, gear mesh frequency, bearing frequency, tower sway, kurtosis, crest factor, and skewness are monitored to generate alarm signals when a faulty condition occurs in the wind turbine drivetrain.

The Condition Diagnostics System developed by Winergy uses magnetostriction-based sensors combined with load sensors to conduct condition monitoring of the shafts in the gearbox of a wind turbine [Kreidler 2010]. The Condition Management System by Moventas utilizes temperature, vibration, torque, oil pressure, speed, and oil particle content to conduct condition monitoring of the rotor, gear, and generator of a wind turbine [Moventas 2012]. The OneProd system developed by ACOEM conducts condition monitoring of the drivetrain and generator of a wind turbine system by means of vibration analysis, electrical current signature analysis, and oil analysis. The system also has the capability to monitor the electrical panels in wind turbines using thermography-based techniques [ACOEM 2012]. Gamesa SMP is a CMS developed by Gamesa that monitors the vibrations in the drivetrain of a wind turbine by means of eight accelerometers [Gamesa 2012]. A Turbine CMS developed by Siemens monitors the main bearing, gearbox, and the generator of a wind turbine by means of accelerometers and Structural Vibration Monitoring (SVM) sensors. SVM sensors monitor the tower sway, rotor imbalance, and tail swing of the wind turbine [Gram 2012].

WindCon, developed by SKF, monitors the condition of the main bearing, transmission, and generator for faulty operating conditions such as unbalanced blades, shaft misalignment, bearing and gear damage, weak foundation, generator stator/rotor problems, and poor lubrication condition [SKF 2012]. WiPro developed by FAG monitors the drivetrain of a wind turbine by means of accelerometers, displacement, speed, force, pressure, and temperature measurements. Determination of oil quality as well as particle counter and torque measurements are also carried out by this system [Schaeffler 2012]. WP4086 CMS is a CMS developed by Mita-Teknik, and it conducts vibration analysis with eight accelerometers. A frequency range of 0.1–5000 Hz is acquired for vibrations of 0.01–5000 *g*. Time domain as well as frequency domain characteristics of the vibration signal are analyzed using FFT and envelope analysis [Mita-Teknik 2012].

BladeControl is a CMS developed for blades and the hub of a wind turbine by Bosch Rexroth Monitoring Systems. This system monitors damage to the rotor blade, icing of the blade, lightning damage, dynamical loads, and rotor imbalance [Bosch 2012]. This is carried out by means of two-axis accelerometers mounted on each rotor blade, a one-axis accelerometer mounted at the shaft, and a rotor blade temperature sensor. FS2500 is a blade-monitoring system developed by FiberSensing that uses sensors based on fiber Bragg grating technology. This system allows for data acquisition at sampling rates of upto 2000 samples per second simultaneously over four sensor channels [FiberSensing 2012]. The Rotor Monitoring System (RMS) is a blade-monitoring system developed by Moog based on a fiber-optic sensing system. This system monitors rotor imbalance, blade damage, and turbine rotor performance [Moog 2012].

10.5.6 Prognostics and Health Management Implementation Strategies

Prognostics and health management-based approaches combine information on the current health condition with historical data along with the physics-of-failure models, or prior information on trends until failure, to predict the future probability of failure. In this method, the failure is defined as a process, whereas failure is an event in condition-based maintenance techniques. This requires the early detection of an anomalous operating condition in a wind turbine to let the operator know the amount of time available for planning necessary maintenance actions before the loss of availability of the wind turbine for power generation.

Prognostics and Health Management (PHM), as discussed in chapter 9, assesses the health of a system and estimates the Remaining Useful Life (RUL) by means of sensor technologies and the interpretation of environmental, operational, and performance-related parameters [Cheng, Azarian, and Pecht 2010]. The first step in the process of implementation of PHM is the identification of the failure precursor parameter. A failure precursor parameter is a feature extracted from monitored sensory data that directly or indirectly indicates impending failure [Vichare and Pecht 2006]. Failure precursor parameter identification requires knowledge of the dominant failure mechanism that can lead to early failure of system components. Failure Modes, Mechanisms, and Effects Analysis (FMMEA) is a systematic approach that can be used for identifying the failure precursor parameters [Mathew et al. 2008]. The implementation of FMMEA at the system level requires the definition of the system under study, which includes the identification of subsystems and their functions. The potential failure modes and causes that can affect the subsystems have to be identified based on prior experience or background knowledge of the physics of operation of the system. This knowledge requires the identification of the potential life-cycle environmental and operational conditions of the system or the subsystem. Identification of failure modes enables identification of the failure mechanisms that govern the failures. Prioritization of these failure mechanisms is then carried out to identify the dominant failure mechanism.

Arabian-Hoseynabadi, Oraee, and Tavner [2010] proposed a method for prioritization of the failure mechanisms in their study. The severity level is classified into four categories, with severity from low to high: "electricity can be generated but urgent repair is required, reduction in ability to generate electricity, loss of ability to generate electricity, and major damage to the turbine as a capital installation." The occurrence level is classified into four categories, from extremely unlikely to frequent: probability of occurrence is less than 0.001, probability of occurrence between 0.001 and 0.01, probability of occurrence between 0.01 and 0.1, and probability of occurrence is more than 0.1. The level of detection of fault is classified into four categories, from almost certain to almost impossible: existing condition monitoring methods always detect the failure, failure can be detected with a good likelihood using the existing condition monitoring methods, failure can be detected with a low likelihood using the existing condition monitoring methods, and no method exists for detection of failure. Values are attributed for each of the cases: failure occurrence level, failure severity level, and detection level of fault. These values are used to calculate the Risk Priority Number (RPN) for each failure mechanism. This number helps in identification of the dominant failure mechanism in a subsystem or at the system level.

Based on the knowledge of the critical failure mechanisms and the associated failure modes for each mechanism, the failure precursor parameters to be monitored are identified, followed by selection of the right sensor for implementation in the system [Ganesan et al. 2005]. The level of detail for carrying out the FMMEA can be classified into four levels: the wind turbine (system level), assembly level such as generator, subassembly level such as stator of generator, and part level such as bearing in the generator. The level of detail used for conducting the FMMEA can lead to a complex analysis and requires in-depth knowledge of the system. The output of the FMMEA step is the identification of the failure precursor parameter and the failure sites in the system. Based on the identified failure precursors, the stress at each failure site can be obtained as a function of loading conditions, geometry, and material properties of the system [Pecht and Jaai 2010].

Damage models can then be used to trend the trajectory of the fault progression and estimate the RUL. These failure models might require additional information, such as material properties, geometry, and environmental and operating loads for the accurate prediction of the RUL. Environmental and operating load data can be obtained from the sensory data obtained from the wind turbine. Some examples of the load-related features required in the models are fatigue cyclic range, mean load, and number of cycles. Data-driven techniques can also be used to predict the trend in the degradation process for the estimation of RUL. These techniques learn the degradation trend from past data or data from laboratory tests for providing estimates of the RUL. These algorithms detect the anomalous operating condition based on the assumption that the statistical characteristics of the system remain relatively unchanged until a fault occurs in the system [Pecht and Jaai 2010]. Anomaly detection is followed by trending of the degradation pattern based on prior learning data to determine the state-of-health of a system. These degradation trends are then used to estimate the time-to-failure of the system once the degradation process crosses a predetermined failure threshold. The advantage of this method is that these algorithms can be used even if the knowledge of the physical degradation process is limited. This flexibility has led to three different categories of data-driven approaches for PHM: supervised learning approach (data representing the healthy and faulty states of the system are available), semi-supervised approach (data for only one class, such as the healthy state of the system, is available), and unsupervised learning approach (no labeled data are available).

Decisions about the system health are typically made using assumptions regarding the system data. It should be noted that employing both the supervised and semi-supervised learning techniques requires reliable training data. This is important, as the classification of incoming data is dependent on the training data, and unreliable training data will lead to errors in detection. In addition to detection, an important aspect of data-driven approaches for PHM is prognostics. The prediction of the degradation process is carried out by techniques such as Markov chains, stochastic processes, and time series analysis. Data-driven approaches can also handle many variables in degradation process prediction by modeling the correlation between parameters and interactions of subsystems as well as the effects of environmental parameters. This is typically carried out by restricting the analyses to parameters that are contributing to anomalous behavior in the system, using methods such as principal components analysis [Pecht and Jaai 2010].

Data-driven approaches require the definition of a healthy baseline of a system. A healthy baseline consists of a collection of system data that represents all the possible variations in the data that represent the healthy operating states of the system [Pecht and Jaai 2010]. This baseline data can be obtained when the system is known to be functioning in a healthy state or can also be obtained from specifications and standards. It is important that the baseline data not contain any operational anomalies that could fail and lead to the misclassification of data. Such misclassification can lead to problems such as false indications of anomalies (false alarms) or the failure to detect anomalous behavior of the system (missed alarms).

Due to the absence of published literature explaining the commercial use of prognostics for predicting the failure of components in a wind turbine, the patent literature pertaining to this aspect is discussed in the following section. The main challenge in the implementation of a prediction process for prognostics of a wind turbine system concerns the unpredictable loading conditions acting on a wind turbine. A patent by General Electric illustrates the use of a rainflow counting algorithm in the calculation of fatigue cycles [Lemieux and Moroz 2007]. Proximity sensors mounted on the shaft as well as vibration and oil condition-monitoring signals are used for the estimation of the remaining cycles under fatigue loading. In a second patent by General Electric, the actual site conditions are estimated based on the data from the CMS, and prediction is carried out based on the risk calculated from physics-based, empirical, and event-based maintenance factors. Uncertainty in prediction is incorporated by using the Dempster-Schafer evidence theory, Bayesian fusion, and fuzzy logic. Rainflow matrices and Miners fatigue damage are used to predict the failure Weibull distributions for each component of a wind turbine [Vittal et al. 2011]. The features that are used to estimate the actual loading conditions are the average wind speed, wind Weibull shape factor, wind turbulence intensity, wind shear, air density, hub height, and rotor diameter. Life prediction is performed using wind speed, generator torque, generator bearing temperature, main bearing temperature, shaft speed, operating hours, total power produced, and the oil debris count.

A prognostic system for the blades of a wind turbine was applied by General Electric [Baller, Sihler, and Breit 2012]. In this system, a blade health indicator is used based on the difference in the actual natural frequency and the healthy state natural frequency for each mode of vibration of the blades. One patent that discusses the algorithms used for prediction of failures in a wind turbine is by Honeywell International [Parthasarathy, Uluyol, and Foslien 2011]. This patent describes the various modules that are required for PHM implementation in a wind turbine. It consists of an anomaly detection module, pattern matching module for fault diagnosis, predictive trending module for estimating future behavior, and a future projection module for predicting time-to-failure. An anomaly detection module consists of a threshold checking system in which changes in the features above a predefined threshold are used for fault detection. The sequential ratio test and principal component analysis are also used for anomaly detection. Fault diagnosis for an anomalous condition is carried out using fuzzy logic techniques, self-organizing feature maps, and least squares estimation.

The existing published literature dealing with research in the implementation of PHM strategies in wind turbines is explained in this section. The existing PHM strategies discussed in the published literature can be classified as follows: techniques based on system identification, state space-based approaches, signal analysis-based approaches, and artificial intelligence-based approaches. System identification-based approaches are model-based approaches in which a system is modeled based on the relationship between the responses of a system to a known input signal to the system. The output of an actual system is monitored in real time and compared with the actual outputs from the developed model. The difference between these values is used as a fault indicator. These methods have also been successfully applied to fault detection in process plants; one example of the implementation of this method is presented by Barszcz

[2009]. State space-based approaches are based on techniques such as Kalman filters or particle filters, which require the definition of models that represent the process dynamics; one such example is given by Butler et al. [2012]. Signal analysis-based approaches are based on fault detection methods that use the time- and frequency-domain characteristics of a signal. This method has been applied successfully in rotating machines such as turbines, electrical machines, gearboxes, and bearings. Various spectral analysis methods, such as cepstrum and envelope analysis, have been used for detailed fault diagnosis as well as remaining lifetime estimates. Artificial Intelligence (AI) based approaches are used when the underlying process in the system is too complex to be modeled analytically. Some examples of this approach are the application of fuzzy techniques or neural networks for modeling the complex relationships in the degradation process of the system.

The definition of failure threshold is an important step in the implementation of PHM in a wind turbine. An example of defining this failure threshold based on the relation between wind speed and active power output of a wind turbine under any operating condition was provided by Caselitz and Giebhardt [2005]. The data values used in their algorithm were the 5-min mean values of wind speed and power output calculated every 3000 samples from data collected at a sampling rate of 10 Hz. During the initial learning phase of this algorithm, the faultless operating condition of the wind turbine is defined, and the mean value and standard deviation for active power and wind speed are calculated. This relationship is called the power characteristic of the wind turbine. As soon as the learning phase is completed, the monitoring phase of the algorithm begins, in which the alarm limits are estimated based on deviations from the faultless condition. Another method for failure threshold definition for the detection of periodic nacelle oscillations using spectral analysis and order analysis with the FFT algorithm was proposed by Caselitz and Giebhardt [2005]. In this method, nacelle oscillations are analyzed using acceleration signals of a variable-speed wind turbine. This algorithm is based on the analysis of samples recorded at equidistant rotational angles instead of the usual method, which is used if there are equidistant time samples. Hence, this algorithm uses the rotor position signal for its analysis. It was observed that the frequencies of nacelle oscillations occur in the range of 0.1–3 Hz for large wind turbines. Amplitude and phase of the 1p nacelle oscillations (oscillations at the rotational frequency of the rotor) are monitored. Trend analysis and definition of the failure threshold are then carried out. Failure threshold is defined as a circle around the tip of the complex mean value of the amplitude and phase values, which represents the faultless condition of the rotor. Changes observed only in phase with constant amplitude are represented by a distance (dX1), and the detection of an amplitude deviation without significant phase shift is represented by a distance (dX2). The two extreme conditions for these distances are defined as failure thresholds.

Wind turbine component failures occur due to fatigue or tensile loading on the metallic supporting structures. Hence, prognostics implementation using a strain monitoring-based algorithm can be promising for the estimation of RUL. Strain loading on a certain body tries to deform or distort the shape due to inherent material properties. The internal resistance of the material to the external load is measured as strain, and monitoring of the strain rate occurring in the structure can provide insights into the deformation process. Any deformation process begins with the initiation of crack due to the strain loads, which can be measured, and monitoring this process can be an indicator of the severity of the impending fault. Strain monitoring can be carried out by means of strain sensors mounted on structures subjected to strains due to the external loads. A methodology for wind turbine blade monitoring using Acoustic Emission (AE) signals for the detection of damage processes in the structure was proposed by Dutton et al. [2003]. The methodology was developed for the detection of peak load events and fatigue loading events. In this method, monitoring is carried out by means of surface-mounted piezoelectric sensors to detect and locate the origin of AE waves within the structure. Features such as amplitude and energy of the AE signal are used to define the damage process occurring in the blade. AE monitoring is used to determine the critical damage parameter using the extracted features for a given load. Pattern recognition was used to analyze and classify AE data obtained from wind turbine blade laboratory tests. The pattern recognition algorithm mathematically separates the AE data based on their AE features and defines a grade for the structural integrity of discrete wind turbine blade sections according to predefined failure criteria using the level of AE activity as well as trends in the AE data. The pattern recognition algorithms are based on mathematical clustering algorithms, which divide each set of AE features into clusters based on the similarity of the features. This method was successfully applied for the detection of failures in blades subjected to fatigue loads.

A method for the prediction of blade failure, proposed by Schulz and Sundaresan [2006], is based on the Structural Neural Systems (SNS). This method performs early detection of damage as well as the detection of multiple damage sites on the blades well before the catastrophic failure of the blade. This detection was made possible by means of strain gauges, which were mounted on the blade and indicated damage at the various locations just before a buckling failure. Continuous sensors mounted on the top and bottom of the blade are also used to record signals independent of the SNS. The continuous sensors were monitored by an AE monitoring system. Early detection of blade damage was achieved using the continuous sensors, followed by tracking of the AE occurrences during the damage growth [Sundaresan et al. 2002]. Another method for blade damage prediction using inertial sensing was proposed by Sørensen et al. [2002]. Inertial sensors detect linear acceleration or angular motion. Because blade damage affects the modal dynamics of a wind turbine blade, inertial sensors can detect blade failures. The placement of the sensors was based on prior experience or determination of the likely positions for acceleration—linear and rotational—which can help in the estimation of modal dynamics. Narrowband signals obtained from the inertial sensors were used for fault detection and prediction. The fault detection logic is based on the underlying assumption that if a fault occurs, both frequency and amplitude will change. A tracking band-pass filter (e.g., implemented with a common phase-locked loop) was used for the estimation of this change. The demodulated signals from the band-pass filter were then correlated with the signal from an anemometer to compensate for variations in wind load. These correlations were used as parameters in a curve-fitting procedure employing a general model for the modal dynamics of the blade. Computations are then carried out utilizing this model for blade fault detection and prediction.

Pitching and yawing systems in a wind turbine use servomotors for changing the position of the wind turbine relative to the wind flow direction. These systems are also prone to frequent failures in a wind turbine. Model-based approaches have been adopted for the condition monitoring of this system since this approach is suitable for non-stationary components. Trend analysis for the prediction process is carried out by a process residual-based method and the model parameter-based method. In the process residual-based method, fault detection and prediction is based on the residual of the process and the estimated output signals. The difference between the output of the actual system being monitored and the output of the model is used for monitoring purposes. Trend analysis of this residue can be used to detect changing characteristics of the system [Verbruggen 2003]. In the model parameters-based approach, continuous estimation of the model parameters is carried out, based on the measured input and output values of the system. These model parameters values are then used to model the trends for prediction. The major disadvantage in both of these methods is that the performance of the prediction depends on the accuracy of the estimation process carried out by the model provided by the user. The number of input and output signals as well as the measurement accuracy of these signals play important roles in the ability to detect changes in trends at an early stage [Verbruggen 2003].

A data-mining-based approach for prediction of failures was proposed by Kusiak and Verma [2011] and was implemented on wind turbine SCADA and status data. The first step in this approach is frequency analysis to identify frequently occurring status patterns. Among the identified status patterns, the rotor and turbine blade based status patterns were found to be the most frequent. It was also observed that the status patterns closely follow a negative binomial distribution. The second step in this approach is Association Rule Mining (ARM), which was used for the identification of relations among parameters in the data set. Reduction in the dimensionality of the data set is the next step in the approach. This step is required to identify the parameters that are significant for the prediction, thereby reducing the dimension of the data under analysis. This dimensionality reduction can be carried out based on some underlying knowledge of the physics of the process to eliminate parameters that are not significant. In addition to these methods, data analysis algorithms such as the boosting tree algorithm, Principal Component Analysis (PCA), and the wrapper approach integrated with the genetic or the best-first search algorithms can also be used for dimensionality reduction. Dimensionality reduction is important for a wind turbine, because the SCADA system in a wind turbine usually records more than 100 parameters of a wind turbine. These recorded parameters can be classified into three categories: system-related data, control parameters, and performance parameters. System-related data include details such as turbine number and index, and other details that are specific to a particular turbine. Hence, this data can be excluded from the prediction process. Control parameters represent the desired values, such as set points on the blade and torque set values. Performance parameters indicate the performance of a wind turbine that enables the identification of the status patterns. The next step is the construction of the prediction model, in which five data-mining algorithms—bagging, ripper, rotation forest, random forest, and k-Nearest Neighbor (k-NN)—were used. Bagging is an ensemble meta-algorithm combining different classifiers. The ripper algorithm reduces classification errors by incremental pruning of an inductive rule algorithm to predict output classes. Rotation forest uses parameter selection to construct a classifier using a base learner and a projection filter. Random forest involves numerous decision trees and can be used for both classification and regression. The k-NN classifies objects based on the training examples in parameter space, whereby objects are classified by the majority of votes from their neighbors. The best-performing algorithm was then used to build the prediction models. The random forest algorithm outperformed the remaining data-mining algorithms. A component performance monitoring scheme was then developed to generate alarm signals based upon the outputs predicted at different time intervals.

Several PHM-based techniques, which also exist in the published literature, pertain to individual components of a wind turbine such as PHM techniques for bearings, gears, and induction motors. The PHM techniques developed for these components have been applied to laboratory-level testing and are yet to find applications in wind turbines. With the continuing advancements in wind turbine technology, such as materials and surface preparation of components, these techniques have to be adapted for the actual usage conditions of wind turbines.

10.5.7 References

ACOEM Group. 2012. "Wind power," http://www.acoemgroup.com/Wind-power-2-14-66-application, Accessed: Dec. 3, 2012.

Arabian-Hoseynabadi, H., H. Oraee, and P. J. Tavner. 2010. "Failure Modes and Effects Analysis (FMEA) for wind turbines," *International Journal of Electrical Power & Energy Systems*, Vol. 32, No. 7, pp. 817–824, Sep. 2010.

Ashley, F., A. G. Briggs, R. J. Cipriano, L. E. Gross, S. Breckenridge, J. Hinkson, and P. A. Lewis. 2007. "Report from the Bethany Wind Turbine Study Committee."

Baines, N. C., and T. Way. 2007. "The development of prognostic / health management (PHM) technologies for wind turbines final report," Great Britain Dept. of Trade and Industry, http://www.dti.gov.uk/files/file37829.pdf, Accessed: Dec. 3, 2012.

Baller, M. K., C. M. Sihler, and M. Breit. 2012. "System and method for monitoring turbine blade," U.S. Patent EP2530448 A205, Dec. 2012.

Barszcz, T. 2009. "Decomposition of Vibration Signals into Deterministic and Nondeterministic Components and its Capabilities of Fault Detection and Identification," *International Journal of Applied Mathematics and Computer Science*, Vol. 19, No. 2, pp. 327–335, Jun. 2009.

Bosch Rexroth. 2012. "Bosch Rexroth Rotor Blade Condition Monitoring Systems."

Brüel & Kjær Vibro. 2012. "Dedicated Monitoring Hardware and Software," http://www.bkvibro.com/en/monitoring-solutions/industries/power-generation/condition-monitoring-of-wind-turbines/dedicated-monitoring-hardware-and-software.html, Dec. 3, 2012.

Butler, S., F. O'Connor, D. Farren, and J. V. Ringwood. 2012. "A feasibility study into prognostics for the main bearing of a wind turbine," in *2012 IEEE International Conference on Control Applications (CCA)*, pp. 1092–1097.

Caithness Windfarm Information Forum. 2013. http://www.caithnesswindfarms.co.uk/, Accessed: Feb. 2, 2013.

Caselitz, P., and J. Giebhardt. 2005. "Rotor condition monitoring for improved operational safety of offshore wind energy converters," ASME, *Journal of Solar Energy Engineering*, Vol. 127, No. 2, pp. 253–261.

Caselitz, P., J. Giebhardt, T. Kruger, and M. Mevenkamp. 1996. "Development of a fault detection system for wind energy converters," in *Proceedings of the EUWEC 1996*, Goteborg, Sweden, pp. 1004–1007.

Chen, Z., and E. Spooner. 2001. "Voltage source inverters for high-power, variable-voltage DC power sources," *Generation, Transmission and Distribution, IEE Proceedings*, Vol. 148, No. 5, pp. 439–447, Sep. 2001.

Chen, Z., J. M. Guerrero, and F. Blaabjerg. 2009. "A Review of the State of the Art of Power Electronics for Wind Turbines," *IEEE Transactions on Power Electronics*, Vol. 24, No. 8, pp. 1859–1875, Aug. 2009.

Cheng, S., M. H. Azarian, and M. G. Pecht. 2010. "Sensor Systems for Prognostics and Health Management," *Sensors*, Vol. 10, No. 6, pp. 5774–5797, Jun. 2010.

Ciang, C. C., J.-R. Lee, and H.-J. Bang. 2008. "Structural health monitoring for a wind turbine system: a review of damage detection methods," *Measurement Science and Technology*, Vol. 19, No. 12, p. 122001, Dec. 2008.

Corten, G. P., and H. F. Veldkamp. 2001. "Insects can halve wind-turbine power," *Nature*, Vol. 412, No. 6842, pp. 41–42.

DOE. 2008. "20% Wind Energy by 2030," DOE/GO-102008-2567, Jul. 2008, http://www.20percentwind.org/20percent_Summary_Presentation.pdf.

Dutton, A. G., M. J. Blanch, P. Vionis, D. Lekou, D. R. V. Van Delft, P. A. Joosse, A. Anastassopoulos, D. Kouroussis, T. Kossivas, and T. P. Philippidis. 2003. "Acoustic emission condition monitoring of wind turbine rotor blades: laboratory certification testing to large scale in-service deployment," in *Proc. European Wind Energy Conf.—EWEC 2003* (Madrid, Spain).

Elsevier. 2005. "Managing the wind: Reducing kilowatt-hour costs with condition monitoring," Elsevier Ltd., *Refocus*, Vol. 6, No. 3, pp. 48–51, May 2005.

Faulstich, S., B. Hahn, P. Lyding, and P. Tavner. 2009. "Reliability of offshore turbines–identifying risks by onshore experience," in *Proc. European Offshore Wind 2009, Conference and Exhibition*, pp. 14–16.

FiberSensing. 2012. "FiberSensing—Bringing light to measurement," http://w3.fibersensing.com/, Accessed: Dec. 3, 2012.

Gamesa. 2012. "Gamesa/Products and services/Operation and maintenance services," http://www.gamesacorp.com/en/products-and-services/operation-and-maintenance-services/predictive-maintenance.html, Accessed: Dec. 3, 2012.

Ganesan, S., V. Eveloy, D. Das, and M. Pecht. 2005. "Identification and utilization of failure mechanisms to enhance FMEA and FMECA," in *Proceedings of the IEEE workshop on accelerated stress testing & reliability (ASTR)*, Austin, Texas.

GE Measurement & Control. 2012. "ADAPT Wind Solution, Wind Turbine Monitor, Wind Condition Monitoring Solution—GE," http://www.ge-mcs.com/en/bently-nevada-monitoring/continuous-online-monitoring/adapt.html, Accessed: Dec. 3, 2012.

GE Measurement and Control. 2012. "Bently Nevada ADAPT Wind Condition Monitoring Solution," http://www.ge-mcs.com/download/monitoring/GEA18079B_AdaptWindBro_r2.pdf, Accessed: Dec. 2, 2012.

Gram & Juhl A/S. 2012. "Gram & Juhl's M-System—Ideal for Industrial Applications," http://www.gramjuhl.dk/index.php?page=m-system, Accessed: Dec. 3, 2012.

Hameed, Z., Y. S. Hong, Y. M. Cho, S. H. Ahn, and C. K. Song. 2009. "Condition monitoring and fault detection of wind turbines and related algorithms: A review," *Renewable and Sustainable Energy Reviews*, Vol. 13, No. 1, pp. 1–39, Jan. 2009.

Harte, R., and G. P. A. G. Van Zijl. 2007. "Structural stability of concrete wind turbines and solar chimney towers exposed to dynamic wind action," *Journal of Wind Engineering and Industrial Aerodynamics*, Vol. 95, No. 9–11, pp. 1079–1096, Oct. 2007.

Hatch, C. 2004. "Improved wind turbine condition monitoring using acceleration enveloping," *Orbit*, Vol. 61, pp. 58–61.

Holst-Jensen, O., and I. Aarhus. 2001. "Advanced condition monitoring of wind turbines," The "Vibration Day 2001" organized by SVIB at the Royal Library in Copenhagen, 21st November 2001.

http://www.boschrexroth.com/en/xc/industries/renewable_energies/wind_energy/products_1/rotor_blade_condition_monitoring_systems_/index, Accessed: Dec. 3, 2012.

Hyers, R. W., J. G. Mcgowan, K. L. Sullivan, J. F. Manwell, and B. C. Syrett. 2006. "Condition monitoring and prognosis of utility scale wind turbines," *Energy Materials: Materials Science and Engineering for Energy Systems*, Vol. 1, No. 3, pp. 187–203, Sep. 2006.

India Environmental Portal. 2012. "Global wind energy outlook 2012 | India Environment Portal," http://www.indiaenvironmentportal.org.in/reports-documents/global-wind-energy-outlook-2012, Accessed: Nov. 30, 2012.

Jeffries, W. Q., J. A. Chambers, and D. G. Infield. 1998. "Experience with bicoherence of electrical power for condition monitoring of wind turbine blades," *Vision, Image and Signal Processing, IEE Proceedings*, Vol. 145, No. 3, pp. 141–148.

Khan, M. M., M. T. Iqbal, and F. Khan. 2004. "Reliability analysis of a horizontal axis wind turbine," in *IEEE 14th NECEC conference* (Newfoundland, Canada).

Kliman, G. B., and J. Stein. 1992. "Methods of Motor Current Signature Analysis," *Electric Machines & Power Systems*, Vol. 20, No. 5, p. 463.

Knudsen, T., T. Bak, and S. M. Tabatabaeipour. 2012. "Detection of Excessive Wind Turbine Tower Oscillations: Fore-Aft and Sideways," in *American Control Conference*, 2012, pp. 5795–5800.

Kreidler, V. 2010. "Online gearbox-load-monitoring to increase gearbox reliability," presented at the International Conference on Renewable Energies and Power Quality (ICREPQ'10), Granada, Spain.

Kusiak, A., and A. Verma. 2011. "Prediction of Status Patterns of Wind Turbines: A Data-Mining Approach," *Journal of Solar Energy Engineering*, Vol. 133, No. 1, p. 011008.

Larsen, F. M., and T. Sorensen. 2003. "New lightning qualification test procedure for large wind turbine blades," in *Proceedings of International Conference on Lightning and Static Electricity*, Blackpool, UK.

Lemieux, D. L., and E. M. Moroz. 2007. "Method and apparatus for condition-based monitoring of wind turbine components," Mar. 7, 2007.

Lu, B., Y. Li, X. Wu, and Z. Yang. 2009. "A review of recent advances in wind turbine condition monitoring and fault diagnosis," in *IEEE Power Electronics and Machines in Wind Applications, 2009*, PEMWA 2009, pp. 1–7.

Manwell, J. F., J. G. McGowan, and A. L. Rogers. 2009. *Wind Energy Explained: Theory, Design and Application*, John Wiley and Sons.

Mathew, S., D. Das, R. Rossenberger, and M. Pecht. 2008. "Failure mechanisms based prognostics," in *International Conference on Prognostics and Health Management. PHM 2008*, pp. 1–6.

Maughan, James. 2007. "Technology and reliability improvements in GE's 1.5 MW wind turbine fleet," in *Proc. 2nd Wind Turbine Reliability Workshop, Sandia National Laboratory, Albuquerque, NM, Sept. 17–18, 2007*, http://windpower.sandia.gov/2007reliability/PDFs/Mon-9-JamesMaughan.pdf.

Mita-Teknik. 2012. "Condition Monitoring System (CMS)," http://www.mita-teknik.com/Solutions/ConditionMonitoringSystem%28CMS%29.aspx, Accessed: Dec. 3, 2012.

MLS Electrosystem. 2007. "Predictive maintenance for blade pitch systems," in *Proc. 2nd NREL Wind Turbine Reliability Workshop, Sept. 17–18, 2007, Sandia National Laboratory, Albuquerque, NM*, http://windpower.sandia.gov/2007reliability/PDFs/Tues-3-B-PaulRowan.pdf.

Moog. 2012. "Moog Rotor Monitoring System," http://www.moog.com/markets/energy/wind-turbines/rotor-monitoring-system/, Accessed: Dec. 3, 2012.

Moventas. 2012. "CMaS Condition Management System," http://www.

moventas.com/what-we-do/service/products-and-solutions/cmas-condition-management-system, Accessed: Dec. 3, 2012.

Musial, W., S. Butterfield, and B. Ram. 2006. "Energy from offshore wind," in *Offshore Technology Conference, Houston, Texas, 2006*, pp. 1888–1898.

Nordex. 2013. "Nordex service and maintenance, plus condition monitoring and n-Mobile," http://www.nordex-online.com/fileadmin/MEDIA/Sonstiges/Nordex_ServiceSuite_EN.pdf, Accessed June 10, 2013.

Parthasarathy, G., O. Uluyol, and W. Foslien. 2011. "Condition-Based Maintenance System for Wind Turbines," U.S. Patent US 2011/0313726 A1.

Pecht, M., and R. Jaai. 2010. "A prognostics and health management roadmap for information and electronics-rich systems," *Microelectronics Reliability*, Vol. 50, No. 3, pp. 317–323, Mar. 2010.

Rademakers, L. W. M. M., T. W. Ver, P. A. van der Werff, and H. Korterink. 2004. "Fiber optic blade monitoring," in *Proceedings of European Wind Energy Conference*, London, UK, November 2004.

Ribrant, J., and L. Bertling. 2007. "Survey of failures in wind power systems with focus on Swedish wind power plants during 1997–2005," in *Power Engineering Society General Meeting, 2007*. IEEE, pp. 1–8.

Rosenbloom, Eric. 2013. "A Problem With Wind Power," AWEO, http://www.aweo.org/problemwithwind.html, Accessed: Feb. 2, 2013.

Schaeffler. 2012. "Schaeffler (UK) Ltd. | Media Library | FAG WiPro Wind Turbine Protection," http://www.schaeffler.co.uk/content.schaeffler.co.uk/en/mediathek/library/library-detail-language.jsp?id=1364147, Accessed: Dec. 3, 2012.

Schulz, M. J., and M. J. Sundaresan. 2006. "Smart Sensor System for Structural Condition Monitoring of Wind Turbines," Subcontract Report, NREL/SR-500-40089.

SKF Group. 2012. "On-line Systems for Wind Segment (WindCon)—KF.com/Products/Condition monitoring products/On-line systems/On-line Systems segment specific," http://www.skf.com/portal/skf/home/products?contentId=884467, Accessed: Dec. 3, 2012.

Sørensen, B. F., E. Jørgensen, C. P. Debel, F. M. Jensen, H. M. Jensen, T. K. Jacobsen, and K. M. Halling. 2004. "Improved design of large wind turbine blade of fibre composites based on studies of scale effects," Danmarks Tekniske Universitet, Risø Nationallaboratoriet for Bæredygtig Energi.

Sørensen, B. F., L. Lading, P. Sendrup, M. McGugan, C. P. Debel, O. J. D. Kristensen, G. Larsen, A. M. Hansen, J. Rheinländer, and J. Rusborg. 2002. "Fundamentals for remote structural health monitoring of wind turbine blades—a preproject," Risø National Laboratory, Risø-R-1336(EN).

Spinato, F., P. J. Tavner, G. J. W. Van Bussel, and E. Koutoulakos. 2009. "Reliability of wind turbine subassemblies." Renewable Power Generation, IET 3, No. 4, pp. 387–401.

Sundaresan, M. J., G. Grandhi, M. J. Schulz, and G. R. Kirikera. 2002. "Embedded continuous sensor for monitoring damage evolution in composite materials," presented at the SPIE conference on smart structures, devices, and systems, RMIT University, Melbourne, Australia, December 16–18, 2002, pp. 222–231.

Sundaresan, M. J., M. J. Schulz, and A. Ghoshal. 2002. "Structural health monitoring static test of a wind turbine blade," *Report, National Renewable Energy Laboratory, NREL/SR-500-28719.*

Thomsen, O. Thybo. 2006. "Sandwich materials for wind turbine blades," pp. 97–114, Jul. 2006.

Verbruggen, T. W. 2003. "Wind turbine operation & maintenance based on condition monitoring," Energy Research Centre of the Netherlands, Petten, The Netherlands, Final report, ECN-C—03-047, Apr. 2003.

Vichare, N. M., and M. G. Pecht. 2006. "Prognostics and health management of electronics," *IEEE Transactions on Components and Packaging Technologies*, Vol. 29, No. 1, pp. 222–229, Mar. 2006.

Vittal, S., S. Nanda, A. Joshi, D. Green, and H. Azzam. 2011. "System and method for wind turbine health management," U.S. Patent 789501622, Feb. 2011.

Walford, C. A. 2006. "Wind turbine reliability: understanding and minimizing wind turbine operation and maintenance costs," Sandia National Laboratories, SAND2006-1100, Mar. 2006.

Watson, S. J., B. J. Xiang, Wenxian Yang, P. J. Tavner, and C. J. Crabtree. 2010. "Condition Monitoring of the Power Output of Wind Turbine Generators Using Wavelets," *IEEE Transactions on Energy Conversion*, Vol. 25, No. 3, pp. 715–721, Sep. 2010.

Wikipedia. 2013. http://en.wikipedia.org/wiki/Vertical_axis_wind_turbine, Accessed: Feb. 15, 2013.

Windmonitor. 2012. http://windmonitor.iwes.fraunhofer.de/windwebdad/www_reisi_page_new.show_page?page_nr=191&lang=en. Accessed: Nov. 30, 2012.

World Wind Energy Association. 2012. "World Wind Energy Report 2011 launched," http://www.wwindea.org/home/index.php?option=com_content&task=view&id=362&Itemid=43, Accessed: Nov. 30, 2012.

CHAPTER 11 Disruptive Technologies

Mike Augustin, *IVHM Inc.*
Keith Jackson and Ravi Rajamani, *Meggitt PLC*
Kirby Keller, *The Boeing Company*
Seth Kessler, *Metis Design Corporation*

> "Whatever will be, will be, but the idea began somewhere."
>
> —Anonymous

11.1 Introduction

New and expanding requirements to implement Integrated Vehicle Health Management (IVHM) systems are gaining momentum in aerospace and other safety- and cost-critical application areas. New technological advances are being investigated to enable and/or enhance health management capabilities, assist in business case convergence, and/or accelerate commercial acceptance into current and future applications. This chapter covers emerging technologies that could have a significant impact on the IVHM field. A team of subject matter experts shares their thoughts regarding new sensors, system concepts, and applications that could impact the future of IVHM.

11.2 Sensor Technologies

11.2.1 MEMS/Digital Sensors

The age of smart sensors is revolutionizing data acquisition and data processing. No longer is the sensor limited to devices with an analog output. Digital technology and the ability to merge mechanical devices into MicroElectroMechanical System packages (MEMS) is leading the way to new architectures, reductions in point-to-point wiring, and importantly, the associated weight reductions for the airborne equipment. In addition, the digital nature of the devices gives designers the ability to reconfigure a network of components and functionality to suit each use. Smart sensors can conceivably operate autonomously, adjusting to the *context* of each vehicle's real-time operations.

The most significant breakthroughs are coming in MEMS packages able to detect motion and/or convert motion into harvested energy. For example, MEMS-based accelerometers are available that could transform Health and Usage Monitoring Systems (HUMS) and other vehicle vibration-monitoring solutions from dedicated hardware solutions to flexible systems of smart, reconfigurable sensors. This type of system has the opportunity to reduce the cost and weight of many vehicle health-monitoring applications significantly.

Motion sensors, which are available from a handful of manufacturers, offer all six Degrees of Freedom (6 DoF), providing linear and angular measurements. These inertial sensors are already used in our phones and gaming devices and are expected to lower the cost and weight of tracking vehicle motions for IVHM systems. Devices that include a magnetometer-based reference system (compass) are also of interest for tracking vehicles with reference to the earth. In essence, it is now possible to have a three-axis accelerometer, a three-axis gyroscope, and a three-axis magnetometer in one device, and at relatively low cost. A single device now provides all of the information needed to support a basic, low-cost Helicopter Flight Data Monitoring (HFDM) system. A complete HUMS system using MEMS technology is discussed by Bechhoefer, Augustin, and Kingsley [2012].

In addition to wired connectivity, the addition of wireless (radios) to MEMS devices has already been announced. These wireless devices could be expected to come in System-In-Package (SIP) configurations. One such offering has an assembly in a 3-mm by 3-mm SIP package (minus the antenna). These are the devices of which an IVHM system designer's dreams are made.

Other technologies that may find their way into sensor applications include direct-write wiring, flexible electronics, and 3-D printing of devices. These technologies have the possibility of supporting the fabrication and integration of embedded sensors, which will be discussed in a later section.

11.2.2 Wireless Sensors/Energy Harvesting

Individual self-powered wireless sensors have been demonstrated successfully in several different aerospace applications for IVHM. They have been used to measure loads in components such as helicopter pitch links, yokes, and rotor blades, as well as the main rotor mast. The form of energy harvesting selected depends on providing the best source of deterministic power for each application. Strain, vibration, solar, and thermal gradient have all been used depending on the application. The capability also exists to power sensors using Radio Frequency (RF) energy. Development kits have been available for several years from a leading technology company. These applications have been documented in a number of technical papers in the rotorcraft or aerospace communities.

Another example of powering a wireless sensor is enabled by an Intel Development kit for Wireless Identification and Sensing Platform (WISP). An example of a successful self-powered wireless sensor (powered by the component's strain) is presented by Arms et al. [2007]. Various forms of wireless sensor using different methods of energy harvesting are also described and demonstrated by Andrews and Augustin [2010].

11.2.3 Integrated Wireless Data Transfer and Power Generation

An alternative to individual wireless sensors is an integrated system that can perform data acquisition and transfer time-synchronized data for a group of sensors installed in a location, which may even be rotating with respect to other systems. These capabilities have been demonstrated in a flight test environment using a variety of technologies. One configuration that was successfully tested involved a Bluetooth "slip ring" with the data acquisition and radios powered by a permanent magnet alternator. Other investigators have proposed lower-power versions implemented via a ZigBee radio. The wireless data transfer could be accomplished after the aircraft had landed, thus eliminating the certification concern of in-flight radio transmission. A full-scale system demonstration illustrating the use of an integrated wireless data transfer system has been documented [Augustin and Yeary 2008; Andrews and Augustin 2010].

11.2.4 RFID

Radio Frequency IDentification (RFID) is now being selected for many aerospace tracking and monitoring applications. Several OEMs, including Boeing and Airbus, collaborated on an SAE International standard for passive RFID for aerospace use. The SAE AS5678 standard for UHF passive RFID was released in 2007. The Boeing 787, for example, was a test bed for RFID deployment. The capability and infrastructure is currently offered by Boeing for implementation and use by the airlines. The technology is currently deployed on required safety items such as flotation equipment and oxygen systems. More recent development work is aimed at harnessing the benefit of semi-passive RFID through RF power recharging methods.

Many aerospace applications would benefit from an autonomic system to track the configuration, usage, and maintenance of installed aerospace components. Semi-passive RFID provides sufficient gain margins to make this possible on an aerospace vehicle. Current research is being conducted to overcome the obstacles for miniaturizing, attaching/embedding, and recharging these devices in-situ on the vehicle. A paper describes the requirements and concept development of an RFID system for on-aircraft parts tracking [Augustin and Yeary 2008].

11.2.5 Noncontact Sensing

A number of sensors are in development to detect loads and/or torque on shafts using noncontact sensor approaches. These efforts allow shaft measurements to be made that would otherwise require mechanical "slip rings" or wireless approaches to acquire the data. Magnetostrictive materials are the primary technology used in these sensors. Hall effect sensors or magnetic coupled pick-up coils can be used as the sensing element. A number of demonstrations have been completed validating the technologies to Technical Readiness Level (TRL) 6 or greater.

11.2.6 Fiber Optics

Fiber-optic sensors of various types and system configurations have been introduced for aerospace applications. The sensor arrays can be multiplexed in several different ways. Time Domain Multiplexing (TDM) and Wavelength Domain Multiplexing (WDM) are two of the most common. Sensing of strain, loads, temperature, and vibration are all feasible using optical sensors. One key reason for the interest in this area is the ability to survive in high-cycle/high-strain applications such as measuring the loads on landing gears and helicopter rotor systems. In addition, the fibers lend themselves to embedding in composite materials or for embedding into mats prepared for installation on metallic or composite component surfaces, easing the placement and installation of the sensor arrays. Fiber interrogators already exist for aerospace applications with reasonable bandwidths (>4 kHz) and relatively low power requirements (< 3 W) compared with previous systems.

Fiber sensors have also been used to detect impact events and locations in challenging applications such as blade strikes along helicopter rotor blades. Determining if the impact level was potentially damaging (or not), identifying which blade(s) were struck, and finding the location of impact on each blade significantly reduces the time, and improves the effectiveness, of NDI evaluations needed to ensure continued airworthiness. The use and demonstration of fiber-optic sensors for load monitoring and impact detection is discussed further in literature [Augustin and Yeary 2008; Andrews and Augustin 2010].

11.2.7 High-Temperature Sensors/Electronics

One person's view of high-temperature sensors and electronics is not necessarily another person's view. Several different limiting factors can come into play. Some of the more frequently encountered are the chip-to-substrate bond, limitations of the packaging (e.g., the use of ABS plastics), or the adhesive between the sensor and point of measurement. Common ABS plastic sensor packaging usually limits sensor applications to 176°F (80°C), so higher-temperature devices require ceramic packaging.

Additionally, there is the traditional limit of 200°C for a typical silicon PN device. Sensor on Chip (SOC) devices are taking those limits higher, as evidenced by work that has been ongoing for a number of years at NASA Glenn Research Center and other venues. Here, the intent is to push the limits to 600°C for engine sensing and control applications. To go higher, one needs to separate the sensing element from the signal conditioning. This is commonly done for monitoring or control of jet engines, or the launch vehicle of a spacecraft. For some extremely high temperatures, fiber-optic sensors are available. Specialized fiber sensors can perform up to 1100°C, as noted by Harper [2010]. Where wireless high-temperature sensors are required, Surface Acoustic Wave (SAW) sensors can be used.

11.2.8 Multifunctional Materials (Structures and Sensing)

In its simplest form, a multifunctional material is one that serves a purpose beyond just being a structural element. Typical vehicles are made of dozens to hundreds of structural elements that can carry load, contain gases or fluids, offer protection, or provide an aerodynamic surface. In principle, because these elements have already "bought their way onto the vehicle" based on their primary function, they could perform a secondary function, at no additional cost or weight penalty. Of course, this is particularly attractive to aerospace applications that are extremely mass sensitive. Some examples of multifunctional capabilities that have been solicited in the past decade, as they relate to IVHM, include energy storage (as in a structural battery) [Thomas et al. 2003], power generation (such as strain energy harvesting) [Priya and Inman 2009], power and data transfer (structural cables) [Jarosz et al. 2011], actuation (such as morphing structures) [Akhilesh and Kudva 2004], self-monitoring and self-healing (discussed in a further section) [Kessler et al. 2010].

In terms of technologies that enable multifunctionality, some have emerged in the past decade. Shape memory alloys can be "programmed" with multiple stable configurations based on thermal or electrical inputs. Carbon fibers, themselves, have reasonable resistance ranges that make them potentially suitable for power and data transfer, as well as structural monitoring via resistivity measurements With carbon-based laminates, the trick then becomes how to isolate various bundles electrically; a complex blend of carbon and glass weaving or layering may be necessary. More recently, many investigators have focused on nanoengineered materials such as Carbon Nanotubes (CNT) for multifunctional purposes, including all of those previously described [Garcia et al. 2008].

The benefit of using materials such as CNT is that they can be readily incorporated into traditional composite laminates, and their properties such as electrical and thermal conductivity can be tailored independently from the host material so as not to impact any designed structural properties. CNT have already been demonstrated as an effective means for power distribution, deicing, and Structural Health Monitoring (SHM) via resistance measurements. Furthermore, CNT are piezoresistive, thus they can record impact events via dynamic strain from the impinging propagating waves. Several universities are presently in development of other forms of nanoengineered materials with properties that can be tailored, which may well be prevalent in structural design for multifunctionality in the coming decades [Méallet-Renault, Denjean, and Pansu 1999].

11.2.9 Embedded Sensors

Second only to inherently multifunctional materials, structures with embedded sensors offer many of the same benefits previously described. The distinction here is that while these sensing elements are inherently part of the structure and thus exposed to the same loading, they are discrete, and only serve to sense (i.e., are not contributing to a structural purpose). The greatest challenges for embedded sensors are:

1. Incorporating them into a structure in a practical and cost-effective manner
2. Assuring the survival of the sensing element through integration and operation
3. Ensuring that the integration does not adversely affect the host structure
4. Externally connecting power and/or transmitting data to these sensors

Conceptually, any sensor could be an embedded sensor; it is just a matter of overcoming these technical challenges.

The most common embedded sensor is the fiber-optic line, described previously, which can be used to measure strain and temperature, as well as for transmission of sensor data [Bocherens et al. 2000]. Many of the challenges previously described have been overcome by decades of research to miniaturize the fiber to the extent that it is finer than the structural fibers in a laminate, and to formulate coatings to protect the fibers from extreme thermal and mechanical loads. While connections still remain a practical challenge, several options have been commercially developed over the years.

Other types of embedded fiber/wire sensors have been demonstrated to a lesser extent for purposes such as integral temperature, pressure, and moisture monitoring [Kim, Ismail, and Springer 1993].

Kapton-based sensors have also been demonstrated as effective embedded sensors by multiple investigators, such as those with metallic traces for strain or crack measurement, or piezoelectric material for elastic wave excitation and measurement [Lin and Chang 2002]. While some of the challenges such as cost-effective and survivable integration have been demonstrated, reductions in static strength and fatigue life have been extensively noted, and practical breakouts have so far not been achieved. Further innovation would be necessary for this style of embedded sensor to be adopted commercially. In general, MEMS sensors, also previously described, have the right scale for embedding. However, they have interconnect challenges even when not inside a structure, and so embedding is likely to further complicate the issue, which is why there has not been a prevalence of MEMS embedded sensors to date [Schoess and Zook 1998].

In general, the greatest advantage of embedded sensors is that they are exposed to the ambient environmental conditions (directly at least), and are thus subjected to less wear and incidental damage. Additionally, they are inherently lightweight and conformal by nature. The greatest disadvantages are that they become impossible to directly inspect, repair, or replace without physically damaging the structure in which they are embedded but, given the "low cost" of installation, this can be mitigated by incorporating additional spare sensors at the outset. Obviously, embedded sensors are also more suited for layered materials, such as composite laminates, as opposed to metal (although this is being investigated with metal additive layer techniques). However, there is the issue of thermal coefficient matching to address so that the sensors do not initiate defects themselves during excursions from ambient conditions.

11.2.10 Deposited Sensors

One way to circumvent some of the toughest challenges of embedded sensors is to deposit them on the surface of a structure, as this greatly simplifies many of the integration and interconnect issues. Furthermore, deposited sensors have significant benefits over simply bonded ones in that they are essentially part of the structural surface, thus conformal, and not susceptible to typical adhesion failures. The downside compared with embedded sensors is that they are exposed to more direct environmental factors.

Two forms of sensor deposition have become commercially available over the past several years, both of which are known as "direct-write" technologies. In the first, conductors and/or insulators are sprayed onto the structure as nano-ink, either over a large area with templates or in fine patterns using "printing" heads [Curtis et al. 2001]. The nano-ink is then cured via an oven, or UV light. The second approach is plasma flame spray, in which the deposited metal is actually liquefied by an extremely large current and then propelled at the structure, where it fuses with the material and cools to form traces [Longtin et al. 2004]. In both technologies, extremely fine trace widths and pitch can be achieved, and they can be applied accurately over large areas using robotic arms. Deposited sensors have been demonstrated commercially for crack detection (simple break trace), and as mature strain gauges and thermocouples.

Several other standard-style sensors have been proposed and demonstrated in prototype by researchers. Furthermore, deposited traces can serve as power and data lines for themselves or for traditional sensors, and have even been demonstrated in impedance controlled formats [Kessler et al. 2009]. Generally, these types of sensors and traces are inherently survivable in extreme environments, and have proven to be quite durable against fatigue loading. The greatest downside is that there is, as yet, no standard, off-the-shelf, direct-write sensor, so generating them still involves large up-front costs.

11.2.11 Self-Aware and Self-Calibrating Sensors

Sensors are an essential aspect for nearly all subsystems of IVHM systems. However, as a consequence, the reliability of sensors dictates the reliability of the overall subsystem and system itself. In fact, sensor reliability is a principal factor that has slowed the introduction of many forms of IVHM, because while the system would have more information available to it, the overall reliability could be reduced by the presence of misinformation. A critical future technology would be truly smart sensors that are self-aware beyond just TEDS (Transducer Electronic Data Sheet as defined in IEEE 1451) information, to the extent that the reliability of their data would be exponentially increased [IEEE 2007]. These are sensors that only provide data that can be substantiated as 99.999% reliable. They would understand normal operating conditions, and either not provide data if outside those conditions, or at least be able to quantify uncertainty under these conditions. They would recognize when they are out of calibration, and potentially be able to periodically self-calibrate. They could identify anomalous data as result of disbond or damage, rather than blindly feed the data to the next system element. In the longer term, they might even be able to repair themselves. Essentially, these are sensors with integral Verification and Validation (V&V) capabilities to greatly enhance their reliability and the integrity of the data they provide.

11.2.12 Self-Reacting and Self-Repairing Components

Beyond smart sensors are intelligent ones: those that are not only self-aware but self-responsive. These are systems that are able to not only identify and validate anomalous conditions in a vehicle, but also take the additional step to mitigate further damage and/or repair the damage. This mitigation essentially entails removing or otherwise circumventing the condition causing the issue. In an electrical system, this could be rerouting current through an alternate path or, similarly, in a mechanical system it could be a redistribution of load through another member. Alternatively, the system could change an operating condition such as turning off nonessential components or slowing speed, rotation, etc. Although such technology and logic is not unrealistic, these systems must be very robust to allow the IVHM sensing system to influence other systems.

Self-repairing systems are even more futuristic. Various prototype systems have been demonstrated with modest success; these include those with embedded spheres filled with epoxy that break upon introduction of damage and can be cured with a network of electrodes [Motuku, Vaidya, and Janowski 1999]. While attractive, these concepts can have an adverse effect on the primary function of the structure, and thus are not ready for commercial application yet.

Other self-assembling methodologies have also been proposed, but at a much more basic research state. Academic institutions have even gone as far as proposing armies of nanobots that would be deployed by a sensing system to seek and repair damaged structures; however, there has not been any evidence to substantiate whether this technology is much more than a concept [Bogue 2008]. While not essential to commercial acceptance of IVHM systems, these intelligent functions would add further value to the system, thus aiding with business case convergence.

11.3 Wireless Data Systems

11.3.1 Introduction

Wireless data communication has played an important role in enhancing aerospace IVHM capability for more than four decades, and its importance will continue to grow in the future. While current systems are quite limited in their data bandwidth capability, this is rapidly changing with the advent of better and cheaper modalities. In this section, the history and the state of the art in wireless communications are examined in relation to aerospace applications, and a brief overview is given as to how wireless communications technology is evolving. Discussion is restricted to commercial applications. It is possible that military wireless communication technology is superior and that in the future the capabilities might be transitioned to the civilian sector, but because there is no access to reliable information on these technologies, the discussion is confined to the commercial aerospace industry.

11.3.2 Legacy

The technology for wireless communications with aircraft using RF has been around for nearly a century, with wireless equipment starting to become standard equipment for commercial aviation even since the 1930s. However, the use of radio to transmit data, especially data relevant to monitoring the health of the aircraft, has only been around since the late 1970s in the commercial arena. The U.S. Air Force (USAF) has employed data communications via RF since the 1950s when it conducted flight tests at Edwards Air Force Base. New jet aircraft pilots were crashing during flight and were not able to debrief the flight test engineers. The USAF developed telemetry systems and remote ground stations to record real-time telemetry to corroborate with the debriefing by the pilots on the new jet aircraft under test. These were later adopted by the space industry and today, telemetry systems are used on every spacecraft and launch vehicle and missile [Losik 2012].

Aeronautical Radio INC. (ARINC) was established in 1929 by a group of airlines as a neutral coordinator of radio communications in the commercial aviation industry. One of its major initiatives was launching a datalink service called Aircraft Communication Addressing and Recording System (ACARS) in 1978. This is still the most commonly used means of transmitting data to and from airborne systems. ARINC is responsible for an extensive set of communication standards, and it provides its customers with a service based on a network of ground-stations for receiving and transmitting data from aircraft. While ARINC is dominant in North America, a competing company called SITA provides customers with similar communication services predominantly in Europe, Asia, and South America [ARINC 2013].

ACARS is limited in scope and is a relatively expensive means of communication. All data must be compressed into reports consisting of a maximum of 220 characters (or bytes), with a message consisting of a maximum of 16 reports (or about 3.5 kB). These are transmitted when the aircraft is within line-of-sight of a ground station through Very High Frequency (VHF) radio communication. In more recent times, this has been enhanced by Satellite Communications (SatCom) to ensure communications while the aircraft is over water or away from a ground station.

Competing services using Inmarsat (geostationary) and Iridium (low-earth-orbit) satellites have also been developed and tested for aircraft communications. They have not been used extensively in IVHM applications so far, but will become increasingly important in the long term. In-Flight Entertainment (IFE) systems have provided much of the impetus for developing these networks and communications systems, and it is conceivable that in the future these two goals will converge to provide airline customers with systems that provide both IFE and IVHM functions.

While the aerospace and naval industries pioneered the use of wireless technology—because of the obvious necessity of communicating with remote assets—its growth in the automotive world has been phenomenal in the last decade or so. This has come about mainly because the automotive industry quickly leveraged the tremendous growth in the ubiquity and coverage of cellular networks around the world. Much has been written about the use of wireless communication in the automotive sector. A good overview of the subject can be found in the introductory chapters of a recent dissertation [Cottingham 2009]. In the transportation industry, in-vehicle wireless signaling has been in use for distributed power (multiple locomotives) control applications since the sixties. In the monitoring and health management area, wireless (cellular modem) technology has gained a lot of ground in recent years [Boughton, Kessner, and Menadue 2006].

11.3.3 Technology

There are several facets to the use of wireless for the purpose of facilitating IVHM functions. Figure 11.1 shows a generic block diagram of an aviation IVHM system with dashed lines representing links that can be replaced by wireless connections.

The data flow follows a standard pattern that is well recognized in the aerospace world (see Figure 1 in [Rajamani and Waters 2011] or the discussion on architecture in chapter 6). The different elements and technologies involved in a wireless IVHM system are described in the following sections.

11.3.3.1 Sensors

While sensors that communicate wirelessly do exist, and growth continues in this area, no known system is yet deployed in a commercial aerospace application. There are several reasons for this: harsh environment for electronics, no ready availability of wireless power, signal interference due to environmental considerations, and the onerous burden of showing that these systems are safe. However, "less-wire" systems have been envisioned and are starting to make their way into this environment. These systems consist of a group of sensors connected to an electronics module that powers the sensors and acts as a data concentrator. The electronics module can process and convert the signals to an appropriate wireless form, which can then be sent to an analysis module (possibly collocated) as well as a transmitter for further relaying to the operator display or a central database. More information about wireless sensors was presented earlier in this chapter.

11.3.3.2 Wireless Networks and Protocols

Local area networks on most platforms are based on standards, often specific to the industry. For example, in the aerospace

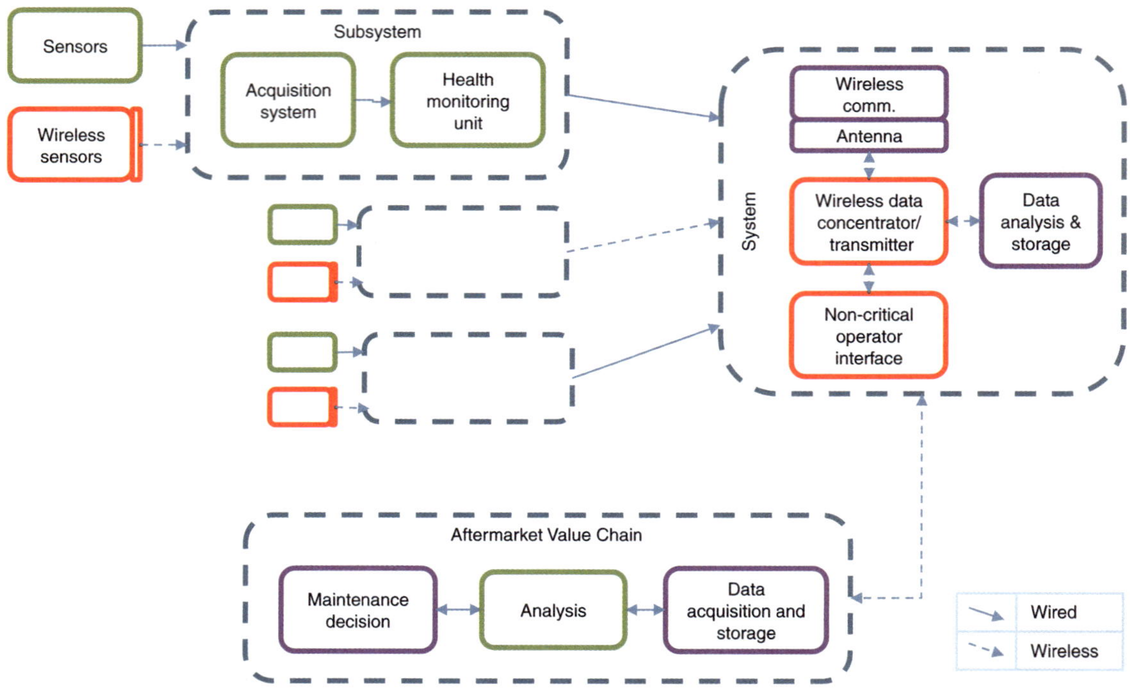

Figure 11.1 System architecture.

world, standards such as ARINC 429, AFDX (ARINC 664), MIL-STD-1553, etc., describe the physical and the logical specifications for wired communications (details can be obtained from the published standards or in reference books [Spitzer 2001]). The Controller Area Network (CAN) serial bus has recently entered the avionics world from the automotive sector to facilitate faster communications over shorter distances [Rao 2009].

It is not easy to translate traffic over these buses to the wireless world, where an entirely different set of communication modes and protocols dominates. Depending on the distance spanned and the power needed, one can consider three sets of modalities: Personal Area Networks (PAN), Wireless Local Area Networks (WLAN), and Wireless Wide Area Networks (WWAN) or cellular networks. The first two categories are covered by IEEE 802.X standards, in which X varies based on the protocol. For example, IEEE 802.11 defines the well-known standards for the standard Wi-Fi systems. Shorter-range communication protocols, such as Bluetooth and ZigBee, are covered by protocols and subcategories under the IEEE 802.15 family of standards [IEEE 2013]. WLAN and PAN networks can help define the architecture of system elements within the confines of an aircraft, with the choice depending on power availability, weight, cost, etc.

Wireless LAN systems can also be used to communicate between the aircraft and either Ground Support Equipment (GSE) or a ground-based access point: say, at the airport gate, where the aircraft is still fairly close at range. In fact, ARINC has developed the Gatelink system, also designated as ARINC 822-1, for Wi-Fi based, parked aircraft-to-airport gate communication [ARINC 2008].

To communicate over a wider area, say from an aircraft on the ground to a remote location, clearly a leading candidate is the cellular modem. While there are restrictions on the use of such devices in the air, connecting and communicating while on the ground is feasible and is already being done. This is a much cheaper solution than ACARS VHF or SatCom. Cellular networks (GSM, 2G/3G/4G, GPRS, etc.) are growing by leaps and bounds, and it will soon be hard to find a populated area of the world, especially near an airport, that is not covered by adequate, if not fast, cellular towers.

This provides the onboard system with a means of communicating with the ground-based system once the aircraft lands. The wireless system can essentially dial up the ground station via a cellular modem and download data. Given the speeds that the newer cell technology affords, during the time the aircraft is taxiing in, parked at the gate, and taxiing out, enough data can be transmitted [ARINC 2008].

In this book, we only deal with wireless networks that are used in commercial aircraft applications. Military and commercial space applications by NASA, USAF, and other space agencies around the world use more sophisticated systems featuring communication standards employing several RF bands and networks of satellites. For example, NASA employs a network of communication satellites and ground stations known as the Tracking and Data Relay Satellite System (TDRSS) for space communications. These communicate in the Ku-, Ka- and S-band RF spectra for space to ground communications. In the future, aircraft communication systems will be able to exploit the infrastructure built up by the space industry to make systems that are more capable and cost effective.

11.3.3.3 Mobility Systems

While slow to catch on in the aerospace world, mobile solutions are nevertheless finding their way into the cockpit and other areas. The first use of these devices within the aerospace world has been in GSE, which can host technical manuals and troubleshooting systems, and in the Electronic Flight Bag (EFB). Technical manuals are rapidly evolving from paper form to electronic form. Interactive Electronic Technical Manuals (IETM) are now fairly mature, with the S100D standard being adopted around the world [S1000D 2013]. Many EFBs are hosted on commercially available mobile devices such as iPads, which are capable of wireless communication but are prevented from utilizing this because of operational prohibitions of using cellular communications [Gondeck 2011]. However, wireless communications are a ready and convenient means of linking to central databases for updating content. This is a critical advantage when compared to their paper-based counterparts. Their acceptance in safety-critical applications will take time and will depend on how well the applicants can demonstrate safety and reliability.

11.3.3.4 Motes

These are small devices that have a very simple structure, communicate via ZigBee protocol, are low powered, and can configure themselves to a local wireless "mesh network" [Richardson 2012]. These are commercially available. A demonstration of the use of these devices in an aircraft interior shows that they can perform quite reliably in such an environment. [Keller et al. 2012].

11.3.3.5 Certification

This is primarily applicable in the aerospace world, although similar standards exist for automotive, energy, medical, and other systems as well. The certification of aerospace IVHM systems is itself a new concept. Only recently has the industry started to take this topic with the seriousness that it deserves. The E-32 committee of SAE International is in the process of publishing guidelines for the selection of assurance levels for engine health management software [SAE 2013]. This is the first document to formally address this topic. Software assurance is so important in the IVHM field, because much of the functionality is deployed via algorithms that are implemented in embedded and application software. Most wireless systems deployed on board are associated with non-safety-critical functions such as IFE and so only need to demonstrate that they do not interfere with critical aircraft functions.

Beyond that, not much needs to be done, since there is no significant consequence of the failure of these systems. As newer wireless systems, more essential to the operation of the aircraft, are deployed, some time needs to be spent considering unique aspects of the wireless systems. The key to any certification exercise is to perform a thorough system safety assessment that clearly identities the consequences of system failures. Existing guidelines for airborne and ground-based systems can be used to guide system designers and reliability and safety experts during both the type certification stage as well as when establishing means of continued air worthiness [SAE 2010].

The cargo Smoke Detection Unit (SDU) is a wireless system that has already been certified to operate on board. This is the first "essential" wireless system to be certified by the authorities. It was developed by Securaplane Technologies and is currently manufactured and sold by Ontic. Details are given in the next subsection along with an example of a more extensive study on the use of wireless systems in a commercial aircraft.

11.3.3.6 Examples of Wireless Systems in the Aviation Industry

The Wireless SDU

The wireless SDU [Blair 2007] system, originally designed and built by Securaplane Technologies and now manufactured by Ontic, is the first essential wireless system certified by the FAA for commercial aviation. The motivation for this system was the ValueJet crash in 1996, after which the FAA mandated that there be a fire detection and suppression system in all airliner cargo bays [Blair 2007; Chilakala 2008]. The wireless SDU was developed around spread spectrum technology to ensure reliability and fail-safe operation. The schematic of this system is shown in Figure 11.2.

Mote Devices

Beyond the local level, wireless technology can provide a means to move data within and from the aircraft to the ground systems at the gate or in flight [Keller et al. 2012]. Wireless enables a plug-and-play network for the entire CBM functional hierarchy. Connectivity tests were conducted on a 737-200 aircraft using commercial off-the-shelf ZigBee motes to determine the level of connectivity between various compartments of the aircraft, as shown in Figure 11.3.

The tests focus on connectivity between likely "gateways" or master nodes located in the wheel wells or the cabin to satellite areas such as the engine nacelles, air conditioning bay, wheels, flap areas, or auxiliary power unit. The connectivity was established and maintained in the majority of locations, showing the potential to support health management communications with a flexible, low-footprint architecture.

11.4 Subsystem Application Technologies

11.4.1 Introduction

IVHM systems are most commonly implemented at the functional or subsystem level due to the close linkage at this level to the most meaningful business case needs. Operational monitoring at this functional level has a more direct relationship to the cost or safety benefits sought by the vehicle health-monitoring system. For example,

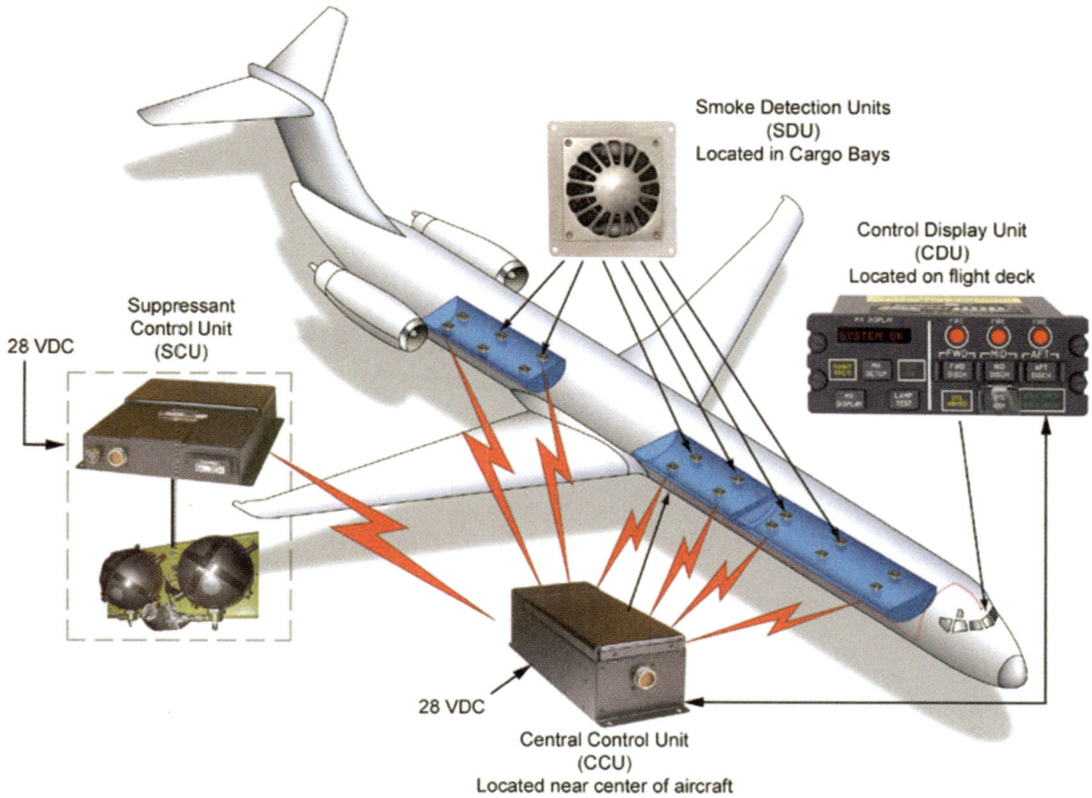

Figure 11.2 Wireless smoke detection system.

Engine Monitoring Systems (EMS) have been aggressively pursued for some time to reduce the operating costs of this critical subsystem. Another example is the pursuit of SHM to mitigate the increased maintenance burden of an aging aircraft fleet. The fact that there are individual SAE International committees working at many of these vehicle functional levels should be of no great surprise. It is of course the integration of all of the vehicle subsystems and conveyance to the stakeholders that ensures the fusion of the benefits that IVHM provides. The following sections and the references cited provide some needed insight into the latest technologies implemented in support of these core business cases that together make up the domain we call IVHM.

11.4.2 Engine Monitoring

Aircraft engine monitoring is one of the more advanced areas of IVHM, with most of the major airline fleets having 24/7 monitoring of their engines. The process to manage their reliability is highly evolved and uses manual and automated (data sensed and transmitted in flight) methods to understand the current condition of the engine. The level of reliability achieved through design and monitoring is now so high that most pilots will not see an in-flight engine shut down in their career. Given this, it is hardly surprising that much effort is put into the management of the fleet using IVHM data, understanding the current condition of the fleet (usage rates), and thereby informing the maintenance

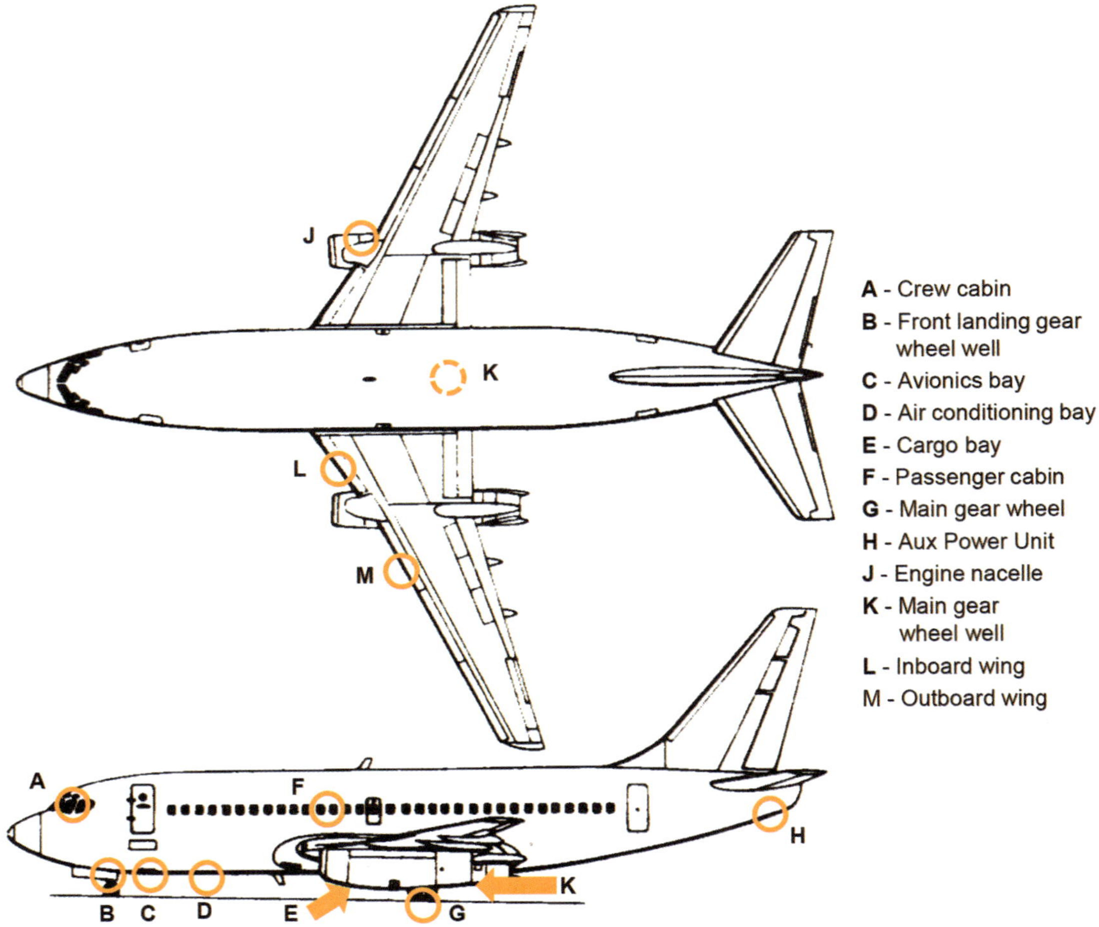

Figure 11.3 B-737 demo.

and supply chain to give the engine users maximum value from their purchase.

However, engines are still subject to damage from foreign objects (birds, debris, and environment) that may damage the blades in an engine. The fan blades (visible at the front of the engine) are made from titanium or carbon fiber. The blades internal to the engine in the compressor and turbine are smaller, run very hot, are subject to heavy loads, and can be subject to cracking or damage caused by foreign objects passing through the engine. Inspection of the blades for damage (which *may* be detected by step changes in the vibration signature in the engine) can be achieved using boroscopes to look inside the engine, rotating the engine and visually looking for damage. There is active research looking at different processing techniques to process engine vibration information and confidently detect if a blade is damaged or cracked with no manual inspection. This saves time (delays to flights) and reduces the risk of secondary damage where part of a blade detaches and causes further downstream damage [Kwapisz, Hafner, and Rajamani 2012].

As the performance requirements for aircraft increase, new engines will be developed that will have additional complexity. Examples are the Pratt and Witney geared turbofan and various open rotor concepts.

Consumer expectation is that new engines will enter service with the same or better reliability than the current generation of engines. While designers will use all the engineering advances (simulation, CFD, etc.) to ensure this, it seems likely that there will be a shortfall between the expectation and reality. Therefore, the IVHM systems will need to be designed to be more interactive, allowing adaption and programming to meet commercial requirements. An example could be the lifing of a component. If it is a safety issue, it will be replaced immediately and is outside the scope of IVHM. If it is a reliability problem, the replacements may be made based on the risk of a delay against the cost to replace. An IVHM system, adapted to gather the appropriate information off the aircraft, combined with a ground-based analysis and planning system can then be used to optimize the cost of delays against the cost of replacement for a given flight schedule. In effect, changes in reliability for a new technology cannot be mitigated, but the impact of any unreliability can.

11.4.3 Structures/Structural Health Monitoring

Perhaps the subsystem with the greatest, but as yet unrealized, cost-savings potential is SHM (or perhaps this is just the subsystem written by the technologist with the greatest bias). While health monitoring of one form or fashion has found its way onto vehicles for most subsystems, fielding SHM has been elusive. Part of the issue is that structures cover the majority of the vehicle, which is of vastly different composition throughout, compared with a few discrete components (e.g., engine, flight controls, APU, etc.) This is not to say that these other subsystems are not of equal or greater complexity, but rather than a single very complex element, structures potentially represent hundreds of different simpler problems. One reason why SHM may be slow to be commercially adopted is the pursuit of the "magic bullet" approach to solve all of these problems. It is quite likely that multiple approaches would need to be choreographed to successfully achieve reliable SHM. However, this also suggests close collaboration between competitors to integrate a system, which can put a strain on the business case.

In general, SHM systems are those that can detect and interpret adverse changes in a structure. These systems can be active, such as guided wave [Kessler, Flynn, and Todd 2011] or eddy current [Goldfine, Schlicker, and Washabaugh 1998] methods that put energy into the structure and sense the response; passive, such as acoustic emission [Ozevin et al. 2007] or fiber optics [Murayama et al. 2003] that monitor for events as a result of outside occurrences; or witness sensors such as strain or crack gauges to measure exceedances or operating conditions to infer damage. For metal structures, key damage modes include fatigue cracks, corrosion, and precursors to fracture. For composite materials, delamination and matrix cracking are the principal concerns. At the substructure level, loosened bolts and/or disbonded ribs are of concern. Levels of SHM characterization range from detection of presence to localization and differentiating type of damage, and finally estimation of size and/or severity.

Requirements for SHM vary greatly depending on the use (and hence business case). To date, the majority of SHM discussion has revolved around AMOC—Alternative Means Of Compliance [A4A 2011]. Here, essentially the SHM system is treated as a conventional Nondestructive Inspection (NDI) tool, where the structure would be inspected on a typical interval, just with an integrated system rather than manually. The future of SHM is Automated Structural Health Monitoring (A-SHM), in which the system would run at preset intervals, without any manual intervention, to enable Condition-Based Maintenance (CBM) of structure rather than preventative maintenance. It is probable that a hybrid form of SHM will eventually be successful, in which the system is allowed to run autonomously to indicate the presence of damage with a high degree of certainty, and with indication of damage type, location, and severity. But, rather than relying on this data, the hybrid system would use it as a guide to perform targeted manual inspection. Periodic, system-wide manual inspections would not be eliminated, but would be spread out much further than presently, as a means to rebaseline and calibrate the SHM to maintain the desired accuracy and reliability levels. By minimizing frequency of tear-down, manual inspection through SHM will not only save life-cycle costs and greatly increase asset availability, but also improve safety (hard to quantify economically) by mitigating human inspector bias and error, particularly for composite structures for which NDI is not as mature for detection of damage beneath the visible surface.

11.4.4 Electrical Power and Wiring

Diagnostics for power and wiring has been the long-term focus of much research. Existing technologies can detect wire chaffing, and advanced technologies can detect defects in live wiring. Other key research is in diagnostics for power generation equipment and batteries. This area is particularly difficult to troubleshoot based on pilot or other reports of functional failures. The use of continuous monitoring and prognostics could have some of the most significant paybacks if flight and mission aborts could be avoided. These technologies are discussed further in literature [Baker, Bordick, et al. 2009; Baker, Thompson, et al., 2009].

Within power systems, the power electronics are a cause of unreliability. In particular, Insulated-Gate Bipolar Transistor (IGBT) devices, the essential building blocks of electrical systems, are prone to failures. Clearly, devices are improving over time but, given the high temperatures and current switching levels at which they operate, they are subject to wearing out. Various strategies can be adopted to combat these failures. Letting them fail and having a backup is one option. Using predictive models and measurements to enable timely replacement is another. Finally, designing systems that have partial redundancy to enable "limp home capability" might be better than a complete backup system, as power electronics are heavy and large.

11.4.5 HUMS/Drivetrain Monitoring

Vibration monitoring systems on vehicles have typically used a collection of analog sensors, primarily accelerometers and 1/Rev interrupters hardwired to a hardware-intensive processing unit. A typical star installation required using twisted shielded point-to-point wiring to each analog sensor. The processor was sized to the needs of the vehicle being monitored and performed all the necessary Time Synchronous Averaging (TSA) and other more advanced diagnostic algorithms for computation of the needed condition and health indicators. Typically, the system operated on a schedule to acquire the necessary data samples under defined torque and or operational conditions (regimes).

MEMS digital sensors and processing can be integrated into stand-alone smart sensors. Systems of this type are already being used to monitor wind energy installations. The use of MEMS sensors in IVHM systems is imminent. The next generation of systems will most likely include these modernizing technologies. Bechhoefer, Augustin, and Kingsley [2012] describe a complete MEMS-based HUMS system. Also, a new means (Polymeric Bead Matrix Technology) [Hedges et al. 2012] of monitoring the quality of fluids used in gearboxes and hydraulic systems may reduce the need and frequency of performing manual oil sample analysis. In the future, laboratory analysis may only have to be performed on-condition.

11.4.6 Rotor System CBM

Applications such as individual aircraft usage monitoring, direct loads monitoring, autonomic configuration identification of critical components, and wireless data transfer (rotating to fixed), are all being addressed at the current time. Monitoring the rotor system in the production environment requires unique technologies and reliability. These technologies were demonstrated in recent U.S. Army CBM Technology Demonstrations at Bell Helicopter and Sikorsky. Bell's work in new sensors and technology for rotor system CBM is described further in references [Augustin and Yeary 2008; Andrews and Augustin 2010].

11.4.7 Platform-Level Diagnostic Fusion

With diagnostic data and analysis available from diverse systems across the platform, it can be expected that fusion of the data can identify relationships, thus enhancing the accuracy and reliability of the diagnostics. In many cases, diagnostics can only detect changes in the measured parameters that could be driven by damage, wear, or maintenance. The integration of ongoing maintenance tasks and adjustments with the health-monitoring data is an essential condition to ensure a robust system with low false alarm rates. A paper describes one approach to platform-level data fusion [Augustin and Hess 2011].

11.5 System-Level Technologies

11.5.1 Unmanned Vehicles

At a system level, an unmanned vehicle needs to exploit the sensing, processing, communications, and actuation technologies that are emerging. Fusing this information and making useful decisions is a complex computing and computer architecture problem. Processing capability will no doubt continue to follow Moore's Law, so the challenges will be around architectures and methods used to process the data.

Unmanned autonomous vehicles have been in operation for decades. One early military example, the V2 rocket, had negligible safety or monitoring capability and resulted in many fatalities, where faults caused the destruction of the launch vehicle or site [King and Kutta 2003].

Today, unmanned aerial vehicles partake in military and surveillance operations with devices ranging from those that are not much more than a smart radio-controlled aircraft through to the General Atomics MQ-1 Predator UAV, which still has "pilots in the loop" flying the aircraft from a remote ground station [DoD 2010].

While details of the IVHM capability of these vehicles tend to be restricted, it can be assumed (because of the time delays in communications) that they have a level of monitoring on the condition of equipment and some degree of control.

IVHM, as applied to commercial aircraft, can still be applied to UAVs. That is, information can be sensed on the UAV, off-loaded at regular intervals, and processed on a ground-based system. Of greater interest to UAVs are the autonomous IVHM capabilities that can be included on the aircraft or vehicle.

Implementation of an IVHM capability on a UAV has three primary functional areas that can be scaled in sophistication depending on the vehicle's requirements and its opportunity to recover from situations.

Monitoring:

Simple monitoring involves sensing parameters and triggering alerts when levels are exceeded. Increased sophistication adds predictive trending of actual or synthetic parameters using single or multivariate models (which may be physical or empirical models). Models may also be adaptive (to varying degrees).

While it might be tempting to apply the most sophisticated techniques, care should be taken to apply the most appropriate monitoring method rather than the most complex.

Cause Isolation:

Where monitoring has indicated "something anomalous," cause isolation is required to find the reason. It may not be necessary to find the root cause, just the general cause. For instance, a noisy electrical power bus may be because of a winding fault in a generator. On the aircraft, all that needs to be isolated is the offending generator so it can be managed.

Cause isolation can be achieved by using "lookup tables" of conditions, but the onboard environment may also allow tests to be run that enable further isolation. For instance, with the previously mentioned electrical bus issue, the vehicle could self-test, switching out individual generators. While cause isolation may be autonomous, it may also be effected using vehicle-to-base communications, which may inform or enact additional tests.

Action:

Where a cause has been fully or partially isolated (a leak may only be partially isolated and may continue to fail), action still needs to be taken. At its lowest level of sophistication, a problem is isolated and the vehicle returns itself to safety. However, judgement is often needed.

In the future, determining the optimal decision is a complex task that may not be isolated to a single vehicle. A group of vehicles may make a better "fleet" decision if the vehicles communicated and exchanged information and potentially replanned their role in a mission. On the other hand, an individual vehicle may make a better decision if it communicated with a base station and people to provide additional information, context, priorities, or recommendations.

As briefly described here, the disruptive technology for UAVS will be the processing and systems integration, which allows data from individual vehicles, other aircraft, and ground systems to be fused to develop "fleet optimized" action plans.

11.5.2 Safety-Critical Applications

Safety-critical systems are developed to an integrity level that allows them to be certified for operation. This process is, quite rightly, very robust, expensive, and hard to change and adapt. IVHM systems need to be the opposite of this. The business case may not support the costs of a safety-critical development and, even if it is affordable, an IVHM system developed to safety-critical standards may not be developed fast enough to capture all of its value (for instance, as new conditions arise in service).

More executive IVHM systems are being developed in which a monitoring system provides information that may be used by an executive control system to change its operation. In this case, the control system is developed to full safety-critical standards, but it can respond to "hints" from the IVHM system, which allows a degree of adaption. Imagine a vehicle that can turn using two different techniques (e.g., steering and wheel speeds). The IVHM system could recommend using wheel speed because the steering may be damaged using steering. The control system is still executive and fully safety-critical, but it is responding to hints and, by so doing, increases the duration of the vehicle.

Looking further into the future, the control systems on UAVs may change. Today's control systems (as part of the certification requirements) are driven by solid, provable logic conditions—e.g., if (this is true) then { do that1 } else { do that2 }. To develop an autonomous vehicle that operates in an open environment (a road or free airspace) will require much more information, from multiple sources, with varying degrees of confidence in the information and adaptive strategies. A driverless train (because of its controlled environment) can move when the signal is green and the doors are closed. A driverless car would make very little progress if it waited for all possible hazards to be classified and removed on the open road.

Therefore, future autonomous vehicles are likely to use probabilistic control methods. In this case, inputs will be vectors of information with indicators of their qualification level. Decisions will be made using decision-making and searching techniques to offer robust solutions to meet the mission objective. For instance, a car detecting a possible object in its path can stop, slow down, sound horn, or ignore. The final decision may be delayed until later as more options may become available [Thrun, Burgard, and Fox 2005].

This is a complex computer science problem to capture, search, and classify vectors or data, so it is not too surprising that one of the most publicised developments in this area is the "Google Driverless Car."

The impact on the changes in the control system on the IVHM system is very significant. The control system has been radically changed to deal with uncertain data, and the IVHM data can now be processed as additional vectors in the control system's information base. For autonomous vehicles, this is likely to be one of the biggest changes for IVHM systems. The IVHM system can be built into the control system's data architecture (reducing the IVHM system's costs), and the control system can be influenced by the IVHM information.

11.5.3 Autonomous/Real-Time Health Monitoring

Providing any form of health monitoring is of very limited utility if there is no way to use the information. The business case for many IVHM applications falls down when the "so what" question is asked: this can be detected, so what can be done about it? If the answer is nothing, then don't bother detecting it.

The control systems described previously and the fusion of vectors of IVHM data into the control systems make autonomous real-time health monitoring possible. The control system will make decisions on how to operate to achieve its mission, but it may also introduce the dimension of preserving its condition into the process. For instance, if a vibration is detected when a vehicle maneuvers in a certain way and the vibration is increasing over time (the fault could be a bearing damage or exterior damage, and it may not need to be fully classified), the vehicle may choose to avoid repeating the action that is increasing the damage (unless forced to do so).

As well as faults, the system may also simply learn to perform better in its environment. A ground-based vehicle may learn which methods of acceleration minimize fuel consumption, vibration, shock, or other parameters. This learning can then be used to generate health parameters for the control systems. For instance, there could be many ways to accelerate a vehicle from 10 to 20 mph but the IVHM system would help to determine which profile would leave the batteries with the best state of charge for the current conditions.

11.5.4 Cloud-Hosted Applications/Push of Software and Configuration Updates

Providing separate IVHM solutions for each system on a vehicle and then integrating them at a higher level is one valid approach to providing a total-vehicle IVHM solution.

However, it may not provide the most cost-effective, lightest, or most functional system. The ability to share data between parts of a vehicle can be used to improve the validation of the data, provide redundancy, or enable additional functions. For instance, a monitoring system detecting unusual power output from a generator may be able to better determine if there is a generator fault if it has information about the loads on the electrical network.

A fully functionally optimized vehicle needs to consider all aspects of the vehicle that are or may be monitored, the required set of sensors to deliver the function, and the information required to deliver each of the functions. The physical implementation of the total-vehicle IVHM system (particularly for an aircraft) is always likely to be driven by the overall weight of the system and the need to find the optimal number of electronics boxes, sensors, and wiring harnesses runs. The logical location of the methods to perform the IVHM analysis may be wherever it is easiest to get access to the correct data and the host processing to deliver the results. Consequently, the logical mapping of the IVHM solution may not be one-to-one onto the physical architecture. The vehicle's IVHM data therefore becomes an "intra-cloud" of data.

While an intra-cloud may provide the most elegant and optimal technical solution, it will not be without its commercial challenges. For instance, the data relating to the performance of a critical item of equipment may no longer be located on that supplier's system, and data privacy has to be considered. Equally, if an item of equipment is being managed and maintained by a party, and a third party hosts the IVHM data and methods, where will responsibility lie in the event of failures in the IVHM system? By virtue of their design, failures in IVHM systems are expensive [Keller et al. 2012].

Looking further than the IVHM intra-cloud, it is well understood that optimal decision making occurs when information from many sources is merged. For instance, when the onset of a fault is detected in an item of equipment (diagnostics), the desire is to provide prognostics by predicting its Remaining Useful Life (RUL) (how long it has left to run). However, providing a useful prediction requires more information than just the state indicated by the current sensors on the equipment. Additional information is likely to include the build specification, overhaul reports, and, most critically, the planned further operation of the equipment (hard use may accelerate the failure). There may even be options to improve or trade the RUL by communication with other vehicles; for instance, a UAV with excessive fuel consumption may change its position in a formation to save fuel.

Therefore, an inter-cloud of IVHM data, in which data are communicated from vehicle to vehicle and between the vehicle and base stations, provides a powerful capability to IVHM systems. While this may seem futuristic, it is already possible to have telephone services that tell you when you are in the vicinity of friends or like-minded social groups and where they are meeting. It is easy to imagine an extension of this where vehicles on the road can broadcast "stopped," "rain," "skid detect," or "sudden move" as a safety warning. It may be that these safety advances are technically enabled by the more commercially viable technology of social media and networking.

What is clear is that cloud-based computing has happened, is growing fast, and many companies are investing considerable sums of money to be the "data hosts" and provide the information services. IVHM systems require the processing and management of disparate and large data sources. Riding the wave of cloud-based computing could provide a step change in IVHM system capability.

11.6 References

A4A. 2011. MSG-3 Operator/Manufacturer Scheduled Maintenance Development, Airlines for America (A4A), 1301 Pennsylvania Avenue, NW, Suite 1100, Washington, DC 20004–1707.

Akhilesh, K. J., and J. N. Kudva. 2004. "Morphing aircraft concepts, classifications, and challenges," *SPIE Proceedings*, Vol. 5388, Smart Structures and Materials 2004, pp. 213–224.

Andrews, James, and Michael Augustin. 2010. "Advanced CBM Technologies for Helicopter Rotor Systems—Full Scale Rotor Demonstration and Test Results," Presented at the American Helicopter Society 66th Annual Forum, Phoenix, Arizona, May 11–13, 2010.

ARINC (Aeronautic Radio Incorporated). 2013. Wikipedia, http://en.wikipedia.org/wiki/ARINC, Accessed June 10, 2013.

ARINC. 2008. ARINC Specification, "822-1 Aircraft/Ground IP Communication," https://www.arinc.com/cf/store/catalog_detail.cfm?item_id=995, Accessed June 10, 2013.

Arms, S., M. Augustin, D. Yeary, P. Darden, and N. Phan. 2007. "Tracking Pitch Link Dynamic Loads with Energy Harvesting Wireless Sensors," Presented at the American Helicopter Society 63rd Annual Forum, Virginia Beach, VA, May 1–3, 2007.

Augustin, Michael, and Andrew Hess. 2011. "Global Data Fusion—Making Diagnostic Data Relevant," Presented at the American Helicopter Society 67th Annual Forum, Virginia Beach, VA, May 3–5, 2011.

Augustin, Michael, and Dave Yeary. 2008. "Advanced Technologies for Rotor System Condition-based Maintenance (CBM)," Presented at the American Helicopter Society 64th Annual Forum, Montréal, Canada, April 29–May 1, 2008.

Baker, T., B. Thompson, C. Ferrie, M. Augustin, and D. Yeary. 2009. "Operations Support and Sustainment Technologies for Current and Future Aircraft," Presented at the American Helicopter Society 65th Annual Forum, Grapevine, Texas, May 27–29, 2009.

Baker, T., N. Bordick, C. Ferrie, M. Augustin, and H. Wilson. 2009. "Operations Support and Sustainment Technology Program, Providing Technology for Condition Based Maintenance," 2009 ISHM Paper given at Stanford University.

Bechhoefer, E., M. Augustin, and M. Kingsley. 2012. "Architecture for a Light Helicopter HUMS," Presented at the American Helicopter Society 68th Annual Forum, Fort Worth, TX, May 1–3, 2012.

Blair, Harvey. 2007. "SecuraNet wireless technology: Intra-aircraft wireless data bus for essential and critical applications," White paper, Securaplane Technologies Inc.

Bocherens, E., S. Bourasseau, V. Dewynter-Marty, S. Py, M. Dupont, P. Ferdinand, and H. Berenger. 2000. "Damage Detection in a Radome Sandwich Material with Embedded Fiber Optic Sensors," *Smart Materials and Structures*, v.9, 310–315.

Bogue, R. 2008. "Swarm intelligence and robotics," *Industrial Robot: An International Journal*, Vol. 35, Is. 6, pp. 488–495.

Boughton, A., J. Kessner, and M. Menadue. 2006. "Communications, an Operator's Perspective," Pacific National Technical Paper, Feb. 2006.

Cottingham, David A. 2009. "Vehicular wireless communication," Technical Report 741, University of Cambridge Computer Laboratory, Jan. 2009.

Curtis, C. J., A. Miedaner, T. Rivkin, J. Alleman, D. L. Schulz, and D. S. Ginley. 2001. "Direct write metallizations for Ag and Al," *Proc. Mater. Res. Soc. Symp.*, Vol. 624, pp. 59–64.

DoD and U.S. Military. 2010. *21st Century Unmanned Aerial Vehicles (UAV) Reliability Study—Predator, Pioneer, Hunter, UAS—Power, Propulsion, Flight Control, Communication, Human Factors*, Progressive Management.

Garcia, E. J., B. L. Wardle, J. A. Hart, and N. Yamamoto. 2008. "Fabrication and multifunctional properties of a hybrid laminate with aligned carbon nanotubes grown *in situ*," *Composite Science and Technology*, v. 68, pp. 2034–41.

Goldfine, N., D. Schlicker, and A. Washabaugh. 1998. "Surface-Mounted Eddy-Current Sensors for On-Line Monitoring of Fatigue Tests and for Aircraft Health Monitoring," *Proceedings of the 2nd Joint NASA/FAA/DoD Conference on Aging Aircraft*.

Gondeck, Matthias. 2011. Electronic Flight Bags (EFBs) in the Airline Industry, Seminar at the Hamburg University of Applied Sciences, September 2011.

Harpin, A. P. R. 2010. "Static And Dynamic Pressure Measurements With Temperature Correction Using High Temperature Optical Pressure Sensors," ASME Turbo Expo 2010: Power for Land, Sea and Air (TE10), June 14–18, 2010, Scottish Exhibition Centre, Glasgow, UK.

Hedges, J., D. Gonzales, J. Everett, B. Hanks, B. Coleman, and M. Augustin. 2012. "Polymeric Bead Matrix Technology for On-Board CBM of Fluid-Lubricated Aircraft Components," Presented at the American Helicopter Society 68th Annual Forum, Fort Worth, TX, May 1–3, 2012.

IEEE 802 Standards. 2013. Wikipedia, http://en.wikipedia.org/wiki/IEEE_802, Accessed June 10, 2013.

IEEE. 2007. IEEE Standard for a Smart Transducer Interface for Sensors and Actuators Wireless Communication Protocols and Transducer Electronic Data Sheet (TEDS) Formats, pp. C1–236.

Jarosz P., C. Schauerman, J. Alvarenga, B. Moses, T. Mastrangelo, R. Reffaelle, R. Ridgley, and B. Landi. 2011. "Carbon nanotube wires and cables: Near-term applications and future perspectives," *Nanoscale*, 2011, 3, 4542–4553.

Keller, K., J. Maggiore, R. Safa-Bakhsh, W. Rhoden et al. 2012. "Sensory Prognostics and Management System (SPMS)," Technical Paper No. 2012-01-2095, SAE International, Warrendale, PA.

Keller, Kirby J., Jeanne Maggiore, R. Safa-Bakhsh, Michael Walz, and William Rhoden. 2013. "Sensory Prognostics and Management System (SPMS)," FAA report DOT/FAA/AR-DTFACT-09-A-80001.

Kessler, S. S., A. Raghavan, C. T. Dunn, S. Wicks, R. Guzman deVilloria, and B. L. Wardle. 2010. "Fabrication of a Multi-Physics Integral Structural Diagnostic System Utilizing Nano-Engineered Materials," *Proceedings of the 2nd Annual Conference of the Prognostics and Health Management Society*, October 10–16, 2010, Portland, OR.

Kessler, S. S., C. T. Dunn, M. Borgen, A. Raghavan, J. Duce, and D. L. Banks. 2009. "A Cable-Free Digital Sensor-Bus for Structural Health Monitoring of Large Area Composite Structures," *Proceedings of the*

Annual Conference of the Prognostics and Health Management Society.

Kessler, S. S., E. B. Flynn, and M. D. Todd. 2011. "Hybrid Coherent/Incoherent Beam Forming Diagnostic Approach to Naval Assets," *Proceedings of the 8th International Workshop on Structural Health Monitoring*, September 12–15, 2011, Stanford University.

Kim, K. S., Y. Ismail, and G. S. Springer. 1993. "Measurement of Strain and Temperature with Embedded Intrinsic Fabry-Perot Optical Fiber Sensors," *Journal of Composite Materials*, v.27, 1663–1677.

King, Benjamin, and Timothy Kutta. 2003. *The History of Germany's V-weapons In World War II (Classic Military History)*, Da Capo Press.

Kwapisz, David, Michael Hafner, and Ravi Rajamani. 2012. "Application of Microwave Sensing to Blade Health Monitoring," European Conference of the Prognostics and Health Management Society.

Lin, M., and F. K. Chang. 2002. "The manufacture of composite structures with a built-in network of piezoceramics," *Composites Science and Technology*, Vol. 62, Issues 7–8, June 2002, pp. 919–939.

Longtin, J., S. Sampath, S. Tankiewicz, R. Gambino, and R. J. Greenlaw. 2004. "Sensors for Harsh Environments by Direct-Write Thermal Spray," *IEEE Sensors Journal*, Vol. 4, No. 1.

Losik, Len. 2012. "Using Analog Telemetry to Measure Spacecraft Equipment Usable Life Upgrading Factory ATP," IEEE Aerospace Conference, Big Sky, MT.

Méallet-Renault, R., P. Denjean, and R. B Pansu. 1999. "Polymer beads as nano-sensors," *Sensors and Actuators B: Chemical*, Volume 59, Issues 2–3, October 19, 1999, pp. 108–112.

Motuku, M., U. K. Vaidya, and G. M. Janowski. 1999. "Parametric studies on self-repairing approaches for resin infused composites subjected to low velocity impact," *Smart Materials and Structures*, 8, p. 623.

Murayama, H., K. Kageyama, H. Naruse, A. Shimada, and K. Uzawa. 2003. "Application of Fiber-Optic Distributed Sensors to Health Monitoring for Full-Scale Composite Structures," *Journal of Intelligent Material Systems and Structures*, January 2003, 14: 3–13.

Ozevin, D., V. Godinez, J. Dong, and M. Carlos. 2007. "Damage Assessment of Gearbox Operating High Noisy Environment Using Waveform Streaming Approach," *Journal of Acoustic Emission*, pp. 355–363.

Priya, S., and D. J. Inman. 2009. *Energy Harvesting Technologies*, Springer.

Rajamani, R., and N. Waters. 2011. "Certification of Engine Health Management Systems: Guidelines for Selecting Software Assurance Levels," SAE Aerotech, Toulouse, France, October 2011.

Rao, T.V. 2009. "CAN Bus in Aviation," *Avionics Today*, May 2009.

Richardson, T. 2012. "Rapid Zigbee Adoption by Utilities & Growth of Zigbee Ecosystem," Wireless Congress, Nov. 2012. (Other publications, presentations, and standards at the Zigbee Alliance website www.zigbee.org.)

S1000D Organization homepage. 2013. http://public.s1000d.org/Pages/Home.aspx, Accessed June 10, 2013.

SAE International. 2010. SAE Aerospace Recommended Practice, "Guidelines for Development of Civil Aircraft and Systems," ARP4754A dated December 2010.

SAE International. 2013. SAE Aerospace Recommended Practice, Establishing software assurance levels for engine health management systems utilized for maintenance credit, ARP5987, under review, April 2013.

Chilakala, Satish Kumar. 2008. "Development and Flight Testing of a Wireless Avionics Network Based on the IEEE 802.11 Protocols,"

MS Dissertation, Kansas University, published by Proquest.

Schoess, J. N., and J. D. Zook. 1998. "Test Results of Resonant Integrated Microbeam Sensor (RIMS) for Acoustic Emission Monitoring," *Proceedings of the SPIE Conference on Smart Electronics and MEMS*, v. 3328, 326–332.

Spitzer, C. R., ed. 2001. *The Avionics Handbook*, Boca Raton, FL: CRC Press.

Thomas, J. P., et al. 2003. "Multifunctional Structure—Battery Materials for Enhanced Performance in Small Unmanned Air Vehicles," Paper #IMECE2003-41512, CD *Proceedings of the ASME International Mechanical Engineering Congress and Exhibition* (New York: ASME, 2003).

Thrun, Sebastian, Wolfram Burgard, and Dieter Fox. 2005. *Probabilistic Robotics (Intelligent Robotics and Autonomous Agents Series)*, The MIT Press.

CHAPTER 12 Summary and Concluding Remarks

Ian K. Jennions, *IVHM Centre, Cranfield University*

> "If you can't explain it simply, you don't understand it well enough."
>
> —Albert Einstein

The scope of this book was to return to the, now familiar, taxonomy (Figure 1.1) and investigate each of the broad theme areas of IVHM in more detail. The aim was to provide a bridge between the introduction to the subject, covered in the first book, and technical papers describing new work. This book is basically in three parts: the Technology chapters (2–9), then the Applications chapter (10), before Disruptive Technologies (11) are discussed. The contributions made in these sections are now summarized in this order, before drawing the book to its conclusion.

- Sensors, Instrumentation, and Signal Processing is where the book, and the IVHM source of data, began. The collection of data from sensors is explored, as is the careful filtering and conditioning of the source signal to remove noise such that the underlying data can be revealed and prepared for further analysis. Feature extraction is dealt with briefly, before a case study brings it all together.
- Anomaly Detection and Diagnosis, the next step in the IVHM chain, is then addressed. It deals with machine learning from data and covers supervised (classification and regression) learning, needed for diagnosis, dealing with random forest and neural networks. Unsupervised learning (e.g., self-organizing maps) is also addressed, along with performance metrics. Examples are used throughout.
- Continuing with algorithms, Prognostics is described next. The taxonomy of prognostic approaches is defined and the model-based and data-driven techniques described. A number of ways to evaluate the accuracy of a prognostic are discussed before giving a detailed case study of an electromechanical actuator, which highlights the depth of knowledge needed about the target system for such an analysis.
- As the subject of IVHM matures, those IVHM Metrics that cater to operators (program managers, plant/fleet managers, owners, and insurers) and engineers (design, researchers, OEMs) have emerged and are discussed. There are many stakeholders in IVHM, all with different objectives, and the use of metrics provides an effective way of brokering consensus. Algorithmic metrics are also covered in detail, as diagnostic and prognostic performance dictates how useful any IVHM implementation will be.
- Architecture provides a way to access the controls and sensor data, route it to the diagnostic and prognostic algorithms, and derive real benefit by placing the data, translated into information, into the hands of those who can make decisions. This chapter provides a background on the evolution of the avionics and IVHM systems, noting that, historically, bandwidth has been a consistent problem for data transmission. It deals with the hardware and software technologies that enable OCBM (Onboard Condition-Based Maintenance), giving examples.
- Feeding from the architecture, the next subject to be discussed is Data Management. Data processing and fusion, combining data from multiple sources to infer information that could not have been gleaned from a single source, is covered before data reduction and compression techniques are discussed. These latter techniques help with the large amounts of data that IVHM produces and needs to move around, addressing the limited bandwidth problem above.
- The next chapter, on Vehicle-Level Reasoning Systems (VLRS), starts by defining how VLRS is fundamentally different from diagnostics and prognostics due to the granularity of the data that each work with. Rather than the component

sensor data of the diagnostic and prognostic algorithms, the VLRS has to deal with pass/fail results or number of exceedances. Subject matter includes architecture, standards, and the technical approaches to reasoning, along with an example of emerging implementation.

- No matter where the data, tools, and techniques come from, they all end up contributing to the system design. The process of starting with stakeholder requirements, developing a framework, and then working through the design considerations to emerge with an IVHM system is discussed. Testing, to include qualification and integration, reveals any mistakes in the design selection and ensures a field-ready system.

The Applications chapter dealt with four different and contrasting examples of IVHM. The SHM (Structural Health Monitoring) and rotating equipment sections dealt with aerospace issues that were touched on in the first book. By contrast, the Motorsport and Wind Turbine sections dealt with IVHM in complementary sectors; indeed, in the case of Motorsport, a sector where the acronym IVHM is not even recognized. This seems to point to a problem of cross-sector identity, a topic for another day. The wind turbine application covers the full scope from installed base and types of failure to the condition monitoring of the vital components and the commercial equipment that is available for this job.

Last, but definitely not least, four subject matter experts share their thoughts on Disruptive Technologies that could impact the future of IVHM. Sensor technologies, that are gradually increasing their TRL (Technology Readiness Level), are explored. Next, the wireless systems that will be available in the future and enable some of these sensors to realize their full potentials, are discussed. Application to legacy product features highly here, along with protocols and standards that will underpin this technology. Moving to a higher level in the plane, subsystem technologies are examined next with requirements to monitor open rotor engines, and IGBTs (Insulated Gate Bipolar Transistors) as the more electric aircraft becomes a nearer reality. The chapter finishes with system-level technologies, examining UAVs and Google's driverless car, to cloud computing and the possibilities that it could unlock.

In closing, the goal of the book was to bring the technology of IVHM to life and explain complex and sometimes daunting concepts in simple-to-understand terms. With this goal in mind, the book has succeeded and produced some remarkable insights into the subject matter, by the foremost exponents in the individual areas, those who truly do understand the subject, to reflect on Einstein's quote.

Appendix: Acronyms

ABCD	Army Bulk CBM Data
AC	Alternating Current
ACARS	Aircraft Communications Addressing and Reporting System
ADC	Analog-to-Digital Converter
AE	Acoustic Emission
AFD	Asynchronous Frequency Domain
AFR	Annualized Failure Rate
AHM	Airplane Health Management
AI	Artificial Intelligence
AIRMAN	AIRcraft Maintenance ANalysis
AIRTAC	AIRbus Technical Aircraft-on-ground Centre
AMISS	Anomaly Monitoring Inductive Software System
AMOC	Alternative Means of Compliance
ANFIS	Adaptive Neuro-Fuzzy Inference System
ANN	Artificial Neural Network
ANOVA	ANalysis Of VAriants
ARINC	Aeronautical Radio INC.
ARM	Association Rule Mining
A-SHM	Automated-Structural Health Monitoring
ASIAS	Aviation Safety Information Analysis and Sharing
ASIP	Aircraft Structural Integrity Program
ATD	Asynchronous Time Domain
AUC	Area Under the Curve
ARMA	AutoRegressive Moving-Average
BIT	Built-In Test
BITE	Built-In Test Equipment
BOLDR	Boeing On-Line Diagnostic Reporting (BOLDR®)
CAN	Controller Area Network
CBA	Cost-Benefit Analysis
CBM	Condition-Based Maintenance
CDF	Common Data Format
CHAPS	CHAfing Protection System
CHAPS-HM	CHAPS-Health Management
CI	Condition Indicator
CMC	Central Maintenance Computer
CMCF	Central Maintenance Computer Function
CMS	Condition-Monitoring System
CNT	Carbon NanoTubes
CFD	Computational Fluid Dynamics
COTS	Commercial Off-The-Shelf
CSV	Comma Separated Variable
DA	Detection Accuracy
DC	Direct Current
DET	DETailed inspection
DFIG	Data Fusion Information Group
DFIG	Doubly Fed Induction Generator

DMS	Database Management System
DOE	Design Of Experiments
DoF	Degrees of Freedom
DStd, DSTD	Dynamic Standard Deviation
EADS	European Aeronautic Defence and Space Company
EASA	European Aviation Safety Agency
ECB	Electronic Circuit Breaker
EFB	Electronic Flight Bag
EKF	Extended-Kalman Filter
ELDK	Embedded Linux Development Kit
EMA	Electro-Mechanical Actuator
emf	electromotive force
EMI	ElectroMagnetic Interference
EMS	Engine Monitoring Systems
EOL	End-Of-Life
EoP	End-of-Prediction
EPDS	Electrical Power Distribution Systems
EPIC	Embeddable Programmable Instrumentation Controller
ERP	Enterprise Resource Planning
FA	Failure Analysis
FAA	Federal Aviation Authority
FBG	Fiber-optic Bragg Grating
FDR	Fault Detection Rate
FFT	Fast Fourier Transform
FIF	Flight Information File
FIS	Fuzzy Inferencing System
FMECA	Failure Modes, Effects, and Criticality Analysis
FMMEA	Failure Modes, Mechanisms, and Effects Analysis
FNR	False Negative Rate
FOD	Foreign Object Damage
FPR	False Positive Rate
FRACAS	Failure Recording, Analysis, and Corrective Action System
GCU	Generator Control Unit
GPRS	General Packet Radio Service
GSE	Ground Support Equipment
GVI	General Visual Inspection
HAWT	Horizontal Axis Wind Turbine
HFDM	Helicopter Flight Data Monitoring
HM	Health Management
HRSG	Heat Recovery Steam Generator
HUMS	Health and Usage Monitoring Systems
ICR	Isolation Classification Rate
IDG	Integrated Drive Generator
IETM	Interactive Electronic Technical Manual
IFE	In-Flight Entertainment
IFSD	In-Flight ShutDown
IGBT	Insulated Gate Bipolar Transistor
IGCT	Integrated Gate Commutated Thyristor
IMR	Isolation Misclassification Rate
IMS	Inductive Monitoring System
ISO	International Standards Organization
ISS	International Space Station
IVHM	Integrated Vehicle Health Management
JDL	Joint Directors of Laboratories

JITP	Just-In-Time Point
KF	Kalman Filter
KR	Kernel Regression
LAN	Local Area Network
LOO	Leave-One-Out
LRU	Line Replaceable Unit
MAA	Military Aviation Authority
MAD	Mean Absolute Deviation
MAE	Mean Absolute Error
MAFT	Major Airframe Fatigue Test
MAPE	Mean Absolute Percentage Error
MCBF	Mean Cycles Between Failures
MCSA	Motor Current Signature Analysis
MEMS	MicroElectroMechanical System
MES	Manufacturing Execution System
MFS	Machine Fault Simulator
MIMOSA	Machinery Information Management Open Systems Alliance
MMBF	Mean Miles Between Failures
MoD	(UK) Ministry of Defence
MOSFET	Metal Oxide Semiconductor Field-Effect Transistor
MRO	Maintenance Repair and Overhaul
MSD	Maintenance Support Device
MSE	Mean Squared Error
MT	Multi-Termination
MTBF	Mean Time Between Failures
MTBMA	Mean Time Between Maintenance Actions
MTBSC	Mean Time Between Service Calls
MTBUR	Mean Time Between Unscheduled Removals
MTBWC	Mean Time Between Warranty Claims
MTB*xx*	Mean Time Between *xx*
NASA	National Aeronautics and Space Administration
NDE	NonDestructive Examination
NDI	NonDestructive Inspection
NFF	No Fault Found
NN	Neural Network
NPV	Net Present Value
O&M	Operation and Maintenance
O&S	Operation and Support
OCBM	Onboard Condition-Based Maintenance
OEM	Original Equipment Manufacturer
OMS	Onboard Maintenance System
OSA-CBM	Open System Architecture for Condition-Based Maintenance
OSA-EAI	Open Systems Architecture for Enterprise Application Integration
OTDR	Optical Time Domain Reflectometer
PAN	Personal Area Networks
PCA	Principal Component Analysis
PDF	Probability Density Function
PF interval	Potential to Functional failure time or interval
PH	Prognostic Horizon

PHM	Prognostics and Health Management
PLM	Product Lifecycle Management
PoF	Physics-of-Failure
PWM	Pulse Width Modulation
QAR	Quick Access Recorder
RCA	Root Cause Analysis
RCM	Reliability-Centered Maintenance
RF	Random Forests
RFID	Radio Frequency IDentification
RGB	Red, Green, and Blue
RMS	Rotor Monitoring System
RMS	Root Mean Squared
RMSPE	Root Mean Squared Percentage Error
ROC	Receiver Operating Characteristic
RoHS	Restriction of Hazardous Substances
ROI	Return On Investment
RM	Rotating Machinery
RPN	Risk Priority Number
RUL	Remaining Useful Life
RUL-OPI	RUL-Online Precision Index
SAMS	Sensor-based Aircraft Maintenance Support
SatCom	Satellite Communications
SAW	Surface Acoustic Wave
SCADA	Supervisory Control And Data Acquisition
SCIG	Squirrel Cage Induction Generators
SCR	Silicon-Controlled Rectifier
SDI	Special Detailed Inspection
SDU	Smoke Detection Unit
SHM	Structural Health Monitoring
SI	Structural Intensity
SIP	System-In-Package
SIR	Sampling Importance Resampling
SNR	Signal-to-Noise Ratio
SNS	Structural Neural Systems
SOA	Service-Oriented Architecture
SOC	Sensor On Chip
SOD	Synchronous Order Domain
SOM	Self-Organizing Map
S-SHM	Scheduled Structural Health Monitoring
STA	Synchronous Time Average
STFT	Short-Time Fourier Transform
SVM	Structural Vibration Monitoring
TAN	Total Acidity Number
TDM	Time Domain Multiplexing
TDRSS	Tracking and Data Relay Satellite System
TEDS	Transducer Electronic Data Sheet
TOC	Total Ownership Cost
TRL	Technology Readiness Level
TSA	Time Synchronous Averaging
TTF	Time To Failure
TWh	Tera Watt hours
UFF	Universal File Formats
UML	Unified Modeling Language
UMTS	Universal Mobile Telecommunications System
USAF	U.S. Air Force

UUT	Unit Under Test
V&V	Verification and Validation
VAWT	Vertical Axis Wind Turbine
VHF	Very High Frequency
VHM	Vehicle Health Management
VIPR	Vehicle Integrated Prognostic Reasoner
VLRS	Vehicle-Level Reasoning System
VQ	Virtual Qualification
WDM	Wavelength Domain Multiplexing
WISP	Wireless Identification and Sensing Platform
WLAN	Wireless Local Area Networks
WWAN	Wireless Wide Area Networks
XML	eXtensible Markup Language

Index

About the Authors

Michael Augustin

Mr. Augustin is president and principal investigator at IVHM Inc., an aerospace consulting business focused on advanced sensors and systems for aerospace applications. He retired from Bell Helicopter in 2011, where he obtained 25 years of experience with advanced systems and sensors, particularly in the HUMS/CBM and enhanced safety areas. Prior to joining Bell Helicopter Textron, he worked in radio frequency communications at Motorola for 17 years. Mr. Augustin has been the chief investigator on a number of research programs dealing with advanced technology. He is the author of numerous papers dealing with the development and implementation of advanced technologies.

Michael H. Azarian

Dr. Michael Azarian is a research scientist and member of the graduate faculty at the Center for Advanced Life Cycle Engineering (CALCE) at the University of Maryland. Prior to joining CALCE, he spent over 13 years in the data storage, advanced materials, and fiber optics industries. His research focuses on the analysis, detection, prediction, and prevention of failures in electronic and electromechanical products. He advises companies on reliability of electronic products, and is the author or co-author of numerous publications. Azarian holds a Masters and PhD in materials science and engineering from Carnegie Mellon University, and a BS in chemical engineering from Princeton University.

Raj Mohan Bharadwaj

Raj Bharadwaj is principal R&D scientist in the vehicle health management group at Honeywell Aerospace Advanced Technologies. His work is centered on Prognostics Health Management (PHM), algorithms, and system design. Prior to joining Honeywell, Bharadwaj was with the General Electric Global Research Center, where he worked on diagnostics for power system equipment and locomotives. He received a PhD in electrical engineering from Texas A&M University.

Chaochao Chen

Dr. Chaochao Chen is a member of the research staff at the Center for Advanced Life Cycle Engineering (CALCE). His research areas include fault diagnosis and failure prognosis, focusing on data-driven approaches such as machine learning and statistical methods; prediction uncertainty management; prognostics and health management (PHM) software implementation, verification, and validation; and fault-tolerant control and their applications to robotics, electronics, batteries, and mechanical systems. Prior to joining CALCE, Dr. Chen was a research fellow at the University of Michigan and Georgia Institute of Technology, working in PHM areas in collaboration with industry and military. He has published numerous technical papers. Chen received his PhD in mechanical engineering from Kochi University of Technology, Japan.

Charlie Dibsdale

Charlie Dibsdale began his engineering career in the Royal Navy, joining the submarine service and moving into operating and maintaining nuclear propulsion and electrical power and distribution systems. The broad range of equipment and the independence on patrols combined to form a solid foundation of operations and maintenance practice. On finishing a full service career, Charlie joined Rolls-Royce Marine with accountability for improving reliability and maintenance; he undertook Reliability-Centered Maintenance (RCM) training and was part of a team that successfully reduced (Trident) submarine maintenance by 20% for no loss of safety or reliability. When OSyS (then known as DS&S) was formed, Charlie moved over. OSyS delivers the predictive maintenance services in all of Rolls-Royce's market sectors in support of the company's product-based services. He has been part of and has influenced the fundamental changes involved as EHM matured and was part of the team that conceptualized the early development of the operations room. Charlie's current roles are as Rolls-Royce global capability owner for Equipment Health Management (EHM), program managing Rolls-Royce IVHM research, and as Chief Engineer for EHM in OSyS, heavily involved with its fitness for purpose and development.

Charlie is active in development of Predictive Maintenance standards and processes, is a member of API 691 defining risk-based maintenance for rotating machinery in the oil and gas sector, and is a member of the SAE International HM1 committee. He has combined his engineering knowledge with a formal education with a BSc (Hons) in Computer Science and an MSc in Information systems.

Neil H. W. Eklund

Dr. Neil Eklund has been a research scientist in the Machine Learning Laboratory of General Electric Global Research since 2002. He has worked on a wide variety of research projects related to remote monitoring, diagnostics, and prognostics for military and commercial aircraft engines, medical devices, wind turbines, manufacturing systems, ground-based gas turbines, and other platforms. Eklund has also been an adjunct professor of electrical engineering and computer science at Union Graduate College since 2005. He was one of the co-founders of the PHM Society and the first editor-in-chief of the *International Journal of Prognostics and Health Management*. He has authored numerous technical publications.

Peter Foote

Peter Foote is professor of sensors for through-life engineering, and head of the Composite Centre in the School of Applied Sciences at Cranfield University, in the UK. He recently joined the university after spending 25 years at BAE Systems developing new concepts for vehicle health management. During that time he spearheaded company efforts on embedded optical fiber sensors for structural monitoring. He is continuing efforts for developing sensors concepts at Cranfield University, and also leads an SAE International Technical Committee developing guidelines for implementation of SHM for the aerospace industry.

Dr. Kai Goebel

Dr. Kai Goebel is a senior scientist working at NASA Ames Research Center where he leads the Prognostics Center of Excellence. He is the Technical Lead for Prognostics and Decision Making in NASA's System-wide Safety and Assurance Technologies Project. Prior to joining NASA in 2006, he was a senior research scientist at General Electric Corporate Research and Development Center since 1997. He has carried out applied research in the areas of real-time monitoring, diagnostics, and prognostics. He has fielded numerous applications for aircraft engines, transportation systems, medical systems, and manufacturing systems. He holds 15 patents and has co-authored more than 200 papers in the field of IVHM. Dr. Goebel was an adjunct professor of the CS Department at Rensselaer Polytechnic Institute (RPI), Troy, NY, between 1998 and 2005 where he taught classes in Soft Computing and Applied Intelligent Reasoning Systems. Dr. Goebel is a member of ASME, AAAI, AIAA, IEEE, VDI, SAE, and ISO. He was the General Chair of the Annual Conference of the PHM Society, 2009 and held numerous chair positions at the PHM conference and the AAAI Annual Meetings series. He is currently a member of the board of directors of the PHM Society and associate editor of the *International Journal of PHM*.

Keith Jackson

Prof. Keith Jackson is the director of technology at Meggitt PLC and a visiting professor at Sheffield University in the UK. He first became involved with SAE International when his team in Cambridge started developing engine control systems for the U.S. heavy-duty diesel market and later for the global automotive market. As the company also provided data-logging equipment for the motorsports industry, it was a natural extension to introduce health-monitoring and data-gathering systems into the road vehicle control systems. Subsequently, he worked for Rolls-Royce PLC, advancing aerospace control and health-monitoring technologies. Jackson is a graduate from University College London.

Ian K. Jennions

Ian K. Jennions is a Professor and Director of the IVHM Centre, Cranfield University, UK. He joined the Centre, which is funded by a number of industrial partners, when it was founded in 2008 and has led its development and growth in research and education over the last three years. Previously, Ian had worked for a number of companies in the gas turbine industry over a 30-year career. He worked for Rolls-Royce, General Electric, and Alstom in a number of technical roles, gaining experience in aerodynamics, heat transfer, fluid systems, mechanical design, combustion and, more recently, IVHM. He has a Mechanical Engineering degree and a Ph.D. in CFD, both from Imperial College, London. He is a Director of the PHM Society, contributing member of the SAE IVHM Steering Group and HM-1 IVHM committee, and a Fellow of IMechE, RAeS, and ASME. He was also the editor of SAE International's *Integrated Vehicle Health Management: Perspectives on an Emerging Field*.

Kirby Keller

Dr. Keller is a Technical Fellow for the Boeing Company. He has over 35 years of experience in the health management, mission systems, and decision-aiding systems for air and space vehicles. His current work includes technical leadership, new business capture, and project management for Integrated Vehicle Health Management (IVHM) programs. He is the principal investigator and system designer for several projects that are focused on the development of architectural concepts to integrate IVHM processing onto new and legacy platforms. He holds a PhD in mathematics from Iowa State University and is the author of numerous technical papers, conference presentations, and technical reports.

Seth S. Kessler

Dr. Kessler is the president and founder of Metis Design Corp. He guest lectures for advanced graduate courses on composite mechanics in the aerospace department at MIT. He has worked at the Lockheed Martin Skunk Works as an advanced concepts engineer on the X33/VentureStar/JSF programs, and was a Draper Fellow working on the DARPA seedling WASP program. Kessler received an SB in aerospace engineering at MIT, studying the effects of a cryogenic environment on composite materials; and an SM in aerospace engineering, creating and validating a design tool to analyze composite structures subjected to extreme inertial loading. He completed his PhD researching structural health monitoring techniques that use piezoelectric devices to detect damage in composites. He has published more than three dozen technical publications.

Ranjith Kumar

Ranjith Kumar received his BE in mechanical engineering from the College of Engineering, Guindy, Anna University, India. He is currently working toward a PhD in mechanical engineering at the University of Maryland, College Park. Kumar is a graduate research assistant with the Center for Advanced Life Cycle Engineering, University of Maryland. His research interests include reliability, dynamics, vibration, tribology, and prognostics and health management.

Robert W. Mah

Robert W. Mah is the Project Scientist for the SSAT Project, one of three projects under the NASA Aviation Safety Program. He has a Ph.D. in Applied Mechanics and Biomedical Engineering from Stanford University. He was PI and head of the Smart Systems Research Laboratory at NASA Ames which developed innovative software solutions and tools for space, aeronautics, and medicine. His experience spans a broad range of research and development work ranging from adaptive control of spacecraft, FT/RM architecture for CEV GN&C, FDIR, and fault tolerant control of the 2.5-m Artificial Gravity Research Centrifuge for the International Space Station, advanced signal processing and precision tools for airborne and ground-based telescopes, high-performance multi-axis motion control capabilities for biological and human research, astronaut space motion sickness research, to novel medical technologies for detection, diagnosis, and prognosis of cancer.

Sony Mathew

Sony Mathew is currently pursuing his PhD at the Center for Advanced Life Cycle Engineering (CALCE) in the mechanical engineering department at the A. James Clark School of Engineering of the University of Maryland. His research areas include reliability, and Prognostics and Health Management (PHM) of electronics. Previously, he managed the activities of the PHM group within CALCE. He developed, executed, and supervised research projects on prognostics of electronics and served as a liaison with CALCE's industry and government partners. Mathew also serves on the editorial board of the *International Journal of Prognostics and Health Management*.

Dinkar Mylaraswamy

Dr. Dinkar Mylaraswamy is the Technology Fellow for condition-based maintenance within Honeywell's Advanced Technology organization. His areas of expertise are fault diagnosis and process monitoring, modeling, and control. In his current role, Dr. Mylaraswamy is responsible for identifying and maturing strategic health management technologies that cut across multiple products and services, providing inputs for strategic technology investments, and mentoring.

Dr. Mylaraswamy joined Honeywell in 1997 after completing his Ph.D. from Purdue University. His Ph.D. thesis on blackboard-based architectures was adopted by the Abnormal Situation Management® Consortium as the basis of an operator tool for addressing the $16B loss suffered by the petrochemical industry from abnormal situations and equipment malfunctions. Dr. Mylaraswamy spent his first six years in Honeywell developing and deploying an Early Abnormal Event Detection application at six refinery sites in North America.

On the Aerospace side, he was the technical lead for Honeywell's Predictive Trend Monitoring program, a web-based application for monitoring aircraft engines. He continues to serve as the technical lead on various health management programs—within Honeywell as well the U.S. Army, NASA, UK-MOD, and Navair—to support the Aerospace Services business within Honeywell.

As the Technology fellow, he routinely works with academic institutes and small businesses, seeking cutting-edge technologies to support the condition-based service business within Honeywell. Dr. Mylaraswamy has authored over 30 papers and holds 14 patents in the area of fault diagnosis and its applications.

Marcos E. Orchard

Dr. Marcos Orchard is associate professor with the department of electrical engineering at Universidad de Chile, associate researcher at the Advanced Mining Technology Center, and project leader at the Lithium Innovation Center. His research interest is the design, implementation, and testing of real-time frameworks for fault diagnosis and failure prognosis, with applications to battery management systems, mining industry, and finance. His fields of expertise include statistical process monitoring, parametric/ nonparametric modeling, and system identification. His research work at the Georgia Institute of Technology was the foundation of novel real-time fault diagnosis and failure prognosis approaches based on particle filtering algorithms. Orchard received a PhD and MS from The Georgia Institute of Technology, as well as a BS and a civil industrial engineering degree with electrical major from Catholic University of Chile. Dr. Orchard has published numerous papers in his areas of expertise.

Michael Pecht

Prof. Michael Pecht is a world-renowned expert in strategic planning, design, test, prognostics, IP, and risk assessment of electronic products and systems. In 2010, he received the IEEE Exceptional Technical Achievement Award for his innovations in the area of prognostics and systems health management. In 2008, he was awarded the highest reliability honor, the IEEE Reliability Society's Lifetime Achievement Award. Prof. Pecht has an MS in Electrical Engineering and an MS and PhD in Engineering Mechanics from the University of Wisconsin at Madison. He is a Professional Engineer, an IEEE Fellow, an ASME Fellow, an SAE Fellow, and an IMAPS Fellow. He has previously received the European Micro and Nano-Reliability Award for outstanding contributions to reliability research, 3M Research Award for electronics packaging, and the IMAPS William D. Ashman Memorial Achievement Award for his contributions in electronics analysis. He served as chief editor of the IEEE Transactions on Reliability for eight years and on the advisory board of IEEE Spectrum. He is chief editor for Microelectronics Reliability and an associate editor for the IEEE Transactions on Components and Packaging Technology. He is the founder and Director of CALCE (Center for Advanced Life Cycle Engineering) at the University of Maryland, which is funded by over 150 of the world's leading electronics companies at more than US$6M/year. The CALCE Center received the NSF Innovation Award in 2009. He is also a Visiting Professor in Electronics Engineering at City University of Hong Kong, and a Chair Professor in Mechanical Engineering and a Professor in Applied Mathematics at the University of Maryland. He has written more than twenty books on product reliability, development, use, and supply chain management and over 400 technical articles. He has also written a series of books on the electronics industry in China, Korea, Japan, and India. He consults for 22 international companies.

Suresh Perinpanayagam

Dr. Suresh Perinpanayagam is a lecturer at the IVHM Centre, Cranfield University. He leads the Electronic Systems' Reliability, Prognostics and Health Management (ERPHM) Group at Cranfield University, in the UK. This group develops data-mining techniques for anomaly detection, diagnostics, prognostics, information fusion, predictive maintenance, and logistical support for electronic systems from the aerospace, rail, automotive, critical infrastructure, and renewable energy sectors. Suresh teaches the sensors and instrumentation module for several MSc courses at Cranfield University and has served in industry as a signal processing expert specializing in vehicle systems applications. He has an ME and PhD from Imperial

College London. He has published widely in academic and professional journals, and has presented many papers at national and international conferences.

Ravi Rajamani

Dr. Ravi Rajamani joined Meggitt in 2011 as an engineering director, after spending nearly 11 years with United Technologies Corporation, first at the Research Center, and then with its Pratt & Whitney division. Before this he was with GE for ten years. His primary focus has been in the area of controls and diagnostics of gas turbines for aerospace and industrial applications. Rajamani has a BTech (ME) from IIT Delhi, an MS (Automation) from IISc, Bangalore, and a PhD (EE) from the University of Minnesota. He also has an MBA from the University of Connecticut. He has published numerous papers in refereed journals and conference proceedings.

Karl Reichard

Dr. Reichard has over 25 years of experience in the design and development of advanced measurement, control, and monitoring systems. He received PhD, MS, and BS degrees in electrical engineering from the Virginia Polytechnic Institute and Virginia Tech. Dr. Reichard is a research associate with the Pennsylvania State University Applied Research Laboratory, and an assistant professor of acoustics with the Penn State graduate program in acoustics. His research experience includes the development of embedded and distributed sensing and control systems for machinery and system health monitoring, acoustic surveillance and detection, active noise and vibration control, and electro-optics. He is the author of numerous papers and articles published in journals and conference proceedings.

Michael J. Roemer

Michael J. Roemer is currently a Technical Fellow with Sikorsky Aircraft with over 20 years' experience in the development of automated health monitoring, diagnostic, and prognostic systems for a wide range of military and commercial applications. He was previously the co-founder and Director of Engineering at Impact Technologies prior to its acquisition by Sikorsky/UTC. His experience includes a wide range of integrated vehicle health management system implementations to detect and predict system faults in real time and perform automated troubleshooting and maintenance planning. He has developed several diagnostic and prognostic capabilities utilizing technologies such as dynamic signature analysis, artificial intelligence, aero-thermal performance monitoring, finite element modeling, probabilistic remaining life analysis, and risk assessment methods. He is the co-founder and Vice President of the PHM Society, Chairman of the SAE HM-1 Integrated Vehicle Health Management Committee, board member and past Chairman of the Machinery Failure Prevention Technology (MFPT) Society, Prognostic Lead for the SAE E-32 Engine Condition Monitoring Committee, Member of the IGTI Marine Committee and ASME Controls and Diagnostics Committee, and Adjunct Professor of Mechanical Engineering at the Rochester Institute of Technology. He is the co-author of a recent book titled *Intelligent Fault Diagnosis and Prognosis for Engineering Systems* and has written or co-authored more than 100 technical papers related to integrated systems health management.

Abhinav Saxena

Abhinav Saxena is a research scientist at the Prognostics Center of Excellence, NASA Ames Research Center. He has been involved in the field of IVHM since 2003 and has published numerous research articles in scientific journals, conference proceedings, and book chapters. His research focuses on developing and evaluating health management algorithms for various engineering systems. He also develops systems engineering processes and methods for verification and validation for health management systems. Mr. Saxena is chief editor of the *International Journal of Prognostics and Health Management*. He has a PhD in electrical and computer engineering from Georgia Institute of Technology. He earned his B.Tech. in 2001 from Indian Institute of Technology, Delhi, and his Masters degree in 2003 from Georgia Tech.

Tarapong Sreenuch

Dr. Tarapong Sreenuch is conducting research at the IVHM Centre at Cranfield University in the UK, focusing on projects related to software architecture and systems integration for IVHM systems. He has many years of experience in engineering software design and distributed systems. He is a lead developer at the IVHM Centre in design and development of IVHM-related data processes and software. Tarapong is an instructor in MSc in IVHM and MSc in autonomous vehicle dynamics and control at Cranfield University. He has published in many academic publications. He has a BEng in electronics and electrical engineering from the University of Surrey, and a PhD in control engineering from Cranfield University—Defence College of Management and Technology.

Ashok N. Srivastava

Ashok N. Srivastava, Ph.D. is the Project Manager for the System-Wide Safety and Assurance Technologies Project at NASA. He is formerly the Principal Investigator for the Integrated Vehicle Health Management research project at NASA. His current research focuses on the development of data mining algorithms for anomaly detection in massive data streams, kernel methods in machine learning, and text mining algorithms.

Dr. Srivastava is also the leader of the Intelligent Data Understanding group at NASA Ames Research Center. The group performs research and development of advanced machine learning and data mining algorithms in support of NASA missions. He performs data mining research in a number of areas in aviation safety and application domains such as earth sciences to study global climate processes and astrophysics to help characterize the large-scale structure of the universe.

Dr. Srivastava is the author of many research articles in data mining, machine learning, and text mining and has edited the book, *Text Mining: Classification, Clustering, and Applications* (with Mehran Sahami, 2009). He is currently editing two more books: *Advances in Machine Learning and Data Mining for Astronomy* (with Kamal Ali, Michael Way, and Jeff Scargle) and *Data Mining in Systems Health Management* (with Jiawei Han).

Dr. Srivastava has given seminars at numerous international conferences. He has a broad range of business experience including serving as Senior Consultant at IBM and Senior Director at Blue Martini Software. In these roles, he led engagements with numerous Fortune Global 500 companies including Bank of America, Chrysler Corporation, Saks 5th Avenue, Sprint, Chevron, and LG Semiconductor.

He has won numerous awards including the IEEE Computer Society Technical Achievement Award for "pioneering work in Intelligent Information Systems," the NASA Exceptional Achievement Medal for contributions to state-of-the-art data mining and analysis, the NASA Distinguished Performance Award, several NASA Group Achievement Awards, the IBM Golden Circle Award, and the Department of Education Merit Fellowship.

Kevin Swearingen

Kevin Swearingen has more than fifteen years of experience developing and integrating vehicle health management technologies in the research and technology division of the Boeing Company. Experience as principal investigator includes developing wireless sensor networking hardware for commercial aircraft, conducting maintenance effectiveness analysis for Army helicopters, sensor data interpretation algorithm development, and airborne parametric data management and analysis software production in support of health management. His expertise includes defining and implementing health management system architectures, data analysis, applications of intelligent software (fuzzy logic, neural network, expert rule-based, and Bayesian constructs), and aircraft hardware and software integration. Swearingen is a PhD candidate at the Missouri University of Science & Technology, and holds MS and BS degrees in electrical engineering from the University of Missouri. He is the author of over a dozen technical papers, conference presentations, and technical reports.

George Vachtsevanos

Dr. Vachtsevanos is serving as Chief Scientist at Impact Technologies and is Professor Emeritus at the Georgia Institute of Technology. He directs the Intelligent Control Systems Laboratory at Georgia Tech, where faculty and students are conducting interdisciplinary research in intelligent control, fault diagnosis, and failure prognosis of complex dynamic systems with emphasis on rotorcraft, and hierarchical / intelligent control of Unmanned Aerial Vehicles. His research in fault diagnosis and prognosis for condition-based maintenance began in 1984 with innovative fault detection and control technologies for the space station program. Under Office of Naval Research (ONR) sponsorship, he developed fault detection and fault-tolerant control systems for a turbojet engine. Jointly with Honeywell, he designed diagnostic and prognostic algorithms for shipboard machinery under ONR sponsorship. More recently, he has been an active participant in DARPA's Prognosis Program, the Aging Aircraft Program, an Advanced Diagnostics Program for U.S. Army vehicles, a U.S. Navy program on Prognostic Enhancements to Diagnostic Systems, an Air Force Space Command Program for CBM Design of Ground Satellite Stations, and other industrial programs. He administers at Georgia Tech and on-site an intensive four-day short course on "Fault Diagnosis and Prognosis for Engineering Systems." He has published over 350 technical papers and is the recipient of the 2002–2003 Georgia Tech School of ECE Distinguished Professor Award and the 2003–2004 Georgia Institute of Technology Outstanding Interdisciplinary Activities Award. He is the lead author of the book, *Intelligent Fault Diagnosis and Prognosis for Engineering Systems*, published by Wiley in 2006.